Lightning

Lightning over Lake Chiem in Bavaria, Germany. Photograph by W. Präpst.
Courtesy, Agfa-Gevaert-Bildarchiv and W. Präpst.

Lightning

Martin A. Uman

Department of Electrical Engineering
University of Florida, Gainesville

DOVER PUBLICATIONS, INC.
NEW YORK

Published in Canada by General Publishing Company, Ltd., 30 Lesmill Road, Don Mills, Toronto, Ontario.

Published in the United Kingdom by Constable and Company, Ltd., 10 Orange Street, London WC2H 7EG.

This Dover edition, first published in 1984, is an enlarged version of the work originally published by McGraw-Hill Book Company, New York, in 1969 in its ''Advanced Physics Monograph Series.'' The Dover edition includes a new preface and a new appendix to the text. This appendix, ''A Review of Natural Lightning: Experimental Data and Modeling,'' originally appeared in *IEEE Transactions on Electromagnetic Compatibility* (Vol. EMC-24, No. 2, May 1982, pp. 79–112) and is reprinted here with the permission of The Institute of Electrical and Electronics Engineers, Inc., 345 East 47 Street, New York, N.Y. 10017.

Manufactured in the United States of America
Dover Publications, Inc., 31 East 2nd Street, Mineola, N.Y. 11501

Library of Congress Cataloging in Publication Data

Uman, Martin A.
  Lightning.

  Previous ed.: New York : McGraw-Hill, 1968, c 1969.
  1. Lightning.   I. Title.
QC966.U4 1984   551.5′63   83-5205
ISBN 0-486-64575-4

To Derek

Preface to the Dover Edition

Fourteen years have passed since "Lightning" was published. These years have been characterized by active research that has resulted in significant progress in our understanding of lightning. Nevertheless, no more up-to-date monograph than "Lightning" has been published. Because the original edition has gone out of print and because there is considerable demand for it, Dover has agreed to republish "Lightning."

The Dover edition is a reprinting of the original "Lightning" with a few minor errors corrected. In addition, Appendix E has been added containing a review of the most important lightning research since the first edition was published. This review is weighted toward the electric and magnetic fields of lightning because it is in this area of research that the most significant advances have occurred.

Although no new lightning monographs have been forthcoming, there have been two recent edited collections of lightning articles—Golde, R. H. (ed.), "Lightning": Vol. 1, "Physics of Lightning"; Vol. 2, "Lightning Protection," Academic Press, N.Y., 1977; and Volland, H. (ed.), "Handbook of Atmospherics," 2 volumes, CRC Press, Boca Raton, Florida, 1982 —and one volume of conference proceedings: Dolezalek, H., and R. Reiter (edd.), "Electrical Processes in Atmospheres," Dr. Dietrich Steinkopff Verlag, Darmstadt, Germany, 1977. These publications survey the more recent work and are essential reading for serious lightning researchers.

Martin A. Uman
December 1, 1982

Preface to the First Edition

"Lightning" was written because there was no up-to-date book or review article on the subject. Two recent books on lightning exist: Malan's "Physics of Lightning" (1963) and Schonland's "Flight of Thunderbolts" (second edition, 1964). Both books are primarily concerned with the results of the studies of Schonland, Malan, and co-workers in South Africa. Neither book covers the important lightning research carried out in the United States, Switzerland, and Japan during the past 10 to 15 years. Neither book contains appreciable references to the literature, and both are often uncritical regarding suggested mechanisms for lightning processes. Perhaps the best available source of information about lightning has been Schonland's 'The Lightning Discharge,' published in 1956 in "Handbuch der Physik." This review article is now somewhat out of date.

The present work represents an attempt to develop a book that would (1) present a balanced, up-to-date coverage of what is known about lightning, (2) separate conjecture regarding physical processes from experimental data, and (3) contain extensive references to the lightning literature. It is hoped that "Lightning" to some extent satisfies these goals.

"Lightning" is written at the level of the advanced undergraduate in physics or engineering. It is primarily intended to serve as a source of information about lightning. However, "Lightning" is written so that it can be read by the nonexpert and hence so that it can be used as a teaching book. In gen-

eral, each chapter builds on information presented in the previous chapter. Chapter 1 presents a general introduction to lightning phenomena and lightning terminology and is intended to serve as background material for the rest of the book. Each chapter except the first and the last contains information about lightning obtained using a particular diagnostic technique. The diagnostic techniques considered are photography (Chapter 2), electric and magnetic field measurements (Chapter 3), electric current measurements (Chapter 4), spectroscopy (Chapter 5), and acoustic measurements (Chapter 6). Chapter 7 is concerned with the theory of the lightning discharge. In organizing the book so that it could be read by the nonexpert and used for teaching, it was necessary to discuss a given topic (for example, the dart leader) in many, if not all, chapters. In order to find all references to a given topic, the reader must use the index. Thus, the index is a very important part of the book. Every possible effort has been made to develop a comprehensive and easy-to-use index.

Rationalized mks units are used throughout the book. If other units are used (for example, the electron volt, the angstrom), those units are stated explicitly.

It is my intention to correct, revise, and bring "Lightning" up to date when such changes are needed. In this regard, I would appreciate receiving reprints of all future lightning articles and having my attention called to past articles that I have not given due consideration.

I wish to express my appreciation to the following gentlemen who have made significant contributions by reading and criticizing part or all the book in its preliminary form: Prof. K. Berger, Prof. Marx Brook, Dr. D. K. Davies, Dr. E. P. Krider, Prof. L. B. Loeb, Mr. F. Myers, Dr. R. E. Orville, Dr. A. V. Phelps, and Dr. Myron F. Uman.

I am indebted to the Westinghouse Research Laboratories, in particular to Dr. A. M. Sletten, Manager of High Voltage and Gas Physics, and to Dr. R. E. Fox, Director of Atomic and Molecular Sciences, for providing the necessary facilities and proper working environment for the preparation of the manuscript. I also wish to thank Mrs. Martha Fischer for her patient and careful typing of the manuscript.

My own research in lightning during the past few years has been supported in part by the Office of Naval Research and by the Federal Aviation Agency. In particular, I owe a debt of gratitude to Mr. James Hughes of the Office of Naval Research for his continuing interest and his moral and financial support of my lightning studies. Without that support this book would probably have not been written.

Martin A. Uman

Contents

Chapter 3   *Electric and Magnetic Field Measurements*   47

Chapter 4   Current Measurements   114

Chapter 5   Lightning Spectroscopy   138

Chapter 6   Thunder   181

Chapter 7   Theory: The Discharge Processes   202

Lightning

1

Introduction to Lightning

1.1 LIGHTNING AND THUNDERCLOUDS

In this introductory chapter we shall briefly survey the information available concerning the lightning discharge. The intent of the survey will be to give the reader a general picture of the overall phenomenon of lightning. This general picture will then serve as the background for the more detailed examination of the lightning discharge to take place in the succeeding chapters. Modern lightning research began in the latter part of the nineteenth century. Most of this book will be concerned with research performed since that time.

Lightning can be defined as a transient, high-current electric discharge whose path length is generally measured in kilometers. Lightning occurs when some region of the atmosphere attains an electric charge sufficiently large that the electric fields associated with the charge cause electrical breakdown of the air. The most common producer of lightning is the thundercloud (cumulonimbus). However, lightning also occurs in snowstorms, sandstorms, and in the clouds over erupting volcanos. Lightning has even been reported to occur in the clear air (McCaughan, 1926; Gisborne, 1928; Myers, 1931; Gifford, 1950; Baskin, 1952), apparently giving rise to the expression "bolt from the blue." In this book we shall be exclusively concerned with the lightning produced by thunderclouds. Such lightning can take place entirely within a cloud (intracloud or cloud discharges), between two clouds (cloud-to-cloud discharges), between a cloud and the earth (cloud-to-ground or ground discharges), or between a cloud and the surrounding air (air discharges).*

Before further discussing the lightning discharge, we will take a brief look at the thundercloud and its electric charges, the sources of lightning. Thunderclouds are formed in an atmosphere containing cold, dense air aloft, and warm, moist air at lower levels. The warm air at low levels rises in strong updrafts to form clouds, and the cold air aloft descends. Such atmospheric conditions occur, for example, when cold polar air masses overrun regions of warmer air or when the earth is strongly heated by the sun and transfers its heat to the air of the lower atmosphere.

As shown in Fig. 1.1, thunderclouds range in size from small clouds, which occur in the semitropics and in which the temperature may everywhere be above freezing, to giant electrical storms, which may have a vertical extent exceeding 20 km. The height of a typical thundercloud is perhaps 8 to 12 km, although, strictly speaking, typical values can only be presented for a given geographic location. Within a typical thundercloud there is a turmoil of wind, water, and ice in the presence of a gravitational field and a temperature gradient. Out of the interaction of these elements, in a way or ways not yet fully understood (Coroniti, 1965), emerge the charged regions of the thundercloud. Typically, the upper part of the thundercloud carries a preponderance of positive charge while the lower part of the cloud carries a net negative charge. Thus the main charge structure of the thundercloud is that of an electric dipole. The charged regions of the dipole are of the order of kilometers in diameter. In addition to the main cloud charges, there may be a small pocket of positive charge at the base of the thunder-

* An interesting class of air discharges is made up of those which extend upward from the tops of thunderclouds (Wright, 1950; Ashmore, 1950; Hoddinott, 1950).

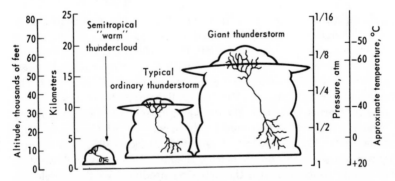

Fig. 1.1 *Comparison of various sizes of convective clouds that produce lightning discharges. (Adapted from Vonnegut (1965).)*

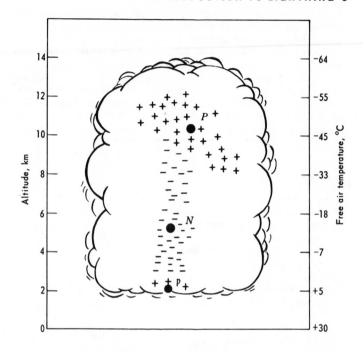

Fig. 1.2 *Probable distribution of the thundercloud charges, P, N and p for a South African thundercloud according to Malan (1952, 1963). Solid black circles indicate locations of effective point charges, typically P = +40 coul, N = −40 coul, and p = +10 coul, to give observed electric field intensity in the vicinity of the thundercloud.*

cloud. Figure 1.2 shows schematically the probable distribution of cloud charge for a typical South African thundercloud according to Malan (1952, 1963). The information needed to construct this cloud-charge picture was obtained from measurements of electric field intensity in the vicinity of thunderclouds. The measured fields are due to both the cloud charges and the induced charges on the earth; or, alternatively, to the cloud charges and their image charges below the conducting earth. Some additional and critical comments regarding the cloud-charge model shown in Fig. 1.2 are presented in Sec. 3.2.

Although the most frequently occurring form of lightning is the intracloud discharge, the greater part of the lightning literature concerns the ground discharge. Cloud-to-ground lightning is sometimes referred to as streaked or forked lightning. A cloud-to-ground lightning discharge is made up of one or more intermittent partial discharges. We shall call

TABLE 1.1

Data for a normal cloud-to-ground lightning discharge bringing negative charge to earth. The values listed are intended to convey a rough feeling for the various physical parameters of lightning. No great accuracy is claimed since the results of different investigators are often not in good agreement. These values may, in fact, depend on the particular environment in which the lightning discharge is generated. The choice of some of the entries in the table is arbitrary. All data in the table are taken from Berger (1967), Berger and Vogelsanger (1965, 1966), Brook and Kitagawa (1960), Brook et al. (1962), Hagenguth and Anderson (1952), Kitagawa et al. (1962), Schonland (1956), Williams and Brook (1963), and Workman et al. (1960) and are discussed in subsequent chapters.

	*Minimum**	*Representative*	*Maximum**
Stepped leader			
Length of step, m	3	50	200
Time interval between steps, μsec	30	50	125
Average velocity of propagation of stepped leader, m/sec†	1.0×10^5	1.5×10^5	2.6×10^6
Charge deposited on stepped-leader channel, coul	3	5	20
Dart leader			
Velocity of propagation, m/sec†	1.0×10^6	2.0×10^6	2.1×10^7
Charge deposited on dart-leader channel, coul	0.2	1	6
Return stroke‡			
Velocity of propagation, m/sec†	2.0×10^7	8.0×10^7	1.6×10^8
Current rate of increase, ka/μsec§	<1	10	>80
Time to peak current, μsec§	<1	2	30
Peak current, ka§		10–20	110
Time to half of peak current, μsec	10	40	250
Charge transferred excluding continuing current, coul	0.2	2.5	20
Channel length, km	2	5	14
Lightning flash			
Number of strokes per flash	1	3–4	26
Time interval between strokes in absence of continuing current, msec	3	40	100
Time duration of flash, sec	10^{-2}	0.2	2¶
Charge transferred including continuing current, coul	3	25	90

* The words maximum and minimum are used in the sense that most measured values fall between these limits.

† Velocities of propagation are generally determined from photographic data and thus represent "two-dimensional" velocities. Since many lightning flashes are not vertical, values stated are probably slight underestimates of actual values. See also comments in Sec. 2.3.1.

‡ First return strokes have slower average velocities of propagation, slower current rates of increase, longer times to current peak, and generally larger charge transfer than subsequent return strokes in a flash.

§ Current measurements are made at the ground.

¶ A lightning flash lasting 15 to 20 sec has been reported by Godlonton (1896).

the total discharge (whose time duration is of the order of 0.2 sec) a *flash;* we shall call each component discharge (whose luminous phase is measured in tenths of milliseconds) a *stroke.* There are usually three or four strokes per flash, the strokes being separated by 40 msec or so. Sometimes lightning as observed by the eye appears to flicker. In these cases the eye is discerning the individual strokes which make up a flash. Cloud-to-ground lightning flashes normally lower tens of coulombs of negative charge from the N region of the cloud (see Fig. 1.2) to the ground.*

In Secs. 1.2 to 1.5 we shall consider the general properties of the usual cloud-to-ground lightning discharge, that is, the discharge that is initiated in the cloud and lowers negative charge to earth. Data concerning this form of lightning are given in Table 1.1. In Secs. 1.6 to 1.9, other forms of the lightning discharge will be considered.

1.2 THE STEPPED LEADER

In the 1930s Schonland, Malan, Collens, and Hodges in South Africa made extensive use of the Boys camera, a camera which provides relative motion between film and lenses, to obtain time resolution of the luminous features of lightning. In addition they measured and interpreted the electric-field-intensity changes occurring during a lightning flash. Based on their observations and on subsequent work, the picture of the luminous features of a cloud-to-ground lightning flash shown in Fig. 1.3 can be constructed. (Streak-camera photographs like that shown diagrammatically in Fig. 1.3 are obtained with a camera composed basically of a stationary lens and a strip of photographic film that is moved horizontally at constant velocity across the image plane.) Each lightning stroke begins with a weakly luminous predischarge, the *leader* process, which propagates from cloud to ground and which is followed immediately by a very luminous *return stroke*. The return stroke propagates from ground to cloud. The cloud-to-ground predischarge preceding the first return stroke in a flash is called the *stepped leader*. The stepped leader is thought by many investigators to begin with a local electrical breakdown between the N and p regions of the thundercloud (see Fig. 1.2). This breakdown would serve to make mobile the electric charges which

---

* This statement and similar ones made throughout this book are not meant to imply that individual charges are transported over the large distance from cloud to ground. Rather the charge transport is an effective one in that the flow of electrons from the cloud into the top of the channel results in the flow of other electrons in other parts of the channel, much as would be the case were the channel a conducting wire. Thus, coulombs of charge can be transferred to ground during the time that an individual electron in the channel moves only a few meters.

previously were attached to ice and water particles. The resulting strong concentration of negative charge within the cloud base would produce electric fields which could then cause a negatively charged column to be propelled downward toward the earth. This column is called the stepped leader because it appears to move downward in luminous steps of typically 50-m length with a pause time between steps of about 50 μsec. During the pause time the stepped-leader channel is not luminous, or, more properly, is not luminous enough to be recorded on photographic film with Boys camera or streak-camera techniques. Each leader step becomes bright and observable in a time less than a microsecond. The faint channel between the top of a step and the bottom of the cloud has been photographed. In Fig. 1.3 the 50-m steps appear as darkened tips on the faintly luminous channel which extends upward into the cloud. A streak-camera photograph of a stepped leader is shown in Fig. 1.4. The typical average velocity of the stepped leader during its trip to the ground is 1.5×10^5 m/sec. It, therefore, takes about 20 msec to traverse a 3-km distance. A typical stepped leader has of the order of 5 coul of negative charge distributed over its length when it is near ground. The average currents which must flow in the stepped leader to deposit this amount of charge in tens of milliseconds are of the order of 10^2 amp. Luminous stepped-leader diameters have been measured photographically to be between 1 and 10 m. It is thought by

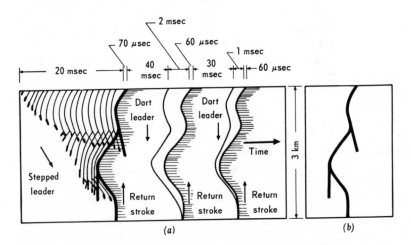

Fig. 1.3 (a) *The luminous features of a lightning flash as would be recorded by a camera with fixed lens and moving film. Increasing time is to the right. For clarity the time scale has been distorted. (b) The same lightning flash as recorded by a camera with stationary film.*

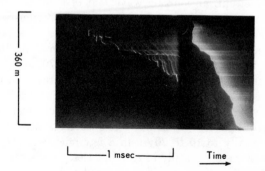

360 m

1 msec   Time

Fig. 1.4 *Photograph of a stepped leader taken with fixed lens and moving film. On the left side of the photograph the intensity of the leader is greatly enhanced. This enhancement was accomplished by varying the exposure in the reproduction process. Photograph is of lightning to Monte San Salvatore, near Lugano, Switz., and was originally published by Berger and Vogelsanger (1966). (Courtesy of K. Berger.)*

most investigators that the stepped-leader current flows down a narrow conducting core at the center of the observed leader, and that the large luminous diameter is due to a corona sheath surrounding the core. Undoubtedly, there are electrical and luminous phenomena occurring in the stepped leader on a time and distance scale much smaller than has been observed. Information regarding the stepped leader is summarized in Table 1.1.

1.3 THE RETURN STROKE

When the stepped leader has lowered a charged column of high negative potential to near the ground, the resulting high electric field at the ground is sufficient to cause upward-moving discharges to be launched from the ground toward the leader tip. When one of these discharges contacts the leader, the bottom of the leader is effectively connected to ground potential while the remainder of the leader is at negative potential and is negatively charged. The situation is somewhat similar to a transmission line charged to a constant potential with a short circuit applied at its end. The leader channel acts like a transmission line (nonlinear and lossy) supporting a very luminous return stroke. The return-stroke wavefront, an ionizing wavefront of high electric field intensity, carries ground potential up the path forged previously by the stepped leader. The return-stroke wavefront propagates at a velocity of typically one-third to one-tenth the speed of light, making the trip between ground and cloud base in a time of the order of 70 μsec. The region between the return-stroke wavefront and ground is traversed by large currents. The excess negative charge deposited on the leader channel is effectively lowered to earth through the highly conducting channel beneath the return-stroke wavefront. The current measured at the ground rises typically to 10 to 20 ka in a few microseconds and falls to one-half of peak value typically in 20 to 60 μsec. Currents of the order of hundreds of amperes may continue to flow for several milliseconds.

The initial gas density in the return-stroke channel is the gas density characteristic of the leader, while the initial temperature in the return-stroke channel is probably much higher than the leader temperature because of the energy delivered to the channel by the return stroke. The channel pressure will therefore exceed the pressure of the surrounding air, and the channel will expand. This expansion apparently takes place with supersonic speed, producing a shock wave which eventually becomes the thunder we hear. The shock-wave phase of the channel expansion is thought to last of the order of 5 to 10 μsec. The gas density in the current-carrying channel behind the shock wave decreases as the shock wave expands. In the latter part of the shock-wave phase the channel temperature as measured by spectroscopic techniques is near 30,000°K. After the shock-wave phase of the channel expansion is completed, the high-temperature, low-density channel approaches, in microseconds or a few tens of microseconds, a state of approximate pressure equilibrium with the surrounding air. When the pressure equilibrium is reached, the channel diameter is probably of the order of a few centimeters.

1.4 THE DART LEADER

After the stroke current has ceased to flow, the lightning flash may be ended. On the other hand, if additional charge is made available to the top of the channel, the flash may contain additional strokes. (The flash is then termed a multiple-stroke flash.) In general, each succeeding stroke appears to drain charge from higher areas in the N region of the cloud. This charge is made available between strokes by the action of the so-called K-streamer and J-streamer processes, electric discharges between the top of the previous return stroke and a higher region of negative charge. If additional charge is made available to the decaying return-stroke channel in a time less than about 100 msec, a continuous or *dart leader* will traverse that return-stroke channel, increasing its degree of ionization, depositing charge along the channel, and carrying the cloud potential earthward once more. The dart leader thus sets the stage for the second (or any subsequent) return stroke. The dart leader appears to be a luminous section of channel about 50 m in length, which travels smoothly earthward at about 2×10^6 m/sec, an order of magnitude faster than the average velocity of the stepped leader. Schematic drawings of streak-camera photographs of dart leaders are shown in Fig. 1.3, and a Boys camera photograph of a dart leader is shown in Fig. 2.6. The dart leader is thought to deposit less charge along its path than does the stepped leader. In addition the dart leader is generally not branched as is the stepped leader. If the current in the previous return stroke has ceased flowing for more than about 100 msec, a dart-stepped leader may occur. The dart-stepped leader begins its path down the decaying channel as a continuous leader, but at some point converts to a stepped leader. If the steps follow the original channel, they are shorter in length and are separated by smaller time intervals than the stepped leader preceding the first return stroke. If the steps do not follow the original channel, they resemble a first-stroke stepped leader. If the current in the previous return stroke has ceased to flow for a relatively long time, say, several hundreds of milliseconds, any subsequent stroke will be preceded by a stepped leader whose path is different from that of the decayed channel. Some data on dart leaders are given in Table 1.1.

1.5 MORE ON STROKES AND FLASHES

The first return stroke in a flash is usually strongly branched downward, as is the preceding stepped leader. Subsequent return strokes following dart leaders show little branching. First return strokes have slower average velocities of propagation, slower rates of increase of current at ground, longer times to peak current, and generally larger charge transfer than do subsequent return strokes.

The time between strokes which follow the same path can be tenths of a second if a small current flows in the channel during the time between the strokes. Apparently a reasonable level of channel conductivity is maintained by the current flow, and the channel is ripe for a dart leader only after this continuing current is terminated. Lightning flashes which contain at least one continuing current interval lower approximately twice as much negative charge to earth as do flashes composed of discrete strokes. Flashes composed of discrete strokes (termed discrete flashes in the literature) and flashes containing continuing-current intervals (termed hybrid flashes) have been reported to occur in New Mexico with approximately the same frequency.

The leader–return-stroke sequence has been observed to occur as many as 26 times in one flash. A channel or part of a channel may, however, suddenly increase in luminous intensity without that increase in luminosity being preceded by a leader process. The channel between a stepped-leader branch point and ground increases in luminosity when the return stroke reaches that branch point. This luminosity increase is known as a branch component. An increase in luminosity of the whole cloud-to-ground channel is called an M component. M components often occur when the channel luminosity is low and hence may be mistaken for discrete strokes. A Boys camera photograph of an M component is shown in Fig. 2.6.

1.6 INTRACLOUD DISCHARGES

Intracloud discharges usually take place between an upper positive charge center and a lower negative charge center. The complete discharge has a duration of the order of 0.2 sec, during which time a continuous low luminosity is observed in the cloud. It is thought that during this period a propagating leader bridges the gap between the two charge centers. Superposed on the continuous luminosity are several relatively bright luminous pulses whose time durations are about 1 msec. It would appear from electric field measurements that these luminous pulses are relatively weak return strokes that occur when the propagating leader contacts a pocket of charge of opposite polarity to that of the leader. The total charge neutralized in an intracloud discharge is probably of the same order of magnitude as the charge transferred in a cloud-to-ground discharge.

1.7 STROKES BRINGING POSITIVE CHARGE TO EARTH

The usual stepped leader lowers negative charge from the cloud toward the earth. Occasionally stepped leaders have been observed that lower positive charge.* Currents due to these "positive" strokes have been measured directly during discharges to instrumented towers. Positive strokes are characterized by a relatively slow rate of rise of current, roughly five times slower than for negative strokes, and a relatively

---

* Since only electrons are sufficiently mobile to constitute the leader current, the leader is charged positively by virtue of the flow of electrons out of the top of the leader into the positive-charge region from which the leader originates.

large charge transfer, roughly three times that of a negative stroke, with a maximum measured value of about 300 coul. Positive discharges are rarely composed of more than one stroke.

1.8 STROKES WITH UPWARD-DEVELOPING LEADERS

Stepped leaders which move upward toward the cloud are often initiated from high structures such as the Empire State Building or towers on mountains. Upward-moving leaders can carry either positive or negative charge. Both types of leaders have been observed. An upward-moving leader carrying positive charge results in the same direction of current flow as does a downward-moving leader carrying negative charge (the usual type of stepped leader). Apparently, when an upward-moving leader carrying positive charge contacts the cloud, no return stroke occurs. Typically, the leader current measured at ground merges smoothly into a more or less continuous current of a few hundred amperes. This initial discharge phase may be followed by one or more of the normal downward-moving dart-leader–upward-moving return-stroke combinations bringing negative charge to ground. It has been reported that upward-moving leaders carrying negative charge may occasionally connect with downward-moving leaders (presumably carrying positive charge), the measured junction points being up to 2 km above the point of origin of the upward-moving leader. Multiple-stroke flashes bringing positive charge to earth are rare, as noted in the preceding section.

1.9 OTHER LIGHTNING FORMS

The word lightning is used in conjunction with various noun-adjectives (the combination is known as a *nominal compound*) such as heat, sheet, rocket, ribbon, bead, and ball. Heat lightning is lightning or lightning-induced cloud illumination not accompanied by thunder. Presumably these discharges are far enough away from the observer so that the thunder cannot be heard. As we shall discuss in Chap. 6, thunder can seldom be heard from lightning discharges over 25 km distant. Sheet lightning is the name given to the lightning-induced cloud illumination which causes a sheetlike section of a cloud or clouds to become luminous. Rocket lightning is the name given to very long air discharges which give the impression of relatively slow progression along their channels. Ribbon lightning occurs when the cloud-to-ground discharge channel appears to be or is shifted (possibly by the wind) in the time between strokes. Each stroke in the flash is then seen separated horizontally in space. To the eye, each ribbon appears to occur simultaneously.

Bead lightning is the name given to that form of lightning in which the channel to ground breaks up, or appears to break up, into luminous fragments generally some tens of meters in length. These beads appear to persist for a longer time than does the normal cloud-to-ground discharge channel. A brief discussion of bead lightning is given in Appendix B. Ball lightning is the name given to the mobile luminous spheres which have been observed during thunderstorms. A typical ball lightning has a diameter of 20 cm and a lifetime of a few seconds. No satisfactory explanation for this phenomenon has been advanced. A brief discussion of ball lightning is given Appendix C.

REFERENCES

1. Ashmore, S. E.: Unusual Lightning, *Weather*, **5**:331 (1950).
2. Baskin, D.: Lightning without Clouds, *Bull. Am. Meteor. Soc.*, **33**:348 (1952).
3. Berger, K.: Novel Observations on Lightning Discharges: Results of Research on Mount San Salvatore, *J. Franklin Inst.*, **283**:478–525 (1967).
4. Berger, K., and E. Vogelsanger: Messungen und Resultate der Blitzforschung der Jahre 1955–1963 auf dem Monte San Salvatore, *Bull. SEV.*, **56**:2–22 (1965).
5. Berger, K., and E. Vogelsanger: Photographische Blitzuntersuchungen der Jahre 1955–1965 auf dem Monte San Salvatore, *Bull SEV.*, **57**:1–22 (1966).
6. Brook, M., and N. Kitagawa: Some Aspects of Lightning Activity and Related Meteorological Conditions, *J. Geophys. Res.*, **65**:1203–1210 (1960).
7. Brook, M., N. Kitagawa, and E. J. Workman: Quantitative Study of Strokes and Continuing Currents in Lightning Discharges to Ground, *J. Geophys. Res.*, **67**:649–659 (1962).
8. Coroniti, S. C. (ed.): Theories of Charge Generation in Thunderstorms, "Problems of Atmospheric and Space Electricity," pp. 237–320, American Elsevier Publishing Company, New York, 1965.
9. Gifford, T.: Aircraft Struck by Lightning, *Meteorol. Mag.*, **79**:121–122 (1950).
10. Gisborne, H. F.: Lightning from Clear Sky, *Monthly Weather Rev.*, **56**:108 (1928).
11. Godlonton, R.: A Remarkable Discharge of Lightning, *Nature*, **53**:272 (1896).
12. Hagenguth, J. W., and J. G. Anderson: Lightning to the Empire State Building, pt. 3, *Trans. AIEE*, **71** (*pt. 3*):641–649 (1952).
13. Hoddinott, M.: Unusual Lightning, *Weather*, **5**:331 (1950).
14. Kitagawa, N., M. Brook, and E. J. Workman: Continuing Currents in Cloud-to-ground Lightning Discharges, *J. Geophys. Res.*, **67**:637–647 (1962).
15. Malan, D. J.: Les décharges dans l'air et la charge inférieure positive d'un nuage orageux, *Ann. Geophys.*, **8**:385–401 (1952).
16. Malan, D. J.: "Physics of Lightning," The English Universities Press Ltd., London, 1963.
17. McCaughan, Z. A.: A Lightning Stroke Far from the Thunderstorm Cloud, *Monthly Weather Rev.*, **54**:344 (1926).
18. Myers, F.: Lightning from a Clear Sky, *Monthly Weather Rev.*, **59**:39–40 (1931).

19. Schonland, B. F. J.: The Lightning Discharge, "Handbuch der Physik," vol. 22, pp. 576–628, Springer-Verlag OHG, Berlin, 1956.

20. Vonnegut, B.: Electrical Behavior of an Airplane in a Thunderstorm, Arthur D. Little, Inc., Cambridge, Mass., February, 1965. (Available from Defense Documentation Center as AD 614 914.)

21. Williams, D. P., and M. Brook: Magnetic Measurement of Thunderstorm Currents, (1) Continuing Currents in Lightning, *J. Geophys. Res.*, **68**:3243–3247 (1963).

22. Workman, E. J., M. Brook, and N. Kitagawa: Lightning and Charge Storage, *J. Geophys. Res.*, **65**:1513–1517 (1960).

23. Wright, J. B.: A Thunderstorm in the Tropics, *Weather*, **5**:230 (1950).

2

Lightning Photography

2.1 EARLY STUDIES

While watching the storm from my house in Ealing, I could in several instances distinctly perceive a flickering appearance in a discharge, and in one particular case the repetitions were at least five or six in number, just sufficiently slow for the eye to detect the variations in brightness without removing the impression of one single flash. . . . the camera was moved in a horizontal plane about the lens as a center at the rate of about once to and fro in three quarters of a second. . . . I hoped, by having the camera moving, to be able to separate the successive components of the flashes, and in this I was fortunately successful.

H. H. Hoffert: Intermittent Lightning Flashes, *Phil. Mag.*, **28**:106–109 (1889).

Hoffert published photographs of several multiple-stroke flashes. He found that each stroke in a given flash was identical in shape, and noted that the first stroke in one 3-stroke flash was branched while the subsequent strokes were not. He reported that in some cases there existed a continuing luminosity between strokes. Hoffert estimated the interval of time between successive strokes to be between $\frac{1}{5}$ and $\frac{1}{10}$ sec, probably somewhat of an overestimation. Weber (1889) also obtained flash-resolved lightning photographs by rapidly moving an ordinary camera back and forth in the horizontal plane. One of the first flash-resolved lightning photographs is due to Kayser (1884), who, using a stationary camera, photographed the spatially-separated strokes of a flash (ribbon lightning). Kayser presented evidence to show that the strokes were separated by the wind.

In the early part of the twentieth century, Walter (1902, 1903, 1910, 1912, 1918) in Germany published the results of a series of careful experiments in which he photographed lightning with a camera whose motion was controlled by a clock-work mechanism. He measured with good accuracy the time between strokes and photographed the leader preceding the first stroke, although his time resolution was insufficient to discover that the first leader was stepped. He noted that the downward branching of the first lightning stroke in a flash was due to the branching of the downward-moving leader.

Further early photographic measurements of lightning were made by Larsen (1905) in the United States. He measured the time between strokes and in one instance recorded a flash which appeared to be composed of 40 strokes. No dart leaders could be resolved so it is likely that some of the apparent "strokes" were increases of channel luminosity due to M components (Malan, 1956; Workman et al., 1960). Larsen reported an average of five to six strokes per flash.

2.2 THE BOYS CAMERA

Coming now to the year 1900, I wished to obtain some experimental evidence, if possible, of the progress of the lightning flash. The impression that there is a downward direction is very common, and occasionally observers believe one has an upward direction. . . . I desired to make a conclusive test. . . . The scheme was to use a pair of identical camera lenses (specially selected for stereoscopic photography) and to mount these on a disk which could be rotated by hand through gearing at any desired speed. In the apparatus I then made I could drive them at any speed up to about forty turns a second. The lenses were four inches apart, center to center; the two images of a lightning flash would then be carried in opposite directions at any speed up to about forty feet per second, and if the flash in each part of its length should be "instantaneous," a difference in time between the two ends of the flash of about 1/40,000 second would be observable. If, for example, the flash were a vertical line and the lenses at the moment were one above the other, one image would be tilted in one direction while the other would be tilted in the other direction, and the more so the greater the duration. If the lightning were not in any part "instantaneous," by which I mean if it lasted long enough for its image to be broadened—that is, 1/40,000 second or more—the two images would fade away, but on opposite sides, and the sharp side would still be available for comparison. . . .

I made this apparatus in 1900, and carried it about with me, for example, to the British Association meeting at Glasgow the following year, and only once obtained a moderately good view of a few flashes, but the developed plates showed nothing at all! Though I have had this now for twenty-six years, I still have not succeeded in obtaining any photographs.

C. V. Boys: Progressive Lightning, *Nature,* **118**:749–750 (1926).

A diagram showing the original Boys camera and illustrating its use is given in Fig. 2.1. A Boys camera photograph is shown in Fig. 2.2.

Eventually, Boys (1928) did obtain photographs with his camera, as did Halliday (1933) a few years later. The conclusions of Boys and Halliday from analyses of their photographs were apparently in error. They found strokes whose luminosity progressed simultaneously from cloud and from ground to a meeting point between cloud and ground. Halliday also found two strokes which began at a point above ground and propagated simultaneously upward and downward. Boys said that he attached no importance to the rough numerical results obtained from his first experiment. What had been shown was that the Boys camera could be a useful tool for lightning research.

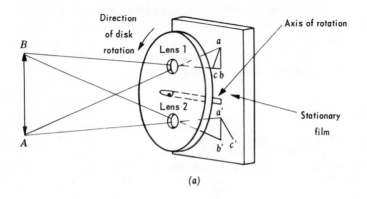

(a)

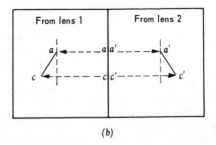

(b)

Fig. 2.1 (a) *Diagram of Boys camera with moving lenses and stationary film. Luminosity progressing from A to B leaves the image ab (a′b′) when the disk carrying the lenses is stationary and ac (a′c′) when the disk is rotating.* (b) *By placing the two photographic images side by side, the time of propagation of luminosity from A to B, (cc′ − aa′)/2v, where v is the velocity of each lens, may be determined.*

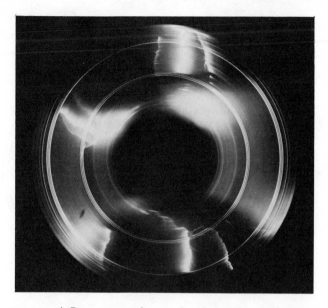

Fig. 2.2 *A Boys camera photograph. Increasing time is in a clockwise direction with 360° being equivalent to 17 msec. Stepped-leader steps are wound all the way around the photograph. A dart leader is apparent in the upper left and lower right hand portions of the photograph. The intensities of the less luminous phenomena have been enhanced using darkroom techniques. The luminous circles are city lights. (Photograph taken near Tucson, Ariz., by E. P. Krider.)*

Boys (1929) published details of an improved version of his camera. The two lenses were mounted one above the other and were stationary; the film, which was mounted on the inner surface of a cylindrical drum, was moved. A drawing of this camera is shown in Fig. 2.3. Many variations of the first Boys cameras have been constructed. Especially of interest are the cameras of Malan (1950, 1957).

The time resolution of a double-lens Boys camera is at best about 1 μsec. This time resolution is needed to observe adequately the high propagation velocity of the return stroke. The leader processes, however, may be observed with a streak camera, that is, a single-lens camera having fast-moving film. With a streak camera, the return-stroke distortion (the visible return stroke is completed in less than 100 μsec) is ignored and the return stroke is used as a reference from which to measure the motion of the slower leaders. (Dart leaders take of the

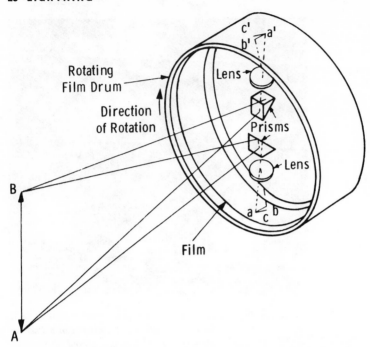

Fig. 2.3 *Diagram of improved Boys camera with moving film and stationary optical system. Luminosity progressing from A to B leaves the image ab (a′b′) when the drum is stationary and ac (a′c′) when the drum is rotating. (Adapted from McEachron (1939).)*

order of 1 msec and stepped leaders of the order of 10 msec to travel from cloud to ground.) Recent measurements made with streak cameras having moving film are in principle little different from the measurements made by Hoffert, Weber, Walters, and Larsen using cameras which were moved by hand or machinery.

2.3 THE SOUTH AFRICAN WORK

2.3.1 Introduction

It is generally believed, though direct evidence has so far been lacking, that the first stage in the preliminaries to spark breakdown is the passage from cathode to anode of an electron avalanche. It has been suggested that the real breakdown occurs after the passage of this avalanche and proceeds from anode to cathode as a thermally ionized tongue.

We have recently obtained what appears to be very clear evidence of the

*reality of these effects in the case of the lightning flash, which we have studied
with a revolving lens camera of the type invented by Prof. C. V. Boys.*

B. F. J. Schonland and H. Collens: Development of the Lightning Discharge,
Nature, **132**:407–408 (1933).

So begins one of the early papers reporting results obtained in South
Africa from Boys-camera and electrical measurements of the lightning
discharge. There are 10 major publications concerning the South
African work (Schonland and Collens, 1934; Schonland et al., 1935;
Malan and Collens, 1937; Schonland, 1938; Schonland et al., 1938a;
Schonland et al., 1938b; Malan and Schonland, 1947; Malan and Schon-
land, 1951a; Malan and Schonland, 1951b; Clarence and Malan, 1957).
In addition, the review paper by Schonland (1956) and the book by
Malan (1963) are of special interest.

In 1933, Schonland and Collens published a preliminary report on
information deduced from photographs of 45 lightning strokes (11
flashes) taken by Collens and 5 strokes taken by Halliday. A total
of 28 downward-moving leaders were identified. Most, if not all, were
associated with unbranched return strokes. Apparently the photographs
were of the dart leader, the leader which precedes strokes subsequent
to the first. Schonland and Collens were the first to present quantitative evidence that return strokes develop in the upward direction, from
ground to cloud. A year later Schonland and Collens published the
detailed results of their analysis of the 45 lightning strokes (11 flashes)
photographed by Collens. The mean straight-line two-dimensional
velocity* found for dart leaders was 1.1×10^7 m/sec and for return
strokes 3.8×10^7 m/sec. Dart lengths were measured to be less than
54 m. In 1935, Schonland, Malan, and Collens reported their observations (the first such observations) of the stepping of the leader preceding
the first stroke in a flash. A diagram of the lightning flash similar to
that shown in Fig. 1.3 was given. The most frequent values for the
effective velocity of the stepped leader was reported to be 1.5×10^5 m/sec.
The existence of dart-stepped leaders were reported. It was shown that
the first stroke in a flash is usually much more intense than the succeeding
strokes. In the 1933–1935 papers are to be found data which still, as
this book is being written, represent a significant fraction of what is
known about lightning. In these papers and in subsequent papers from
South Africa a great deal of detailed lightning data obtained by photography was presented and discussed. We review that data now.

* A velocity measurement made from a streak or Boys camera photograph is only a
true representation of the actual velocity if the channel is straight and is vertical.
If the channel exhibits small-scale tortuosity, the actual velocity may be considerably greater than the straight-line velocity. If the channel is not vertical, the
actual velocity may be of the order of 10 percent greater than the measured velocity.

2.3.2 The stepped leader

Stepped leaders can be divided on the basis of step length and average earthward velocity into two classes, α and β (Schonland et al., 1938a,b). Type α stepped leaders exhibit a low and relatively uniform average earthward velocity ($\sim 1.0 \times 10^5$ m/sec) throughout their trip from cloud base to earth. The α steps are generally shorter than the β steps and do not vary appreciably in length or brightness. Type α leaders are weakly luminous compared to type β leaders. Schonland (1956) states that the majority of the leaders photographed in South Africa appear to be type α. However, electrical measurements indicate that type β are in the majority. Type β leaders exhibit relatively long, fairly bright steps and high average earthward velocity (usually 8×10^5 to 2.4×10^6 m/sec). They are also characterized by extensive branching beneath the cloud base. As a type β leader approaches the earth, its velocity decreases, and it becomes similar to a type α leader. Diagrams of streak-camera photographs of α and β leaders are shown in Fig. 2.4. Table 2.1 gives average downward velocities for a number of α and β leaders.

There are two important variations of the type β leader (Schonland et al., 1938b). These are termed β_1 and β_2. Type β_1 is different from a normal β leader in that it exhibits a *marked* discontinuity in downward velocity at some point in its earthward trip. Before the discontinuity the β_1 is heavily branched and exhibits an average velocity of about 0.6 to 2.6×10^6 m/sec. After the discontinuity the leader becomes essentially a type α leader with few branches, weak luminosity, and average velocity of about 0.7 to 3.2×10^5 m/sec. Table 2.2 gives examples of β_1 leader properties.

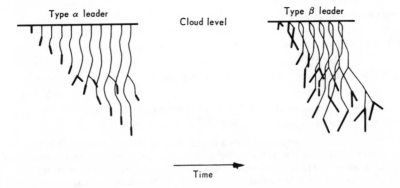

Fig. 2.4 *Diagrams of streak-camera photographs of type α and type β stepped leaders.*

TABLE 2.1

Average downward velocities of type α and type β stepped leaders in South Africa. Adapted from Schonland (1956).

Velocity, m/sec	No. of type α	No. of type β and air discharge
1.0– 2.0 $\times 10^5$	27	
2.0– 3.0	9	
3.0– 4.0	13	
4.0– 5.0	4	
5.0– 6.0	3	1
6.0– 7.0	2	1
7.0– 8.0	2	1
8.0– 9.0		3
9.0–10.0		2
10.0–11.0		2
11.0–12.0		2
12.0–13.0		1
17.0–18.0		1
20.0–21.0		1
23.0–24.0		1
25.0–26.0		1

TABLE 2.2

Properties of type β_1 leaders.   The average downward velocity of the leader in the first stage, near the cloud, is v_1; in the second stage, near the earth, v_2. Adapted from Schonland et al. (1938b).

Flash designation	First stage		Second stage	
	Length, km	v_1, m/sec	Length, km	v_2, m/sec
92	1.12	11.2 $\times 10^5$	0.64	0.8 $\times 10^5$
77	0.90	10.0	1.00	1.9
43	0.48	16.0	1.44	1.1
140	1.90	6.3	1.22	1.3
N2	1.00	7.2	1.54	3.2
139	1.90	10.0	1.26	1.4
40	1.15	26.0	1.90	10.5
135	1.20	>10.0	1.20	0.9
112	0.20	. . . .	1.20	0.7
98	0.40	. . . .	0.70	1.0

Type β_2 leaders are different from normal type β leaders in that during the slow second stage, near earth, the β_2 leader channel is illuminated by a dart leader passing quickly from cloud base to leader tip. Profuse leader branching is observed at the earthward end of the dart. A number of these dart leaders may overtake the slow-moving stepped leader as shown diagrammatically in Fig. 2.5. The time interval between the darts is of the order of 0.01 sec, as is the time interval between the appearance of the first stage of the type β_2 leader and the first dart. The second stage of the β_2 leader is the slowest and least luminous form of leader process. There is not much difference in the leader velocity before or after the appearance of a dart. In Table 2.3 examples of β_2 leader properties are given. Workman et al. (1936) have apparently photographed a type β_2 leader which appeared to travel to ground in four successive large darts. The leader steps were unresolved on the photograph.

There is necessarily some variation in the photographically measured step length due to variations in the component of the steps along the line of sight. Measured step lengths varied from about 10 to 200 m. Out of 21 different first leaders whose photographs were suitable for accurate step-length measurements, 15 were found to have step lengths between 20 and 85 m (Schonland, 1956). The step length and step brightness are usually found to increase with increased leader average velocity. Both α and β leaders exhibit an increased average velocity and increased step brightness as they approach very close to the earth.

The pause time between leader steps for 21 first leaders varied between 37 μsec and 124 μsec (Schonland, 1956). Nineteen of the twenty-one

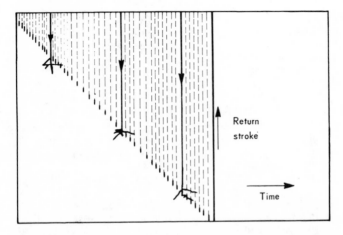

Fig. 2.5 *Diagram of a streak-camera photograph of a type β_2 stepped leader.*

TABLE 2.3

Properties of type β_2 leaders. The average downward velocity of the leader in the first stage, near the cloud, is v_1; in the second stage, near the ground, before the dart, v_2; after the dart, v_2'. Adapted from Schonland et al. (1938b).

	First stage			*Second stage*			
Flash designation	*Interval before dart, sec*	*Length, km*	v_1, *m/sec*	*Length, km*	v_2, *m/sec*	*Length, km*	v_2', *m/sec*
102	0.0082	1.2	2.0×10^6	0.72	9.0×10^4	0.48	1.3×10^5
92	0.0090	0.5	9.0×10^5	0.80	9.2×10^4	0.64	7.0×10^4
92	0.0067	1.6	2.6×10^6	0.31	5.0×10^4	0.61	2.0×10^5
BX	0.0032	0.9	$>1.0 \times 10^6$	0.40	1.2×10^5	0.40	1.2×10^5

leaders whose photographs were found suitable for measurement had pause times between 37 and 92 μsec. The average downward velocity of the stepped leader is found by dividing the step length by the pause time, since the steps become luminous in a time which is short compared to the pause time. For the 21 first leaders discussed above, the average velocity varied between 2.6×10^5 and 2.1×10^6 m/sec. In general, the longer step lengths were associated with longer pause times.

The velocity of propagation of luminosity along a step cannot be measured with the Boys camera because the step length is short and the time resolution of the camera is at best of the order of 1 μsec. The steps apparently become bright in a time less than 1 μsec. Thus for a 50-m step the velocity of propagation of luminosity along the step must exceed about 5×10^7 m/sec. Schonland et al. (1935) report that the upper part of the leader step is broadened on Boys camera photographs and suggest that this effect indicates a downward-traveling luminosity.

Lightning discharges which fail to reach the ground are known as air discharges. Most air discharges are profusely branched type β leaders. Thus air discharges and type β leaders appear in the same column in Table 2.1. Some air discharges are type β_2 leaders, and it has been suggested by Sourdillon (1952) that these air discharges are what is known as rocket lightning (see Sec. 1.9).

Luminous stepped-leader radii have been reported by Schonland (1953) to be between 0.5 and 5.0 m. These measurements were made from Boys camera photographs, and hence the possibility exists that the streaking of the leader on the film resulted in an enlarged image.

Schonland (1953) argues that this is not the case since no stepped leader radius was found to be smaller than 0.5 m even when the streak direction was parallel to the steps. It is unlikely that a photographic measurement of the stepped-leader radius would result in an underestimation of that radius (Evans and Walker, 1963).

2.3.3 Dart and dart-stepped leaders

The dart leader would appear on a still photograph taken with fast shutter speed as a thin luminous line extending downward from the

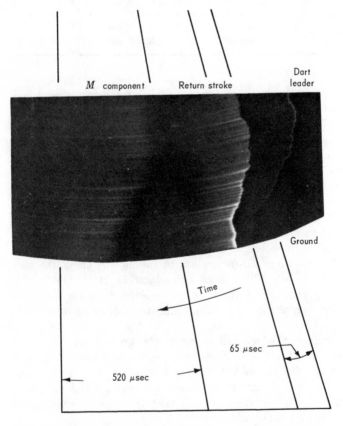

Fig. 2.6 *A Boys camera photograph showing dart leader, return stroke, and M component. The intensities of the dart leader and M component have been enhanced with respect to the return stroke using darkroom techniques. (Photograph taken near Tucson, Ariz., by E. P. Krider.)*

TABLE 2.4

Distribution of dart-leader velocities. Adapted from Schonland
et al. (1935).

Velocity range, m/sec	No. of dart leaders
$1.0–3.0 \times 10^6$	17
$3.0–5.0$	8
$5.0–7.0$	6
$7.0–9.0$	5
$9.0–11.0$	5
$11.0–13.0$	2
$13.0–15.0$	6
$15.0–17.0$	1
$17.0–19.0$	2
$19.0–21.0$	2
$21.0–23.0$	1

cloud base with a bright tip about 50 m long. To the Boys camera the dart-leader tip appears to move smoothly earthward, usually without branching. A Boys camera picture of a dart-leader–return-stroke combination is shown in Fig. 2.6. Dart-leader velocities were measured to be from 1.0×10^6 to 2.3×10^7 m/sec with a most frequent value of 2.0×10^6 m/sec (see Table 2.4). Dart-leader velocities often decrease as the dart leader approaches the ground. Further, the dart-leader velocity is related to the time interval between the previous stroke and the appearance of the dart leader. High velocities are associated with short intervals and low velocities with long intervals. It has been suggested (Schonland et al., 1935) that this effect is due to the decrease in channel conductivity with time, lower conductivity yielding a slower dart leader. The change in the channel radius with time and the increase in mass density as the temperature decreases may also contribute to the observed effect (see Secs. 7.5 and 7.7). In general, the longer the time interval between strokes, the more intense the subsequent return stroke. These effects are depicted in Table 2.5. Data obtained by Brook and Kitagawa on the relation between dart-leader velocity and interstroke time interval are considered in Sec. 2.5.3 (Fig. 2.10).

When the time interval between strokes is long, the dart leader may change from a continuous moving leader to a stepped leader of high velocity, short step length, and short time interval between steps. The rapid stepping follows the channel of the previous stroke. Properties of dart-stepped leaders are given in Table 2.6. Sometimes the dart

TABLE 2.5

Properties of dart leaders and subsequent return strokes. First stroke intensity is unity. Adapted from Schonland (1956) and Schonland et al. (1935).

No. of dart leaders	Time interval from previous strokes, sec	Mean interval, sec	Velocity m/sec	Mean velocity m/sec	Intensity of return stroke	Mean intensity
5	0.005–0.12	0.044	15–22×10^6	19×10^6	0.3–0.8	0.46
5	0.07–0.48	0.17	1.7–2.8×10^6	2×10^6	1.2–5.0	2.2

leader or dart-stepped leader will revert to a normal stepped leader. In these cases the leader does not necessarily follow the channel of the previous strokes.

TABLE 2.6

Average velocity, average step length, and average time interval between steps for dart-stepped leaders preceding second stroke of multiple-stroke flash. Adapted from Schonland (1956).

Flash designation	Velocity, m/sec	Step length, m	Time interval, μsec
67	1.2×10^6	9.0	7.4
64	1.1	10.0	9.0
75	1.0	7.4	7.4
130	1.7	25.0	15.0
657	1.7	13.0	7.8
X7	0.48	12.0	25.0

2.3.4 First return strokes

Stepped leaders are branched downward and the first return stroke illuminates these branches as well as the main channel. The first return-stroke luminosity travels upward from the ground decreasing abruptly in intensity as each branch point is reached. The velocity of the return stroke along the main channel, in general, also decreases as each branch point is reached. Velocity as a function of channel height for one return stroke is given in Table 2.7. Return-stroke velocity at the base of the channel is typically of the order of 1×10^8 m/sec, and at

TABLE 2.7

A first return-stroke timetable. Velocity values are accurate to within 20 percent. Adapted from Schonland et al. (1935).

Section of channel	Return-stroke velocity, m/sec
Earth to second branch from bottom; lower $\frac{1}{3}$ of channel	1.6×10^8
Branch 2 to branch 3; next $\frac{1}{6}$ of channel	2.1
Branch 3 to branch 5; next $\frac{1}{4}$ of channel	0.97
Branch 5 to cloud base; next $\frac{1}{4}$ of channel	0.55

the top of the channel is typically of the order of 4×10^7 m/sec. The return stroke travels out along the branches with velocities which range from 1.5×10^7 to 1.2×10^8 m/sec. The return-stroke velocity along a branch tends to be higher than the velocity on the main channel near the branch point.

Schonland (1937) has photographically measured return-stroke-channel radii ranging from 7.5 to 11.5 cm. He found branch radii from 5.5 to 8.5 cm. The measurements were made from a still-camera photograph of close lightning showing the various strokes of a multiple-stroke flash separated in space by the wind (ribbon lightning). It is unlikely that a channel-radius measurement made from a photograph would result in an underestimation of that radius (Evans and Walker, 1963). Results of channel radius measurements made in Arizona using time-resolved photography will be given in Sec. 2.5.2.

2.3.5 Subsequent return strokes

The dart leader preceding a second stroke travels down the remnants of the defunct first-stroke channel illuminating only very prominent first-stroke branches. Often no branches are illuminated. Thus primarily the main channel is conditioned for supporting the second return stroke in its trip from earth to cloud. Return strokes subsequent to the first usually do not, therefore, show much branching. Branches present in a second stroke will almost always disappear by the third stroke. The propagation velocity of a given return stroke subsequent to the first is relatively uniform throughout the trip from earth to cloud. The velocity range for strokes subsequent to the first is from 2.4×10^7 to 1.1×10^8 m/sec. The luminosity of subsequent return strokes tends to decrease as the stroke is propagated upward, but this decrease is not pronounced.

2.3.6 Branch components and M components

The properties of branch components and M components have been considered by Malan and Collens (1937) and Malan and Schonland (1947). These references contain many interesting photographs and drawings.

When a first return stroke reaches a branch, the whole channel between branch point and ground increases in luminosity. The velocity of propagation of the branch-component luminosity exceeds 10^8 m/sec. A determination of the exact velocity and of the direction of propagation has apparently not been made. For short-length branch components the time resolution of the Boys camera is inadequate and for long branch components the edge of the luminosity on the photograph is too diffuse to enable a measurement to be made.

After the return stroke enters the cloud, additional increases of channel luminosity along the total visible channel may be observed. These are termed M components. A few may be associated with branching of the channel inside the cloud; most are not. M components occur while the channel is faintly luminous and apparently while it still carries some current. The time intervals between M components as measured by Malan and Schonland (1947) are usually between 0.3 and 3.0 msec. The time duration of M components was found by the same authors to increase with each new M component for a given stroke.

Working in New Mexico (see Sec. 2.5.3) Kitagawa et al. (1962) found that the most frequent time between M components was about 6 msec and that during the first 15 msec after a return stroke the time interval between M components tended to increase with elapsed time. The earliest M components were separated in time by less than 1 msec. After 40 msec the time intervals showed no dependence on the elapsed time. The data of Kitagawa et al. (1962) apparently refer primarily to M components occurring during the continuing current phase which follows some strokes, whereas the data of Malan and Schonland (1947) apparently refer primarily to M components occurring during the low-current "tail" which may well follow most strokes. An M component is visible on the Boys camera photograph given in Fig. 2.6.

2.3.7 Miscellaneous

Malan (1955, 1963) has developed a special camera for photographing luminous phenomena that produce a general diffuse illumination of the cloud. Malan's camera records the luminosity coming from only a small portion of the cloud and streaks this luminosity on film. The problem of separating overlapping images from a large source is thus resolved. The camera records, in addition, the cloud-to-ground region. Malan has found that luminosity occurs within the cloud during the period between cloud-to-ground strokes (supporting the idea of the J process). He has also studied intracloud lightning and found the most frequent duration of an intracloud discharge to be 0.25 sec. Cloud discharges were found to exhibit continuous luminosity punctuated by intermittent intensity increases at time intervals similar to those between cloud-to-ground strokes. Photoelectric measurements of cloud discharge luminosity will be considered in Sec. 3.8.

Luminosity fronts which move inward from the tips of branches to the main channel of a return stroke or air discharge have been sometimes observed. These have been called *recoil streamers*. In return strokes the recoil luminosity is observed after the return stroke has reached the end of the branch. The recoil luminosity has a relatively low propagation velocity, of the order of 6×10^5 m/sec (Schonland, 1956). Photographs of recoil streamers in air discharges are given by Ogawa and Brook (1964). The relation of the recoil streamers to the intracloud discharge is considered in Sec. 3.8.

Schonland and Malan (as reported by Schonland, 1956) have determined the number of strokes per flash in South Africa for 1,800 flashes.

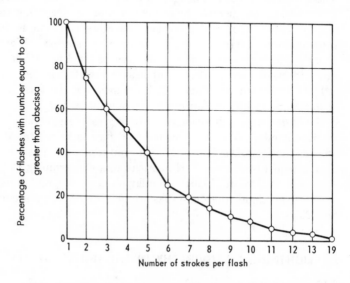

Fig. 2.7 *Cumulative frequency distribution of the number of strokes per flash in South Africa. (Adapted from Schonland (1956).)*

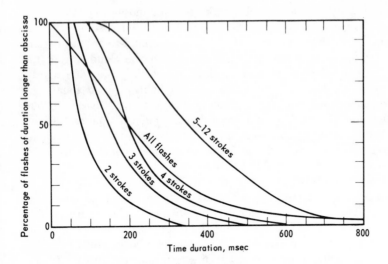

Fig. 2.8 *Cumulative frequency distribution of total duration of flashes to ground in South Africa. Curve for all flashes includes one-stroke flashes. (Adapted from Malan (1956).)*

These data are given in Fig. 2.7. Fifty percent of the flashes have four or more strokes, ten percent have nine or more. According to Schonland (1956) flashes with many strokes are more common in large frontal thunderstorms than in minor convective storms. Wormell (1953) reports that the most common flash measured in England consists of only one stroke although 62 percent of flashes have two or more strokes.

Data presented by Malan (1956) concerning the total time duration of flashes to ground in South Africa are given in Fig. 2.8. Fifty percent of the flashes have a duration equal to or greater than 0.2 sec. According to Schonland (1956), 20 percent of all flashes in South Africa contain one or more strokes followed by a continuing current whose duration may extend to 200 msec. Numerical data concerning this phenomenon are given by Malan (1956). Similar observations regarding continuing currents in cloud-to-ground flashes in New Mexico are discussed in Sec. 2.5.3.

2.4 LIGHTNING PHOTOGRAPHY IN THE UNITED STATES: 1933–1949

2.4.1 Introduction

The first photographic study of lightning undertaken in the United States was that of Larsen (1905) (see Sec. 2.1). Jensen (1932, 1933) published the results of photographic and electrical studies of lightning and, in particular, of the branching properties of lightning. From his studies Jensen concluded that lightning is usually downward branched from a negatively charged cloud. We shall consider Jensen's electrical measurements in Chap. 3. Soon after the first South African lightning stroke measurements were made, Lloyd and McMorris (1934) in the United States obtained photographs of dart leaders using a Boys camera. Their work, which included the measurement of the dart-leader and return-stroke velocities, was also reported by McEachron (1934).

2.4.2 The Empire State Building study

The Empire State Building in New York City is a steel frame building whose height is about 410 m above street level. At its peak is an antenna structure and above that a lightning rod. From 1935 to 1949, excluding the years 1942 to 1946, extensive measurements were made by personnel of the General Electric Company of lightning strikes to the building. Several different types of cameras (Hagenguth, 1940; Kettler, 1940), including Boys cameras, were operated at a distance of 836 m from the top of the Empire State Building between 1935 and 1941, and at a distance of 885 m from 1947 through 1949. Photographs of lightning strikes to the building were made simultaneously with oscilloscopic recording of measurements of lightning currents. The measurements of lightning current will be described in Chap. 4.

During the time that the Empire State Building study was in progress, there were an average of 22.6 discharges per year to the building. The pertinent results of the Empire State Building study are given in papers by McEachron (1939, 1941) and Hagenguth and Anderson (1952). We will consider now some of these results.

In 1939 McEachron reported discovery of (1) upward-moving stepped leaders and (2) continuous lightning currents and resultant continuous luminosities lasting for times of the order of 0.5 sec and more. Subsequently, statistics on these two phenomena were accumulated. Of 135 flashes studied with current-recording oscillograph and camera, about 50 percent had no pronounced luminosity peak or current peak at all. Measurements showed that the majority of the flashes to the Empire State Building began with an upward-moving stepped leader. The stepped leader was not followed by a return stroke, but rather the leader current merged smoothly into a continuous current flow with an average amplitude of 250 amp. In about half of the flashes initiated by upward-moving stepped leaders, the continuous current flow was interrupted one or more times (average 2.3) by a downward-moving dart leader followed by an upward-moving return stroke (the normal cloud-to-ground dart-leader–return-stroke combination). A schematic diagram

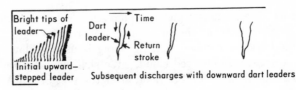

Still—camera photograph of flash

High—speed photograph of the same flash

Fig. 2.9 *Diagram showing usual lightning between cloud and Empire State Building. (Adapted from McEachron (1939).)*

of the usual flash to the Empire State Building is shown in Fig. 2.9. The upward-moving stepped leader is apparently characteristic of the Empire State Building and other tall structures. Berger and Vogelsanger (1966) (see Sec. 2.5.4) report upward-moving stepped leaders from structures on mountains. On occasion, the normal cloud-to-ground stepped-leader–return-stroke combination was also observed between cloud and Empire State Building.

It was reported by Flowers (1944) that for slowly varying stroke properties the density of a stroke image on photographic film was proportional to the stroke current as measured on an oscillograph. The experimental correlation was made on a millisecond time scale during periods of continous current (up to 0.5 sec). It would therefore appear possible, with proper calibration, to measure slowly varying lightning currents by photographic or photoelectric techniques.

McEachron (1939) reported that upward-moving stepped leaders have an average step length of about 8.2 m, the range being 6.2 to 23 m. Most of the step lengths were near the average. He stated that there may be considerable error in these measurements due to the smallness of the image on the photographs. These step lengths are considerably smaller than those measured in South Africa for downward-moving stepped leaders (see also Sec. 2.5.4 and Table 2.8). The small steps may be due to the effect of the building on the discharge. The time interval between steps varied from 20 to 100 μsec, the average being 30 μsec. Most of the time intervals were close to the average. Velocities of 20 upward-moving stepped leaders were measured. These ranged from 4.7×10^4 to 6.4×10^5 m/sec with an average of 2.6×10^5 m/sec. The data on time intervals and stepped-leader velocities are similar to those found in South Africa for downward-moving stepped leaders.

McEachron (1939) reported dart-leader velocities ranging from 5.8×10^5 to 3.9×10^7 m/sec, with an average of about 1.2×10^7 m/sec. Return stroke velocities following dart leaders ranged from 3×10^7 to

8.5 × 10⁷ m/sec, with an average of 6 × 10⁷ m/sec. The data on dart leaders and return strokes are in reasonable agreement with the South African data.

A flash having a duration of 1.5 sec was recorded, this being the maximum duration. Half of the flashes had a time duration in excess of 0.27 sec.

2.4.3 The Pittsfield study

A photographic investigation of lightning was started in Pittsfield, Massachusetts, in 1935 at the site of the high-voltage engineering laboratory of the General Electric Company. The purpose of the investigation was (1) to obtain a large amount of statistical data concerning lightning to normal terrain and (2) to compare these data to the lightning data obtained from the Empire State Building study. The investigation was carried on for seven consecutive years. The main results of the study are reported by Hagenguth (1947).

To determine statistics concerning the number of luminosity peaks (apparently both return strokes and M components) per flash, 619 lightning photographs were analyzed. It was found that 40 percent of the flashes had two or more luminosity peaks and that 10 percent had six or more peaks. Hagenguth (1947) states that if the data are adjusted for flash visibility during the storm and distance from the flash to the observatory, 50 percent of the flashes will have three or more luminosity peaks. The maximum number of peaks recorded in a flash was 26.

The maximum flash duration was found to be 0.8 sec. The photographic technique necessarily provides an underestimation of the actual value. The duration of a flash was found to be largely independent of the number of luminosity peaks.

Fifty percent of the time intervals between luminosity peaks were 15 msec or greater. A maximum of 0.5 sec was recorded. A large number of luminosity peaks occurred at time intervals of 1 msec or less. These were probably M components (Malan, 1956; Workman et al., 1960). Thirty percent of the luminosity peaks had a persistence of at least 100 μsec, five percent of at least 10 msec.

Hagenguth (1947) reports that the change in luminous intensity of the channel when the first return stroke reaches a branch, the branch component, is not nearly so pronounced in the Pittsfield photographs as it is in the photographs of Malan and Collens (1937). The differences may be due to the types of films used and the exposures obtained.

A number of still photographs showed apparent upward branching, mostly near the cloud. Similar observations had been made in previous studies (e.g., Halliday, 1933; Schonland et al., 1938b). In those cases for which time-resolved photographs were available, no true upward branching was found. The apparent branching resulted from a change in position within the cloud of the downward leader from one stroke to the next or from the union of a cloud discharge with a cloud-to-ground stroke. Sourdillon (1952) has photographed the latter effect in his study of the relation between air discharges and cloud-to-ground discharges.

Some cloud-to-cloud discharges* were photographed. In most cases these were of the continuing type although in one case a distinct number of weak luminosity peaks were noted. A few cloud-to-cloud leaders were observed with estimated velocity of propagation of the order of the cloud-to-ground stepped leaders. The velocity estimate should not be considered too reliable because of the uncertainty in stroke distance.

A relation of about 1 to 10 was found to exist between the stepped-leader average velocity and the dart-leader average velocity. Of 45 dart-leader velocities measured, 5 percent exceeded 9×10^6 m/sec, 50 percent exceeded 3×10^6 m/sec, and 80 percent exceeded 1.5×10^6 m/sec. For nine stepped-leader average velocities determined, the median value was about 3×10^5 m/sec. For four return-stroke velocities measured, the average was 6.4×10^7 m/sec. All of this quantitative data is in reasonably good agreement with the data from South Africa.

2.5 RECENT STUDIES

2.5.1 Introduction

Significant photographic studies of lightning have been carried out recently in Arizona, in New Mexico, and in Switzerland. The lightning investigations at these sites are still in progress as this book is being written. Near Tucson, Arizona, and near Socorro, New Mexico, photographic, electrical, and spectroscopic studies of lightning have been made. Lightning strikes to two towers near Lugano, Switzerland, have been photographed and their currents measured. We consider now the significant results of the photographic portions of the investigations carried out of these three sites.

2.5.2 Lightning photography in Arizona

Evans and Walker (1963) photographed lightning from a distance of about 110 m. They used a high-speed Beckman and Whitley Dynafax framing camera, capable of taking 224 frames at a frame rate of 26,000

---

* It is not clear whether these were all cloud-to-cloud flashes as was stated by Hagenguth or whether some or all were intracloud flashes as is likely from the discussion of their properties (see Sec. 3.8).

frames per second, to photograph lightning striking a television transmission tower atop Mount Bigelow near Tucson. Exposure times of 2 μsec were used. The time interval between frames was 77 μsec. The area of field at the top of the tower as viewed by the camera was about 8 $\times$ 11 m. Analysis of the photographs has yielded determinations of the lightning-channel radius, luminosity vs. time curves for the channel, and data concerning changes in the geometry of the channel within individual return strokes and from stroke to stroke within a single flash. Since the camera was continuous running and the time spacing between frames was 77 μsec, the time of the occurrence of the first frame relative to the initiation of the stroke could not be determined (within 77 μsec). It is not known whether the strokes photographed were initiated by upward-moving leaders or by downward-moving leaders, or by both.

Photographs of three flashes are discussed. One flash contained four strokes with time durations of luminosity of 2.3, 1.1, 1.0, and 0.5 msec, respectively. The second flash contained six strokes with shorter durations of luminosity, 0.2 to 0.7 msec. The third flash contained only one stroke whose luminosity lasted about 0.6 msec. For most strokes recorded, the first frame was the brightest and the intensity fell off monotonically thereafter. In one stroke two intensity peaks were measured. Channels appeared to have thick regions and thin regions, especially in the brighter frames. In the less bright frames, the channel showed a great deal of small-scale (order of centimeters) twisting and bending, "small-scale embroidery." The shape of the channel was found to remain essentially the same during a stroke, but to change somewhat from one stroke to the next within a flash. That the channel shape changes from stroke to stroke was also noted by Schonland (1937).

Lightning channel radii between 1.5 and 6 cm were measured. These values should be compared with the radii of from 7.5 to 11.5 cm measured photographically by Schonland (1937) (Sec. 2.3.4). Uman (1964 and unpublished) has measured the holes melted in Fiberglas screens when lightning passed through those screens. Hole radii from a few millimeters to a few centimeters have been usual. This technique for measuring channel radii may yield smaller values than the photographic technique because the outer regions of the channel, although luminous, may not be sufficiently hot for a sufficiently long time to melt the Fiberglas. Hill (1963), on the basis of the observation of the effects of lightning on metal electrodes, concludes that the channel radius is of the order of millimeters. Any deduction of the channel radius from an observed electrode phenomenon must be considered unreliable. The electrical conductivity of the mixture of metal vapor and air near an electrode will be higher than the conductivity of the air away from the electrode, and thus a smaller channel radius is required near an electrode to carry a given current. In addition, electrode phenomena in high-current arcs are complex and not well understood. Additional discussion regarding the lightning-channel radius is given in Sec. 7.6.

2.5.3 Lightning photography in New Mexico

Workman et al. (1960) photographed a lightning flash which consisted of 54 luminosity peaks of which 26 were leader–return-stroke combinations. The flash had a time duration of 2 sec. In the typical multiple-stroke flash the stroke luminosity decreases with stroke number, the first stroke being the brightest. In the flash observed such regularity was not apparent. However the 26 strokes could be divided into groups, each of which exhibited the typical behavior. Strokes 6 to 14, 15 to 19, 20 to 22, and 23 to 25 each appeared to constitute a group in which the luminosity decreased with stroke number. The duration of each of these groups was found to be of the order of a typical cloud-to-ground flash. Other lightning flashes photographed in the same storm were also characterized by a relatively large number of strokes, several exceeding 20. Workman et al. (1960) conclude that they observed unusual thundercloud conditions in which the succeeding strokes had access to four to six times the volume of charge normally available in a thundercloud. It is postulated that this charge resided in four to six separate cells, each possessing a charge structure somewhat as shown in Fig. 1.2, located side by side. Presumably all cells were ripe for electrical activity at the same time, and there was considerable horizontal movement of charge during the flash.

The analysis of photographic records of 193 lightning flashes is reported by Kitagawa et al. (1962) and Brook et al. (1962). Electrical measurements were also reported and will be discussed in the next chapter. It was found that 50 percent of the multiple-stroke flashes (constituting about 90 percent of cloud-to-ground flashes) contained at least one stroke which was followed by a very long continuing luminosity lasting from 40 to 500 msec, with an average of 180 msec. These flashes are termed hybrid flashes to distinguish them from flashes with only discrete strokes. The continuing luminosity was found to contain M components (Sec. 2.3.6). Evidence is presented to show that when a stroke follows a previous stroke channel after a time of from 100 to 500 msec the major portion of this time interval is taken by the continuing luminosity of the preceding stroke. The nonluminous portion of the time interval is comparable to the maximum time interval between strokes without continuing luminosity.

For multiple-stroke flashes with no long-continuing (greater than 40 msec) luminosity, longer interstroke time intervals were found to occur after short-continuing (less than 40 msec) luminosity involving M components than after true discrete strokes. Kitagawa et al. (1962) suggest that after the current in the channel diminishes to a value below which no luminosity is recorded, the conductivity of the channel decays more slowly following a short-continuing stroke than following a discrete stroke. After the channel loses its luminosity, a subsequent stroke can follow the same channel if dart-leader initiation occurs within about 100 msec. For times longer than 100 msec a subsequent stroke takes a different channel with a new stepped leader.

As mentioned, single-stroke flashes were found to occur infrequently. Of the single-stroke flashes only about 1 percent had continuing luminosity. Additional and conflicting data regarding continuing current flow following single-stroke flashes are given in Sec. 3.7.5. No first strokes in multiple-stroke flashes were found to have continuing luminosity. It was found that after the occurrence of the return stroke in a single-stroke flash, there usually followed a period during which several leaders tried unsuccessfully to reach the ground; they terminated in the air beneath the cloud.

Kitagawa et al. (1962) summarize the results of their photographic study in seven tables: classification of 193 flashes to ground, analysis of 11 discrete single flashes, analysis of 36 discrete multiple flashes, analysis of 36 hybrid flashes, comparison of single flashes from three different storms, comparison of discrete multiple flashes in three different storms, and comparison of hybrid multiple flashes in three different storms. The reader is referred to these tables for a great deal of detailed information.

Brook and Kitagawa (as reported by Winn, 1965) have measured dart-leader velocity as a function of interstroke time interval (see Fig. 2.10). Slower dart leaders are associated with longer time intervals. As discussed (see Table 2.5), Schonland et al. (1935) first related dart-leader velocity to interstroke time interval.

Brook and Vonnegut (1960) report visual observation of the J-streamer process, the postulated process by which electric discharges move upward during the interstroke interval of a multiple-stroke discharge, connecting the channel to ground with new sources of charge:

> The uppermost region from which the return stroke originated was observed to move (in steps reminiscent of darts) upward and outward, illuminating new regions of cloud, before a new section was added to the channel of the previous return stroke and a new stroke occurred. Some discharges to ground appeared to originate from a vertical column, but by far the greater number were seen to progress horizontally or inclined to about 30° to the horizontal. . . . The manner of horizontal spread of the junction streamer is also of interest. It appeared sometimes to progress horizontally in divergent direc-

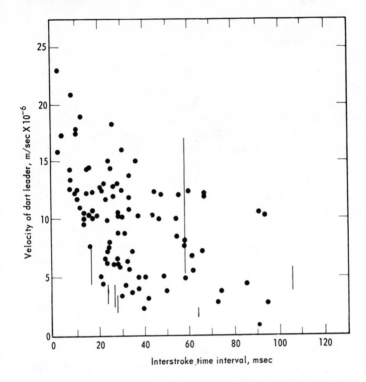

Fig. 2.10 *Data of M. Brook and N. Kitagawa showing how dart*
leaders tend to travel more slowly down older channels.
Vertical lines indicate velocity range of those dart leaders
whose velocities were not constant. Shortest interstroke
time is 3 msec. (Adapted from Winn (1965).)

tions, and, although the in-cloud paths were distinct and separate, and often
quite long, the return strokes were seen to follow the same below-cloud path to
ground. This observation is important because it shows that lightning
strokes may often bridge individual convective cells when many thunderstorms
are simultaneously active along a line.

This conclusion is supported by the measurements of Workman et al.
(1960), previously discussed. Similar visual observations were reported
by Schonland et al. (1935). A photograph showing this phenomenon is
reproduced in the frontispiece.

2.5.4 Lightning photography in Switzerland

Berger and Vogelsanger (1966) (see also the review paper by Berger,
1967) report on photographs taken of lightning discharges to towers on

two mountains near Lugano, Switzerland, and on photographs of discharges to earth in the area surrounding the towers. The summit of Monte San Salvatore is 640 m above the level of Lake Lugano and 915 m above sea level. The tower on Monte San Salvatore is about 55 m higher than the summit. The top of the tower on Monte San Carlo is about 650 m above the lake. The two towers are separated by 365 m horizontal distance. Still cameras and streak cameras (maximum time resolution 5 μsec) were located so that strikes to the towers could be photographed from distances of about 350 m and 3.2 km. In addition, the lightning current to the two towers was recorded.

The following new and significant results were reported. (1) Four different types of stepped leaders were identified: downward-moving negatively charged (the usual stepped leader), downward-moving positively charged, upward-moving negatively charged, and upward-moving positively charged. Distinct differences were found between positively charged and negatively charged leaders. (2) Evidence was presented to show that upward-moving leaders carrying negative charge occasionally connect with downward-moving leaders, the junction points being up to 2 km above the point of initiation of the upward-moving leader. (3) A weakly luminous corona discharge was photographed in advance of the bright tips of two negatively charged, upward-moving stepped leaders. We will now discuss these and other results.

We have already seen (Sec. 2.4.2) that the Empire State Building is the source of upward-moving stepped leaders. The sign of the charge carried by these stepped leaders was not discussed in the papers of McEachron (1939), McEachron (1944), and Hagenguth and Anderson (1952). Upward-moving stepped leaders initiate about 75 percent of the flashes to the measuring towers on Monte San Salvatore and Monte San Carlo. Strokes resulting from upward-moving stepped leaders are easily identified because they are branched upward, just as strokes from downward-moving stepped leaders are branched downward. Sixteen upward strokes have been photographed in the area surrounding the measuring towers (excluding discharges to the towers), thirteen to exposed points such as buildings, towers, or poles on other mountains, two to forested mountain tops, and one to a church tower located a mere 200 m above the level of Lake Lugano. Photographs of an upward-moving stepped leader carrying negative charge and the resultant flash are shown in Fig. 2.11. In general, it appears that upward strokes occur mainly near the end phase of a thunderstorm and that they are coincident in time with long horizontal strokes between clouds. Upward strokes thus appear dependent on cloud-to-cloud discharges for the high local electric fields at the earth's surface needed for their initiation. Strongly branched upward strokes were found from tower measurements of current to carry large charge magnitudes, of the order of 100 coul.

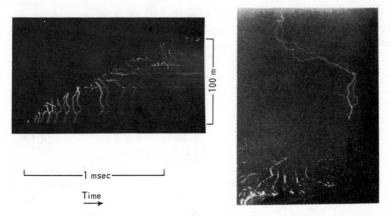

Fig. 2.11 *An upward-moving stepped leader carrying negative charge and the resultant flash. Leader is from tower on Monte San Salvatore, near Lugano, Switz. (Photograph from Berger and Vogelsanger (1966); courtesy of K. Berger.)*

The primary visual differences between first leaders carrying negative charge and first leaders carrying positive charge are summarized below, the charge sign being determined by current measurements. Negative stepped leaders (stepped leaders carrying negative charge) whether upward moving or downward moving exhibit regular, distinct steps throughout the length of the leader channel. Positive stepped leaders (stepped leaders carrying positive charge) exhibit steps only over a small portion of the channel length. The remainder of the channel length is occupied by an apparently continuous leader which exhibits periodic or irregular intensity variations. The leader carrying positive charge emits considerably less light than the negative leader. Leaders could be identified in only 7 of the 46 photographed upward, positive strokes from the measuring towers. In these 7 cases, the positive leader could not be seen until the channel was at least 40 m long (at which time its current had reached several hundred amperes). It then exhibited upward steps for 60 to 150 m vertical distance. Above about 200 m of the tower top, the positive leader exhibited the continuous form described above. Photographs of an upward-moving stepped leader carrying positive charge and the resultant flash are shown in Fig. 2.12. The upward-moving stepped leader from the Empire State Building, whose photograph is reproduced by Hagenguth and Anderson (1952), exhibits the typical properties of a upward-moving stepped leader carrying positive charge.

Some of the data accumulated by Berger and Vogelsanger (1966) from analysis of lightning photographs are given in Table 2.8. The downward-moving, negatively charged stepped leaders to the towers (row 1) resulted in strokes whose peak currents were between 16 and 53 ka. The downward-moving, negatively charged stepped leaders to earth (row 2) exhibited high average velocity and long step length near the cloud. The step length and average velocity decreased as the leader moved earthward. Berger and Vogelsanger (1966) observed short vertical discharges from a given tower occurring simultaneously with upward-moving negative leaders and subsequent high current flow from the other tower. The short discharges (row 4) ranged in total length between 10 and 55 m. It is reasonable to assume that discharges from both towers carried the same sign of charge. Currents in the short discharges were less than 20 amp, the measuring threshold. Only one downward-moving, positive stepped leader (row 7) was observed. Since this stroke did not hit a measuring tower, no measurement could be made of its charge, and hence its identification is solely due to the similarity of its luminous properties with those of upward-moving stepped leaders

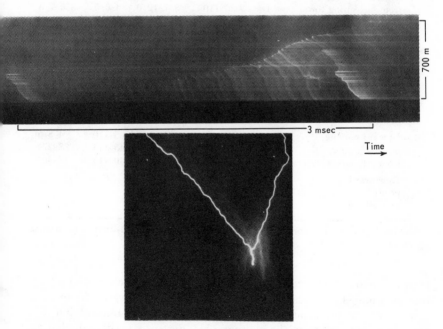

Fig. 2.12   *An upward-moving stepped leader carrying positive charge and the result-ant flash. Leader is from tower on Monte San Salvatore, near Lugano, Switzerland. (Photograph from Berger and Vogelsanger (1966); courtesy of K. Berger.)*

TABLE 2.8

Stepped-leader properties as determined by photography. Unless otherwise noted, pause time, step length, and leader average velocity represent averages over about 5 to 30 steps. Recorded lengths are two dimensional. Adapted from Berger and Vogelsanger (1966).

Leader direction, charge, and description	No. of strokes	Observed height above tower or earth, m	Leader average velocity, 10^5 m/sec	Pause time between steps, μsec	Step length, m
1. Downward moving, negatively charged, to measuring tower	4	0–100	1.8–2.2	40–52	8–10
2. Downward moving, negatively charged, to earth	14	0–1,300	0.85–4.4	29–47	3–17
	1	0–1,750	0.65	} 41	} 29
		1,750–2,000	} 7.0	} 47	} 50
		2,000–2,350	11.0		
3. Upward moving, negatively charged, from measuring tower	8	0–110	1.2–1.9	33–50	4.5–8
	3	250–1,200	1.1–4.5	40–47	5–18
	1	20–110	8.7–12.0	4–6.5	3.5–7.5
4. Upward moving, negatively charged, short discharges from measuring tower	6	0–55	0.85–1.4	34–47	3–6
5. Upward moving, negatively charged, from Mt. Sighignola (4 steps)	1	540–900	22	55	120
6. Upward moving, positively charged, from measuring tower	4	40–110	0.40–0.75	65–110	4–8
	7	110–500	1.3–4.9	45–115*	8–27*
	7	500–1,150	1.1–9.7	40–115*	12–40*
7. Downward moving, positively charged, to earth	1	320–920	} 24		
		920–1,660	17		
		1,660–1,870	3.6		

* Above 150 m, values represent displacement and time of intensity maxima during continuous phase of the stepped leaders.

carrying positive charge. The resultant stroke was characterized by intense thunder. A photograph of a positive, downward stroke to a tower with simultaneous current measurement was obtained but the leader was not visible on the photograph. Forty-six positive discharges (current measurements) have been reported by Berger and Vogelsanger (1965) in a study which did not include photographic measurements. Of these 46, 18 were apparently due to downward-moving, positive leaders as determined from the measurement of current as a function of time.

In general, step lengths of leaders photographed at heights up to a few hundred meters above the measuring towers were of the order of 10 m, in agreement with the step lengths found in the Empire State Building study (Sec. 2.4.2). Longer step lengths, in better agreement with the South African data, are probably more common at greater heights, perhaps above the height at which the grounded structure introduces significant distortion into the electric field pattern due to the cloud.

Multiple-stroke flashes were found to be frequent following both downward-moving and upward-moving stepped leaders for discharges between negative cloud charge and positive earth (downward-moving, negative leaders; upward-moving, positive leaders). Multiple-stroke flashes between positive cloud charge and negative earth were rare. Normal dart leaders moving from cloud to ground preceding strokes subsequent to the first were observed. For 80 dart leaders, an average dart leader velocity of 9×10^6 m/sec was found. Dart-stepped leaders, similar to those observed in South Africa, were also photographed.

Evidence is presented to show that an upward-growing channel from a negative tower may sometimes connect with a downward-growing channel from the cloud. In the five of these cases recorded, in 3 to 14 msec after the initiation of the upward negative leader impulse currents of 22 to 106 ka flowed to the tower. Unfortunately, the junction points were above the field of view of the cameras or were, for other reasons, not recorded photographically. The upward-growing channels or connecting discharges were presumed to be between 500 and 1,800 m in length. Current rise times in strokes bringing positive charge to earth are lower than in strokes bringing negative charge to earth (Berger and Vogelsanger, 1965). Data are presented to show that the slower rise time in positive strokes is because of the effect of the length of the connecting discharge and that among positive strokes the ones with the faster current rise times have the shorter connecting discharges.

For strokes bringing negative charge to earth the connecting discharges, if they exist, are much shorter in length. From the photographs of tower impacts one sees that the tower and the top of the last leader step are connected by the main discharge when the leader tip is 20 to 70 m above the tower top. An upward-moving discharge from the tower would carry positive charge. This type of discharge is very weakly luminous near the tower and hence would be difficult to photograph. Evidence for the existence of connecting discharges to downward negative strokes is found in photographs obtained by Berger and coworkers of short channels extending upward which occur when there is lightning to ground nearby. Further photographic evidence for the existence of positively charged connecting discharges is presented in Sec. 7.6. Evidence is also provided (Berger and Vogelsanger, 1965) by the current oscillograms which indicate that the current due to a downward negative stroke may reach a few hundred amperes in a few tenths of milliseconds before it is transformed into the usual microsecond rise time lightning wavefront. The time scale is approximately right for the propagation of an upward-moving leader for about 40 m with a velocity of about 10^5 m/sec.

Berger and Vogelsanger (1966) give photographs of two upward-moving, negatively charged stepped leaders that show a faint corona discharge in advance of the bright leader tip. The corona discharge extends upward for about one step length. It does not develop continuously between two steps but rather is impressed on the film in a time less than 5 μsec, the time resolution of the camera. The emission of a new corona discharge appears to be simultaneous with the conversion of the previous corona discharge into the luminous leader head. The observation of corona in advance of the leader tip is significant in that it represents information regarding the electric field in advance of the leader, information important in developing an adequate theory of the stepped leader.

REFERENCES

1. Berger, K.: Novel Observations on Lightning Discharges: Results of Research on Mount San Salvatore, *J. Franklin Inst.*, **283**:478–525 (1967).

2. Berger, K., and E. Vogelsanger: Messungen und Resultate der Blitzforschung der Jahre 1955–1963 auf dem Monte San Salvatore, *Bull. SEV*, **56**:2–22 (1965).

3. Berger, K., and E. Vogelsanger: Photographische Blitzuntersuchungen der Jahre 1955–1965 auf dem Monte San Salvatore, *Bull. SEV.*, **57**:1–22 (1966).

4. Boys, C. V.: Progressive Lightning, *Nature*, **118**:749–750 (1926).

5. Boys, C. V.: Progressive Lightning, *Nature*, **122**:310–311 (1928).

6. Boys, C. V.: Progressive Lightning, *Nature*, **124**:54–55 (1929).

7. Brook, M., N. Kitagawa, and E. J. Workman: Quantitative Study of Strokes and Continuing Currents in Lightning Discharges to Ground, *J. Geophys. Res.*, **67**:649–659 (1962).

8. Brook, M., and B. Vonnegut: Visual Confirmation of the Junction Process in Lightning Discharges, *J. Geophys. Res.*, **65**:1302–1303 (1960).

9. Clarence, N. D., and D. J. Malan: Preliminary Discharge Processes in Lightning Flashes to Ground, *Quart. J. Roy. Meteorol. Soc.*, **83**:161–172 (1957).

10. Evans, W. H., and R. L. Walker: High Speed Photographs of Lightning at Close Range, *J. Geophys. Res.*, **68**:4455–4461 (1963).

11. Flowers, J. W.: Lightning . . . Measuring Lightning Currents Photographically. Equipment, Procedure, and Examples of Results, *Gen. Elec. Rev.*, **47**(4):9–15 (1944).

12. Hagenguth, J. H.: Lightning Recording Instruments, *Gen. Elec. Rev.*, **43**(5,6): 195–201, 248–255 (1940).

13. Hagenguth, J. H.: Photographic Studies of Lightning, *Trans. AIEE*, **66**:577–585 (1947).

14. Hagenguth, J. H., and J. G. Anderson: Lightning to the Empire State Building, pt. 3, *Trans. AIEE*, **71**(*pt. 3*):641–649 (1952).

15. Halliday, E. C.: On the Propagation of a Lightning Discharge through the Atmosphere, *Phil. Mag.*, **15**:409–420 (1933).

16. Hill, R. D.: Determination of Charges Conducted in Lightning Strokes, *J. Geophys. Res.*, **68**:1365–1375 (1963).

17. Hoffert, H. H.: Intermittent Lightning Flashes, *Phil. Mag.*, **28**:106–109 (1889); also in *Proc. Phys. Soc.*, **10**:176–180 (1888–1890).

18. Jensen, J. C.: The Relation of Lightning Discharges to Changes in the Electrical Field of Thunderstorms, *Trans. Am. Geophys. Union*, **13**:190–191 (1932).

19. Jensen, J. C.: The Branching of Lightning and the Polarity of Thunderclouds, *J. Franklin Inst.*, **216**:707–747 (1933).

20. Kayser, H.: Über Blitzphotographien, *Ber. Königliche Akad. Berlin*, 611–615 (1884).

21. Kettler, C. J.: Cameras Designed for Lightning Studies, *Photo Technique*, 38–43, (May, 1940).

22. Kitagawa, N., M. Brook, and E. J. Workman: Continuing Currents in Cloud-to-Ground Lightning Discharges, *J. Geophys. Res.*, **67**:637–647 (1962).

23. Larsen, A.: Photographing Lightning with a Moving Camera, *Ann. Rept. Smithsonian Inst.*, 119–127 (1905).

24. Lloyd, W. L., Jr., and W. A. McMorris: Strikes Twice and Even Ten Times, *Gen. Elec. Rev.*, **37**:349–350 (1934).

25. Malan, D. J.: Apparel de grand rendement pour la chronophotographie des éclairs, *Rev. Opt.*, **29**:513–523 (1950).

26. Malan, D. J.: Les décharges lumineuses dans les nuages orageux, *Ann. Geophys.*, **11**:427–434 (1955).

27. Malan, D. J.: The Relation between the Number of Strokes, Stroke Intervals, and the Total Durations of Lightning Discharges, *Geofis. Pura Appl.*, **34**:224–230 (1956).

28. Malan, D. J.: The Theory of Lightning Photography and a Camera of New Design, *Geofis. Pura Appl.*, **38**:250–260 (1957).

29. Malan, D. J.: "Physics of Lightning," The English Universities Press Ltd., London, 1963.

30. Malan, D. J., and H. Collens: Progressive Lightning, pt. 3, The Fine Structure of Lightning Return Strokes, *Proc. Roy. Soc. (London)*, **A162**:175–203 (1937).

31. Malan, D. J., and B. F. J. Schonland: Progressive Lightning, pt. 7, Directly Correlated Photographic and Electrical Studies of Lightning from Near Thunderstorms, *Proc. Roy. Soc. (London)*, **A191**:513–523 (1947).

32. Malan, D. J., and B. F. J. Schonland: The Electrical Processes between the Strokes of a Lightning Discharge, *Proc. Roy. Soc. (London)*, **A206**:145–163 (1951a).

33. Malan, D. J., and B. F. J. Schonland: The Distribution of Electricity in Thunderclouds, *Proc. Roy. Soc. (London)*, **A209**:158–177 (1951b).

34. McEachron, K. B.: New England Checks African Studies, *Elec. World*, 15–16, (July 7, 1934).

35. McEachron, K. B.: Lightning to the Empire State Building, *J. Franklin Inst.*, **227**:149–217 (1939).

36. McEachron, K. B.: Lightning to the Enpire State Building, *Trans. AIEE*, **60**:885–889 (1941).

37. Ogawa, T., and M. Brook: The Mechanism of Intracloud Lightning Discharge, *J. Geophys. Res.*, **69**:5141–5150 (1964).

38. Schonland, B. F. J.: The Diameter of the Lightning Channel, *Phil. Mag.*, **23**: 503–508 (1937).

39. Schonland, B. F. J.: Progressive Lightning, pt. 4, The Discharge Mechanism, *Proc. Roy. Soc. (London)*, **A164**:132–150 (1938).

40. Schonland, B. F. J.: The Pilot Streamer in Lightning and the Long Spark, *Proc. Roy. Soc. (London)*, **A220**:25–38 (1953).

41. Schonland, B. F. J.: The Lightning Discharge, "Handbuch der Physik," vol. 22, pp. 576–628, Springer-Verlag OHG, Berlin, 1956.

42. Schonland, B. F. J., and H. Collens: Development of the Lightning Discharge, *Nature*, **132**:407–408 (1933).

43. Schonland, B. F. J., and H. Collens: Progressive Lightning, *Proc. Roy. Soc. (London)*, **A143**:654–674 (1934).

44. Schonland, B. F. J., D. B. Hodges, and H. Collens: Progressive Lightning, pt. 5, A Comparison of Photographic and Electrical Studies of the Discharge Process, *Proc. Roy. Soc. (London)*, **A166**:56–75 (1938a).

45. Schonland, B. F. J., D. J. Malan, and H. Collens: Progressive Lightning, pt. 2, *Proc. Roy. Soc. (London)*. **A152**:595–625 (1935).

46. Schonland, B. F. J., D. J. Malan, and H. Collens: Progressive Lightning, pt. 6, *Proc. Roy. Soc. (London)*, **A168**:455–469 (1938b).

47. Sourdillon, M.: Étude a la chambre de Boys de "l'éclair dans l'air" et du "coup de foudre a cime horizontale," *Ann. Geophys.*, **8**:349–354 (1952).

48. Uman, M. A.: The Diameter of Lightning, *J. Geophys. Res.*, **69**:583–585 (1964).

49. Walter, B.: Ein photographischer Apparat zur genaueren Analyse des Blitzes, *Physik. Z.*, **3**:168–172 (1902): Über die Entstehungsweise des Blitzes, *Ann. Physik*, **10**:393–407 (1903); Über Doppelaufnahmen von Blitzen . . . , *Jahrbuch Hamb. Wiss. Anst.*, **27**(beihefte 5):81–118 (1910); Stereoskopische Blitzaufnahmen, *Physik. Z.*, **13**:1082–1084 (1912); Über die Ermittelung der zeitlichen Aufeinanderfolge zusammengehöriger Blitze sowie über ein bemerkenswertes Beispiel dieser Art von Entladungen, *Physik. Z.*, **19**:273–279 (1918).

50. Weber, L.: Über Blitzphotographieen, *Ber. Königliche Akad. Berlin*, 781–784 (1889).

51. Winn, W. P.: A Laboratory Analog to the Dart Leader and Return Stroke of Lightning, *J. Geophys. Res.*, **70**:3265–3270 (1965).

52. Workman, E. J., J. W. Beams, and L. B. Snoddy: Photographic Study of Lightning, *Physics*, **7**:345–379 (1936).

53. Workman, E. J., M. Brook, and N. Kitagawa: Lightning and Charge Storage, *J. Geophys. Res.*, **65**:1513–1517 (1960).

54. Wormell, T. W.: Lightning, *Quart. J. Roy. Meteorol. Soc.*, **79**:474–489 (1953).

3

Electric and Magnetic
Field Measurements

3.1 INTRODUCTION

The photographic method of studying the lightning discharge by means of the Boys camera has the unique advantage of giving direct information concerning events in the discharge in two dimensions of space and one of time and could be extended if necessary to include the third space dimension. The luminous events which it records are, however, secondary processes, and the primary movements of electrical charge which cause them can only be inferred by an application of ideas gained from the laboratory study of the passage of electricity through gases.

The direct study of these primary electrical processes involves the observation of the electric field during the discharge by means of a cathode-ray oscillograph. . . . The method gives information concerning the total electric moment of the cloud charges and requires to be compared with the photographic data before its results can be interpreted in terms of the charges themselves and their movements.

B. F. J. Schonland, D. B. Hodges, and H. Collens: Progressive Lightning, pt. 5, A Comparison of Photographic and Electrical Studies of the Discharge Process, *Proc. Roy. Soc. (London)*, **A166**:56–75 (1938).

In this chapter we shall first examine the theory and the experimental techniques used to infer lightning properties from measurement of the electric and magnetic fields associated with lightning. We shall then examine the results of electric and magnetic field studies and of correlated electric and magnetic field studies and optical studies of the lightning discharge.

3.2 ELECTROSTATICS

The electric field intensity **E** at a distance r from a positive point charge in air or vacuum is

$$\mathbf{E} = \frac{Q\mathbf{a}_r}{4\pi\epsilon_0 r^2} \qquad \text{volts/m} \tag{3.1}$$

where $\mathbf{a}_r$ is a unit vector directed along r in the direction away from the charge and ϵ_0 is the permittivity of vacuum (which is essentially equal to the permittivity of atmospheric air). In mks units $(4\pi\epsilon_0)^{-1} \cong 9 \times 10^9$. If the source of the field were a negative point charge, the field would be directed toward the charge and Eq. (3.1) would have a minus sign in front of its right-hand side. Equation (3.1) also describes the electric field intensity outside of a spherically symmetric charge distribution of total charge Q (Sec. 7.3.1).

It has been found by measurement that the fine-weather electric field vector above the earth is directed toward the center of the earth. That is, the earth is negatively charged, the atmosphere above the earth is positively charged. The fine-weather electric field intensity at the ground is of the order of 100 volts/meter. In most of the literature the fine-weather electric field is termed a positive electric field. We will therefore adopt the following convention: An electric field at the ground is defined as positive if it is due to positive charge above ground level, that is, if the vector field is directed toward the earth; an electric field at the ground is defined as negative if it is due to negative charge above ground level, that is, if the vector field is directed away from the earth. This terminology will be used to discuss the electric fields due to thunderstorm charges.

The charge configuration we consider now will serve as the basis for the calculation of the electric field at the ground due to thunderstorm charges above the earth. In the model to be used the earth will be treated as a flat conducting plane and the thunderstorm charge centers as point charges or as spherically symmetric charge distributions. We wish now to calculate the electric field intensity due to a positive point charge $+Q$ located a distance H above a conducting plane. Such a configuration is shown in Fig. 3.1. By the method of electrical images, the effect of the charges induced on the conducting plane can be reproduced by eliminating the plane and replacing it with a negative image charge $-Q$ located a distance H below the plane. The magnitude of the electric field intensity

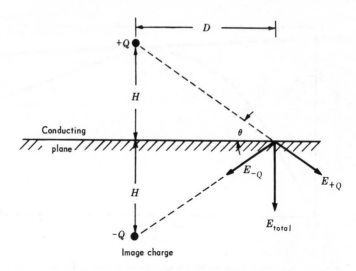

Fig. 3.1 *Diagram for the calculation of the electric field intensity at D due to a positive point charge +Q at height H above a conducting plane.*

at the level of the plane and a distance D down the plane due to *each* charge is, by Eq. (3.1),

$$E = \frac{Q}{4\pi\epsilon_0(H^2 + D^2)} \tag{3.2}$$

The electric field vector resulting from each charge is however pointing in a different direction. The total electric field is obtained by vector addition. Both electric fields can be decomposed into components parallel to the plane and components perpendicular to the plane. The parallel components are of equal magnitude but of opposite direction and thus add to zero. Therefore there is no horizontal electric field at the plane. (The electric field at any conducting surface is always perpendicular to that surface.) The perpendicular components are both positive, in the sense discussed in the preceding paragraph, and thus can be added directly to obtain the total electric field intensity. The perpendicular field component due to either point charge is found by multiplying the total field due to that charge by $\sin \theta = H/(H^2 + D^2)^{1/2}$. Thus the total electric field magnitude is

$$E_{\text{total}} = \frac{2QH}{4\pi\epsilon_0(H^2 + D^2)^{3/2}} \tag{3.3}$$

with a direction perpendicular to the plane and positive.

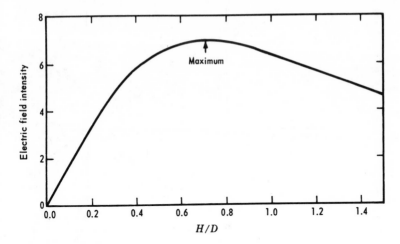

Fig. 3.2 *Electric field intensity for the charge configuration given in Fig. 3.1.
To obtain electric field in volts per meter multiply value given in
graph times $Q \times 10^9/D^2$.*

If the height H of the charge $+Q$ is varied, the field at D will pass
through a maximum as shown in Fig. 3.2. We can explain this field
variation from a physical point of view. If H is very small, the electric
field at D will have little vertical component (the vertical component is
the total field) because θ is small. As H increases, the vertical component
increases. The field will decrease for larger H because the distance from
the charge to the observation point increases. We can determine the
value of H/D for which the field is maximum by taking the derivative
of E_{total} with respect to H and setting that derivative equal to zero.
The result is $H/D = 1/\sqrt{2}$.

If D is much greater than H, Eq. (3.3) can be approximated as

$$E_{\text{total}} \cong \frac{2QH}{4\pi\epsilon_0 D^3} = \frac{M}{4\pi\epsilon_0 D^3} \tag{3.4}$$

where $M = 2QH$ is termed the electric dipole moment of the charge $+Q$
and its image. If electric field measurements are made far from a
thunderstorm, the measurement of the change in electric field due to the
destruction of thunderstorm charge (by, for example, its flowing to ground
in a lightning discharge), a knowledge of D, and the use of Eq. (3.4)
allow a calculation of the change in the dipole moment.

Using Eq. (3.1), we can calculate the electric field intensity at the
ground due to the three regions of charge, P, N, and p, within a model
thundercloud. The results of such a calculation, for several values of

p and with the values of P and N suggested by Malan (1952, 1963), are shown in Fig. 3.3. The analytical expression for the electric field intensity with the charges and charge heights given in Fig. 3.3 is

$$E = 1.8 \times 10^{10} \left[\frac{2 \times 10^3 p}{(4 \times 10^6 + D^2)^{3/2}} - \frac{2 \times 10^5}{(2.5 \times 10^7 + D^2)^{3/2}} + \frac{4 \times 10^5}{(10^8 + D^2)^{3/2}} \right] \quad \text{volts/m} \quad (3.5)$$

where p is the lower positive charge. Because of its proximity to the ground the lower positive charge significantly affects the electric field at small D. Note that in the cloud-charge model used the charged regions are treated as points or as spherically symmetric charge distributions and are assumed to be above one another. The cloud-charge model we have employed is useful for describing the relationship between the cloud and lightning phenomena. It should not necessarily be considered

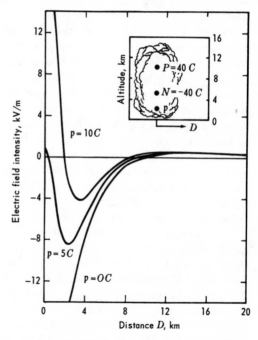

Fig. 3.3 *Electric field intensity at the ground vs. distance for $P = 40$ coul at 10-km height, $N = -40$ coul at 5-km height, and three values of p at 2-km height. See Fig. 1.2. (Adapted from Malan (1952, 1963).)*

an accurate representation of the charge distribution or the charge magnitudes within a real thundercloud. The electric field values for the model thundercloud found from Eq. (3.5) do not include the effects of space charge between the cloud and ground, around the cloud, or between the cloud and ionosphere. An actual measurement will give the resultant electric field from both the cloud charges and the space charge. In general, values of cloud charges are deduced from electric field measurements assuming that space charge effects are negligible. It is not known whether this is a valid assumption. Kasemir (1965) has reviewed the data available regarding the magnitude of the cloud charges. He suggests that cloud-charge models such as Malan's (1952, 1963) derived from electric field measurements are in error due to the neglect of space-charge shielding effects. In the cloud-charge model advanced by Kasemir (1965) continuity of current and an assumed form for the electric conductivity are used to calculate the electric field in and around the cloud. From the divergence of the electric field the charge density is obtained and hence the net charge in a region of the cloud is also obtained. Kasemir (1965) proposes a cloud model consisting of an upper positive charge of 60 coul, a lower negative charge of −340 coul, and a positive charge in the cloud base of 50 coul. The type of model for thunderstorm charges advanced by Kasemir, a model taking into account current flow, was first proposed by Holzer and Saxon (1952).

In general, the exact value of the steady-state cloud charges will not be of concern to us in our study of the lightning process. During a lightning flash charge is moved or destroyed and this *change* in the charge configuration reflects itself in a change in the measured electric field. The change in electric field, since it occurs more quickly than space charge or cloud charge not related to the flash can rearrange itself appreciably, is essentially independent of the magnitudes of the space charge or cloud charge. Some additional consideration of the steady-state electric fields of the thundercloud is given in Sec. 3.6.

We shall now examine a model for the leader process from which we can calculate the electric field at the ground due to the leader. We idealize the leader as a vertical line charge (or equivalently as a charge distribution with cylindrical symmetry). In Fig. 3.4 a positive line charge of length $y - x$ and charge per unit length ρ is shown oriented vertically above a conducting plane. We consider now the electric field at D due to the line charge and its electrical image. The electric field due to a small element of charge $\rho \, dz$ within the line charge is identical to the field from a small point charge. Thus the magnitude of the field due to the charge within dz is

$$dE_{+\rho \, dz} = \frac{\rho \, dz}{4\pi\epsilon_0(z^2 + D^2)} \tag{3.6}$$

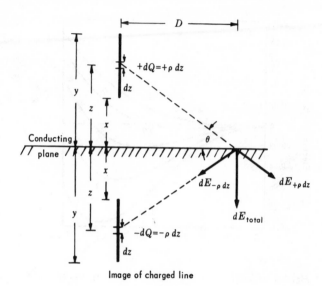

Image of charged line

Fig. 3.4 *Diagram for the calculation of the electric field intensity at D due to a vertical charged line of length y − x with positive charge per unit length ρ located above a conducting plane.*

The total field due to this charge element and its image is

$$dE_{\text{total}} = \frac{2\rho z \, dz}{4\pi\epsilon_0 (z^2 + D^2)^{3/2}} \tag{3.7}$$

where the field direction is perpendicular to the plane and positive. The field from all the charge in the line is found by integrating Eq. (3.7) from $z = x$ to $z = y$

$$E_{\text{total}} = \frac{2\rho}{4\pi\epsilon_0} \left[\frac{1}{(D^2 + x^2)^{1/2}} - \frac{1}{(D^2 + y^2)^{1/2}} \right] \tag{3.8}$$

where we have assumed ρ to be constant. If the line charge were negative, the field at the ground would be negative, and Eq. (3.8) would have a minus sign in front of its right-hand side.

If a positively charged leader emerges from a volume of positive charge, the charging of the leader as its length is extended results in a decrease of charge in the source volume. If the leader length is ℓ, the height of the source charge is H, and the charge source can be treated as a point charge or as a spherically symmetric distribution, the field *change* at the ground due to the decrease in the source charge is

$$\Delta E_S = - \frac{2\rho \ell H}{4\pi\epsilon_0 (H^2 + D^2)^{3/2}} \tag{3.9}$$

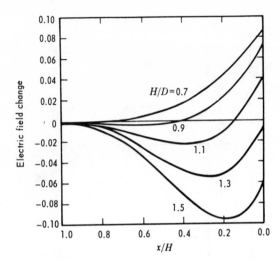

Fig. 3.5 *Electric field change for a negative leader moving downward from a negative charge center at height H. The factor x represents the height of the leader tip above the ground. Multiply field change given in graph times $\rho/2\pi\epsilon_0 D$ to obtain mks units of volts per meter.*

since $\rho\ell$ is the amount of charge lost by the source volume. The source field has become less positive since positive charge is removed from the source volume; hence the negative sign on the right-hand side of Eq. (3.9). If the source charge and the leader charge were negative, the field *change* due to the decrease of source charge would be positive.

We now consider the case of a negatively charged leader moving downward from a spherically symmetric, negatively charged volume. This situation approximates that existing in the dart leader and to some extent in the stepped leader. The upper end of the leader is at H, in the center of the source charge. Hence in Eq. (3.8) $y = H$. In the case under consideration $\ell = H - x$. The total field change due to the leader extension and the source charge decrease can be written

$$\Delta E = -\frac{2\rho}{4\pi\epsilon_0 D}\left[\frac{1}{(1 + x^2/D^2)^{\frac{1}{2}}} - \frac{1}{(1 + H^2/D^2)^{\frac{1}{2}}} - \frac{H - x}{D}\frac{H}{D}\frac{1}{(1 + H^2/D^2)^{\frac{3}{2}}}\right] \quad (3.10)$$

where $x = H$ at $t = 0$, the time of leader initiation. As time increases, x decreases until the time the leader touches the ground, $x = 0$. The field change calculated from Eq. (3.10) is given in Fig. 3.5 for several values of H/D. If the velocity v of the leader tip is constant, the variable x in Fig. 3.5 and in Eq. (3.10) can be replaced by $H - vt$; that is, $\ell = vt$. Measurements made close to the leader should yield initially a negative field change; measurements made at large distance should yield a positive field change. The field change when the leader touches the ground, $x = 0$, should be zero for $H/D = 1.27$, negative for larger values of H/D, and positive for smaller values of H/D. The field change measured as a function of time due to a close leader should exhibit a characteristic hook shape. Leaders for which $H/D \ll 1$ should be characterized by field changes which are parabolic as a function of time if the leader tip velocity is constant. To see that this is so, we expand Eq. (3.10) assuming that $x = H - \ell$ is small compared to D and that $\ell = vt$. If all terms of order greater than $(H/D)^2$ are ignored, we find

$$\Delta E \cong \frac{\rho \ell^2}{4\pi\epsilon_0 D^3} = \frac{\rho v^2 t^2}{4\pi\epsilon_0 D^3} \tag{3.11}$$

We consider now the electric field change associated with a downward-moving, positively charged leader. Positive leaders may possibly move downward from the P region toward the N region, initiating an intracloud discharge, or may possibly move from either the P region or the p region toward ground initiating a positive discharge to ground. The field change is that given in Eq. (3.10) but with opposite sign

$$\Delta E = + \frac{2\rho}{4\pi\epsilon_0 D} \left[\frac{1}{(1 + x^2/D^2)^{1/2}} - \frac{1}{(1 + H^2/D^2)^{1/2}} - \frac{H - x}{D} \frac{H}{D} \frac{1}{(1 + H^2/D^2)^{3/2}} \right] \tag{3.12}$$

Plots of Eq. (3.12) for various values of H/D are given in Fig. 3.6.

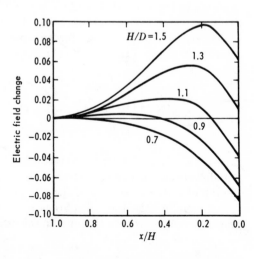

Fig. 3.6 *Electric field change for a positive leader moving downward from a positive charge center at height H. The factor x represents the height of the leader tip above the ground. Multiply field change given in graph times $\rho/2\pi\epsilon_0 D$ to obtain mks units of volts per meter.*

The field changes shown in Fig. 3.6 for the lowering of positive charge are of the type we would expect from an examination of Fig. 3.2. In Fig. 3.2, as H/D decreases from a value greater than about 0.7, as a point or spherically symmetric positive charge is lowered and this process observed at close range, the field increases, reaches a maximum, and then decreases. In Fig. 3.6, the lowering of positive charge observed at close range results in a field increase to maximum and then a subsequent decrease. In Fig. 3.2, as H/D decreases from a value less than about 0.7, as a point or spherically symmetric positive charge is lowered and this process observed at a considerable distance, the field monotonically decreases. In Fig. 3.6, the lowering of positive charge observed at a considerable distance results in a monotonic field decrease. We conclude, then, that although there are some quantitative differences between the electric field change due to the motion of a point charge and the electric field change due to the motion of a line charge out of a point-charge source, the qualitative features of the field changes are the same.

We consider now a negative leader moving upward from a negative-charge center. A possible example of this type of discharge might be a leader moving upward from the N toward the P region initiating an intra-cloud discharge. For this situation $x = H$ and $\ell = y - H$ in Eq. (3.8) and Eq. (3.9). The field change is given by

$$\Delta E = -\frac{2\rho}{4\pi\epsilon_0 D}\left[\frac{1}{(1 + H^2/D^2)^{1\!/\!2}} - \frac{1}{(1 + y^2/D^2)^{1\!/\!2}} - \frac{y - H}{D}\frac{H}{D}\frac{1}{(1 + H^2/D^2)^{3\!/\!2}}\right] \quad (3.13)$$

where $y = H$ at $t = 0$, and y increases with time. At close distances the initial field change is positive; far away the initial field change is negative, as shown in Fig. 3.7. The field changes shown in Fig. 3.7 can be qualitatively predicted from an examination of Fig. 3.2.

The expected field change for an upward-moving, positive leader is shown in Fig. 3.8. The analytical expression for the electric field change is

$$\Delta E = +\frac{2\rho}{4\pi\epsilon_0 D}\left[\frac{1}{(1 + H^2/D^2)^{1\!/\!2}} - \frac{1}{(1 + y^2/D^2)^{1\!/\!2}} - \frac{y - H}{D}\frac{H}{D}\frac{1}{(1 + H^2/D^2)^{3\!/\!2}}\right] \quad (3.14)$$

A positive, upward leader might possibly precede a discharge between the p and N regions, or between the P region and the ionosphere, or between the top of a return stroke and the N region.

It is useful to form the ratio of the total field change due to the cloud-to-ground leader process to that due to the return stroke. We will consider

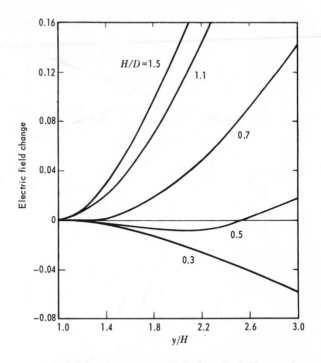

Fig. 3.7 *Electric field change for a negative leader moving upward from a negative charge center at height H. The length of the leader above the charge center is y − H. Electric field changes which are negative for small y/H become positive for large y/H. Multiply field change given in graph times ρ/2πε₀D to obtain mks units of volts per meter.*

the downward-moving, negatively charged (and, according to our model, uniformly charged) leader whose total field change is given by Eq. (3.10) with $x = 0$ and $y = H$. The return stroke removes the negative charge on the leader and hence causes a positive field change. The leader field which is destroyed by the return stroke is given by Eq. (3.8) with $x = 0$ and $y = H$. The ratio of these two field changes is plotted in Fig. 3.9 as a function of H/D. Far away from the discharge the ratio approaches one. When $H/D \ll 1$, the leader field change is essentially equal to the field change caused by moving a point charge (equal the leader charge) from H to the average height of the leader charge ($H/2$ for a uniformly charged leader). The return-stroke field change is equal to the field change caused by moving the point charge from the average height to ground.

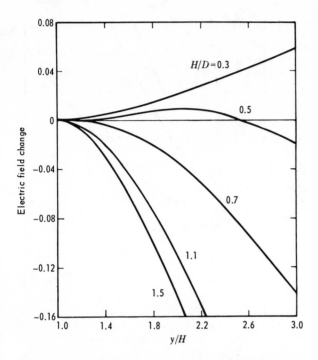

Fig. 3.8 *Electric field change for a positive leader moving upward from a positive charge center at height H. The length of the leader above the charge center is $y - H$. Electric field changes which are positive for small y/H become negative for large y/H. Multiply field change given in graph times $\rho/2\pi\epsilon_0 D$ to obtain mks units of volts per meter.*

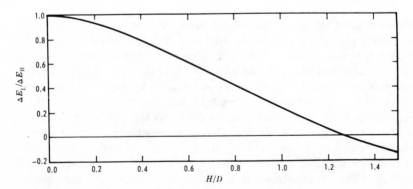

Fig. 3.9 *Ratio of electric field change due to leader to that due to return stroke vs. ratio of height of charge center to horizontal distance from charge center.*

Since the ratio of leader-process field change to return-stroke field change is only a function of D and H, a measurement of the ratio and a knowledge of D make possible a calculation of H, the height of the charge center.

3.3 MAGNETOSTATICS

The motion of charges within a cloud or between a cloud and ground constitutes an electric current. With this current is associated a magnetic field which may be measured at the ground. We shall consider in this section a simple model to describe the magnetostatic effects due to current flow during a lightning discharge. We idealize the current flow as vertical and focus on a current-carrying line (and its image below the ground plane) as shown in Fig. 3.10.

The magnetic flux density $d\mathbf{B}$ at a distance r from a short element of length dz carrying current I is

$$d\mathbf{B} = \frac{\mu_0 I \, dz}{4\pi r^2} \, (\mathbf{a}_I \times \mathbf{a}_r) \tag{3.15}$$

where $\mathbf{a}_r$ is a unit vector directed outward along r, $\mathbf{a}_I$ a unit vector in the direction of the current flow through dz, and μ_0 the permeability of vacuum (which is essentially equal to the permeability of atmospheric

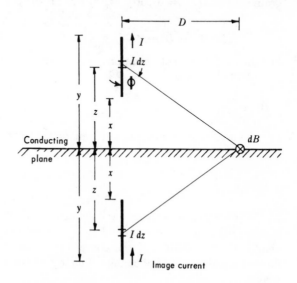

Fig. 3.10 *Diagram for the calculation of the magnetic flux density at D due to a vertical line of length $y - x$ carrying current I above a conducting plane.*

air). In mks units $\mu_0/4\pi \cong 10^{-7}$. If, as shown in Fig. 3.10, the current is flowing vertically upward, then the vector magnetic flux density at D points into the page. The magnitude of the magnetic flux density at D due to $I\,dz$ is

$$dB = \frac{\mu_0 I\,dz}{4\pi}\,\frac{D}{(z^2 + D^2)^{3/2}} \tag{3.16}$$

since

$$|(\mathbf{a}_I \times \mathbf{a}_r)| = \sin\phi = \frac{D}{(z^2 + D^2)^{1/2}} \tag{3.17}$$

To find the magnitude of the total magnetic flux density at D due to the current flow in the vertical line of length $y - x$, we integrate Eq. (3.16) from $z = x$ to $z = y$ and multiply by 2 to take account of the image current with the result that

$$B = \frac{\mu_0 I}{2\pi D}\left[\frac{y}{(y^2 + D^2)^{1/2}} - \frac{x}{(x^2 + D^2)^{1/2}}\right] \tag{3.18}$$

The vector field points into the page. If the current direction were reversed, the vector magnetic flux density would be directed out of the page.

If the current flow is between a charge center at height H and the ground ($y = H$, $x = 0$), Eq. (3.18) can be written

$$B = \frac{\mu_0 I}{2\pi D}\,\frac{H}{(H^2 + D^2)^{1/2}} \tag{3.19}$$

For the case that an observation is made very close to a discharge, $H \gg D$, Eq. (3.19) becomes

$$B \cong \frac{\mu_0 I}{2\pi D} \tag{3.20a}$$

For the case that an observation is made far from the discharge, $H \ll D$, Eq. (3.19) becomes

$$B \cong \frac{\mu_0 I H}{2\pi D^2} \tag{3.20b}$$

In the preceding section we have defined the electric dipole moment of a charge $+Q$ located a distance H above a conducting plane and its image charge as $M = 2QH$. If the charge $+Q$ flows to ground, the current flowing equals the rate of change of charge at the source, and thus we can write

$$\frac{dM}{dt} = 2IH \tag{3.21}$$

If we combine Eq. (3.21) with Eq. (3.20b), we find that

$$B = \frac{\mu_0}{4\pi D^2} \frac{dM}{dt} \tag{3.22}$$

a result that we will consider again in the next section.

3.4 ELECTROMAGNETICS

In the preceding two sections we have examined the electric and magnetic fields due to static or slowly varying distributions of thunderstorm charges and currents. In a lightning discharge there will be a considerable variation in charge and current in a relatively short time interval. We therefore consider now the time-varying electromagnetic fields associated with the lightning discharge.

We can assign to the charge distribution present within a thundercloud a total electric dipole moment. That is, we can sum the dipole moments due to each individual charge or charge conglomeration and its image to obtain a total dipole moment

$$M = 2 \sum_i Q_i H_i \tag{3.23}$$

where the sum is over every charge or charge conglomeration above the ground and terms involving negative charge are negative in sign. The electric and magnetic fields on the ground at D due to the dipole moment and its time variation can be calculated in a straightforward manner if a number of simplifying assumptions are made (see, for example, Jones, 1964). Subject to these assumptions, which will be listed in a moment, the field magnitudes calculated are

$$E = \frac{[M]}{4\pi\epsilon_0 D^3} + \frac{1}{4\pi\epsilon_0 c D^2}\left[\frac{dM}{dt}\right] + \frac{1}{4\pi\epsilon_0 c^2 D}\left[\frac{d^2M}{dt^2}\right] \tag{3.24}$$

$$B = \frac{\mu_0}{4\pi D^2}\left[\frac{dM}{dt}\right] + \frac{\mu_0}{4\pi c D}\left[\frac{d^2M}{dt^2}\right] \tag{3.25}$$

where c is the velocity of light and the values of the quantities in brackets are their retarded values obtained at time $(t - D/c)$. For Eq. (3.24) and Eq. (3.25) to be valid the following conditions must be satisfied: (1) $D \gg H_i$, (2) the magnitude and the phase of the current resulting from the change in M must be constant along the path of the current, and (3) the H_i's must be constant. If these conditions are met to some approximation, the problem of calculating the fields reduces to the problem of calculating the total dipole moment as a function of time.

The first term on the right-hand side of Eq. (3.24) is called the electrostatic term. It is essentially the same as Eq. (3.4) corrected for the field propagation velocity. The second term on the right-hand side of Eq. (3.24) is called the intermediate or induction-field term. This term is proportional to the current, represents a reactive energy storage, and, since it is proportional to D^{-2}, maintains a significant magnitude to larger values of D than does the electrostatic term. The last terms on the right of Eq. (3.24) and of Eq. (3.25) are the radiation terms. They represent energy propagated away from the source at the speed of light and are proportional to the time rate of change of current. The radiation terms are significant at large values of D, at which the remaining terms in Eq. (3.24) and Eq. (3.25) can be ignored. The first term on the right-hand side of Eq. (3.25) is called the magnetostatic term. It is essentially the same as Eq. (3.22) corrected for the field propagation velocity. Of all the terms given in Eq. (3.24) and Eq. (3.25) only the electrostatic term has a nonzero value before and after the lightning discharge.

The electrostatic field is the dominant electric field at about 10 km from a discharge if the effective frequency of the field variation under study does not exceed a few kHz. If measurements are made closer than about 10 km, condition (1) is not satisfied and thus Eq. (3.24) is not valid. The pertinent equations of Sec. 3.2 should be used although these also may be

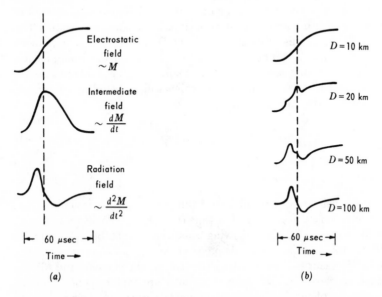

Fig. 3.11 *(a) Electric field components vs. time for a return stroke. (b) Total electric field intensity vs. time at several distances for a return stroke.*

inaccurate if the charge motion is rapid. Similarly, Eq. (3.25) is not valid for measurements made close to a discharge, and the magnetostatic approximation of Sec. 3.3 may not be valid if current changes are rapid. At distances above 100 km the radiation field is dominant. At intermediate distances all field components may contribute significantly to the total field. In Fig. 3.11 the expected qualitative form of the three electric field components are shown for a return stroke bringing negative charge to earth. The return stroke causes a monotonic increase in M. Also shown in Fig. 3.11 is the expected total electric field due to a return stroke as a function of distance.

3.5 FIELD-MEASURING APPARATUS

3.5.1 Electronic measurement of electric field intensity

The electric field intensity at ground can be measured as a function of time by measuring the voltage between an antenna and ground. We will discuss now the pertinent features of such a measurement. We consider a flat-plate antenna as shown in Fig. 3.12. The plane of the antenna is oriented perpendicular to the electric field vector, parallel to the ground, that is, along an equipotential surface. The electric field is assumed to be uniform. The antenna is a height h above the ground. In the absence of any loading of the antenna (Fig. 3.12a) the electric field near the antenna is E (the value that would exist in the absence of the antenna), and the potential difference between ground and antenna is $V_g = Eh$. The stray capacitance between antenna and cloud is C_c, between antenna and ground, C_g, where $C_g \gg C_c$. The potential difference between cloud and ground is V. By *cloud* we mean the effective charge center which is the origin of E. The cloud-to-ground potential difference is divided across the two capacitances C_c and C_g. The potential difference across C_g is

$$V_g = V \frac{C_c}{C_g + C_c} \tag{3.26}$$

which, since $V_g = Eh$, can be rewritten

$$V = Eh \frac{C_c + C_g}{C_c} \tag{3.27}$$

When the measuring circuit shown in Fig. 3.12b is attached to the antenna, a potential v is measured which is less than V_g. This is so because the RC circuit attached to the antenna loads the antenna. We assume that R is a very large impedance compared to C, so that we need only consider the effects of C on the determination of v. Since C and C_g

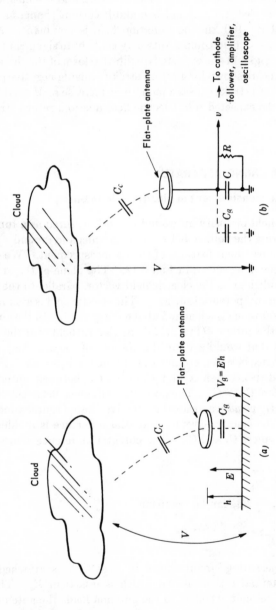

Fig. 3.12 (a) *Flat-plate antenna not attached to electronics.* (b) *Flat-plate antenna with associated electronics.*

are in parallel, the voltage v is given by

$$v = V \frac{C_c}{C_g + C_c + C} \tag{3.28}$$

We can eliminate the unknown V by substituting Eq. (3.27) into Eq. (3.28)

$$v = Eh \frac{C_c + C_g}{C_g + C_c + C} \tag{3.29}$$

Since $C_g \gg C_c$, Eq. (3.29) is to good approximation

$$v = Eh \frac{C_g}{C_g + C} \tag{3.30}$$

The measured voltage is proportional to the electric field E. The proportionality constants may be measured or calculated. In practice $C > C_g$, and thus C can be used to control the magnitude of the measured voltage. The effect of R is to allow the voltage v to decay with a time constant $R(C + C_g)$, or RC if $C \gg C_g$. If RC is made large compared with the times of interest, the effects of R on the measurements can be neglected.

Kitagawa and Brook (1960) have described two measuring circuits for use with an antenna. The first, which they call "electric-field meter (slow antenna)," has $RC = 4$ sec. The frequency response extends from dc to better than 20 kHz.* Five different values of C are used to vary the voltage gain over a range of 80 db. Values of R used vary between 10^7 and 10^{11} ohms. The output voltage is displayed on an oscilloscope which is allowed to free-run, each sweep being slightly displaced vertically from the previous sweep. The sweep rate used is usually 50 msec per sweep, so that several seconds can be easily displayed on the oscilloscope screen. The time resolution is a fraction of a millisecond. The second circuit, which Kitagawa and Brook call "electric-field-change meter (fast antenna)," has $RC = 70$ μsec. The upper frequency response extends above 1 MHz. When combined with a film recording system, Kitagawa and Brook obtain a time resolution of 10 μsec. Higher time resolution can be obtained using a tape recording system. The fast antenna circuit allows very rapid field changes to be delineated, although accurate field values can only be obtained for times much less than 70 μsec due to the exponential signal decay through R. Simultaneous use of the electric-field meter and the electric-field-change meter allows the determination of electric field data over the time of a complete lightning flash with good time resolution of the rapidly occurring field variation.

---

* For later studies (e.g., Ogawa and Brook, 1964), the electric-field meter was modified to have $RC = 10$ sec and an upper frequency response of 1 MHz.

3.5.2 The field mill or electrostatic fluxmeter

The length of time over which the electric field intensity can be adequately monitored electronically is a function of the effective RC time constant of the electronic system. We have seen in the previous section that times of the order of seconds are practical. For the measurement of electric field intensity over longer periods of time a continuous-running mechanical instrument known as the field mill or electrostatic fluxmeter may be used. The field mill can be constructed to have a time resolution of about a millisecond. The information provided by the field mill is thus similar to that provided by the electric-field meter described in the previous section.

The field mill consists of an exposed conductor which is periodically shielded from the ambient electric field by another conductor. The former conductor is connected to ground through a resistor; the shielding conductor is grounded. The charge induced on the exposed conductor by the ambient field flows to ground through the resistor during the shielding period. The magnitude of the current flow is a measure of the charge induced on the exposed conductor which is in turn a measure of the ambient field. The output signal is oscillatory with a period equal to the time taken to shield and then unshield the exposed conductor. The output signal after amplification may be stored on magnetic tape or may be displayed directly on an oscilloscope screen and photographed. The field mill is calibrated in the laboratory by exposing it to known electric fields. Details regarding the design and operation of field mills for lightning measurements are given by Malan and Schonland (1950), by Schonland (1956), and by Malan (1963).

3.5.3 Miscellaneous electric field measuring techniques

Of historical interest is the field-measuring technique used by Wilson (1916, 1920) and several other early investigators. They employed a mercury–sulfuric-acid capillary electrometer attached to an antenna to measure the electric field intensity with a time resolution of about 0.1 sec. The charge induced on the antenna caused the motion of a drop of sulfuric acid within a capillary tube filled with mercury. The displacement of the acid drop was proportional to the field. The time resolution was determined by the inertia of the acid and mercury.

Other types of electrometers were also employed by early investigators for measuring thundercloud electric fields and field changes due to lightning. For example, a gold leaf electrometer was used by Wilson (1916) and a string electrometer was used by Appleton, Watson-Watt, and Herd (1926). To measure the *steady* electric field of a thundercloud using an electrometer attached to an antenna, it is necessary to shield and then unshield the antenna, as is done mechanically in the field mill.

If a grounded sharp conductor is placed in a strong electric field, corona current will flow. Wormell (1927) first used corona-current measurements to obtain electric field data under and near thunderstorms. According to Simpson and Scrase (1937) (who refer to measurements made by Whipple and Scrase) the corona current is given by

$$I = a(E^2 - M^2) \tag{3.31}$$

where a and M are constants and E must be greater than M before there is current flow. Chapman (1958) has discussed the validity of Eq. (3.31) and has suggested that under certain conditions the corona current should vary linearly with electric field. Chapman (1958) gives numerous references to the literature on corona discharges from points in thundercloud fields. Additional information regarding point discharge current is given by Ette (1966). A variation of the corona-current technique has been used by Simpson and Scrase (1937) and by Simpson and Robinson (1941) to measure and record electric fields in and near thunderclouds. Their device, called an *alti-electrograph*, was flown on balloons into thunderclouds. The alti-electrograph recorded the corona current between two electrodes separated by about 3 mm. One electrode was connected to the case of the instrument, the other to an insulated wire about 25 m long trailing beneath the instrument. The two electrodes rested on a disk of pole-finding paper which was rotated by a small clock. The pole-finding paper was stained by the current flowing from the positive needle, the width of the stain being roughly related to the electric field intensity in the vicinity of the alti-electrograph. The alti-electrograph was also equipped to provide a reading of atmospheric pressure. Simpson and Scrase (1937) were the first to observe the small, concentrated positive charge pocket at the base of the thundercloud, the p region. In addition, they verified the existence of the N-P dipole structure first suggested by Wilson (1916, 1920) (see Sec. 3.6). A survey of information concerning the p region is given by Williams (1958).

3.5.4 Magnetic field measurements

The voltage induced in a loop antenna of area A oriented with its plane perpendicular to a magnetic flux density B due to the rate of change of the magnetic flux density is

$$V_{\text{antenna}} = A \frac{dB}{dt} \tag{3.32}$$

where it is assumed that A is constant and B is essentially uniform over the area A. Thus the rate of change of magnetic flux density can be measured by appropriate connection of the loop antenna to an oscilloscope. In order to measure directly the magnetic flux density, the antenna is connected in series with a resistance R and a capacitance C. The equation for the current i flowing in the series circuit is

$$A \frac{dB}{dt} = Ri + \frac{1}{C} \int_0^t i \, dt' \tag{3.33}$$

where the capacitance is assumed to be initially uncharged. We consider the case in which R and C are chosen such that

$$Ri \gg \frac{1}{C} \int_0^t i \, dt' \tag{3.34}$$

that is, the impedance $1/\omega C$ to a sine wave signal is small compared to R for all effective angular frequencies ω present due to the changing magnetic field. Using Eq. (3.34) we can solve Eq. (3.33) for the current

$$i \cong \frac{A}{R} \frac{dB}{dt} \tag{3.35}$$

If the output signal is the voltage across the capacitance, that signal is

$$V_C = \frac{A}{RC} \int_0^t \frac{dB}{dt'} \, dt' = \frac{A}{RC} B(t) \tag{3.36}$$

Thus the magnetic flux density can be measured as a function of time. The measured magnetic flux density from Eq. (3.36) can be used in Eq. (3.19) to determine lightning current if H and D are known and the assumptions necessary for the validity of Eq. (3.19) are satisfied. The magnetic loop antenna must be well shielded from electric fields if the measured signal is to be meaningful.

In addition to the magnetic loop antenna and its associated circuitry, a number of magnetometers (instruments to measure magnetic flux density) are commercially available. The commercial instruments are available with a wide range of sensitivities and time responses. Of special interest is the fluxgate magnetometer used by Williams and Brook (1963) to measure continuing currents and leader currents. The instrument is sensitive to changes in the magnetic field of 1γ (10^{-9} webers/m^2). It has a response time of 20 msec. Williams and Brook (1963) report that they have built a fluxgate magnetometer with a response time of 1 msec.

3.6 EARLY ELECTROSTATIC FIELD MEASUREMENTS: THE ELECTRIC-DIPOLE CLOUD MODEL

The pioneering measurements of the electrostatic fields of thunderclouds and of the electrostatic field changes due to lightning were made by

Wilson (1916, 1920), by Appleton, Watson-Watt, and Herd (1920), and by Schonland and Craib (1927). Among the conclusions of these studies were (1) that the thundercloud is essentially an electric dipole, with a net positive charge located above a net negative charge, and (2) that the usual cloud-to-ground lightning flash lowers negative charge from the cloud to the ground, resulting on the average in the destruction of an electric dipole moment of about 100 coul-km. The experiments of Wilson and of Schonland and Craib were performed using antennas connected to electrometers of either the gold leaf (Wilson's earlier work) or the capillary variety. Appleton, Watson-Watt, and Herd used an antenna attached to an oscilloscope in addition to using equipment similar to that of Wilson. We consider now the theory and the measurements that led to the two conclusions listed above.

That the thundercloud is essentially a vertical electric dipole with the positive charge on top (a positive dipole) can be deduced from measurements of the steady electrostatic field of the thundercloud as a function of D or can be deduced from field *change* measurements. As can be seen from Fig. 3.3, far from a thundercloud (in which the magnitude of the N and P charges are shown equal, although this equality has not been established with certainty) the steady electric field of that thundercloud is positive; at a value of D of about 8 km the field undergoes a reversal in sign and at smaller D is negative. The reversal of the steady field was predicted by Wilson (1920) on the basis of his field change measurements, and the first evidence of such an effect was reported by Schonland and Craib (1927).

We consider now the electrostatic field changes to be expected if (1) a lightning flash lowers negative charge to ground, (2) a lightning flash lowers positive charge to ground, and (3) an intracloud discharge partially destroys a positive electric dipole. Recall that a positive electric field at ground is defined as due to positive charge above ground. If positive charge is destroyed, the field *change* is negative, that is, the positive field decreases. Similarly, if negative charge above ground is destroyed, the field change is positive; that is, the negative field decreases. The electrostatic field change measured on the ground at D due to the discharge of ΔQ_N coul of negative charge from height H_N in the cloud to the ground is, from Eq. (3.3),

$$\Delta E_{NG} = + \frac{2}{4\pi\epsilon_0} \frac{\Delta Q_N H_N}{(H_N{}^2 + D^2)^{3/2}} \tag{3.37}$$

It is positive for all values of D. If $D \gg H_N$, the resultant dipole moment change can be written

$$\Delta M_{NG} = +2\Delta Q_N H_N = 4\pi\epsilon_0 \Delta E_{NG} D^3 \tag{3.38}$$

The electrostatic field change measured on the ground at D due to the discharge of ΔQ_P coul of positive charge from height H_P in the cloud to the ground is, from Eq. (3.3)

$$\Delta E_{PG} = -\frac{2}{4\pi\epsilon_0} \frac{\Delta Q_P\, H_P}{(H_P{}^2 + D^2)^{\frac{3}{2}}} \tag{3.39}$$

It is negative for all values of D. If $D \gg H_P$, the resultant dipole moment change can be written

$$\Delta M_{PG} = -2\Delta Q_P\, H_P = 4\pi\epsilon_0\, \Delta E_{PG}\, D^3 \tag{3.40}$$

The electrostatic field change measured on the ground at D due to an intracloud discharge destroying a portion of a vertically oriented positive electrical dipole can be calculated by summing Eq. (3.37) and Eq. (3.39) with $\Delta Q_N = \Delta Q_P = \Delta Q$

$$\Delta E_{PN} = -\frac{2\Delta Q}{4\pi\epsilon_0}\left[\frac{H_P}{(H_P{}^2 + D^2)^{\frac{3}{2}}} - \frac{H_N}{(H_N{}^2 + D^2)^{\frac{3}{2}}}\right] \tag{3.41}$$

The field change suffers a reversal in sign with increasing D: at small D the field change is positive; at large D the field change is negative. That this is the case can also be seen from Fig. 3.3. If a portion of the dipole is destroyed, the positive steady field at large distances will decrease, a negative field change, while the negative steady field at close distances will also decrease, a positive field change. If the magnitudes of the N and P charges are not equal, the field-change reversal distance will differ from the steady-field reversal distance. The reversal distances will also differ if the positions of the upper and lower ends of the discharge are different from the effective centers of the two charged regions supporting the discharge. If $D \gg H_P$, the electric field change given in Eq. (3.41) can be written

$$\Delta E_{PN} = -\frac{2}{4\pi\epsilon_0} \frac{\Delta Q\,(H_P - H_N)}{D^3} \tag{3.42}$$

and the resultant dipole moment change is

$$\Delta M_{PN} = -2\Delta Q\,(H_P - H_N) = 4\pi\epsilon_0\, \Delta E_{PN}\, D^3 \tag{3.43}$$

In summary, negative flashes to ground exhibit positive field changes, positive flashes to ground exhibit negative changes, and intracloud discharges neutralizing a portion of a positive dipole exhibit positive field changes at small D and negative field changes at large D.

We look now at the measurements. Wilson (1916, 1920) working in England, where according to Wormell (1953) cloud-to-ground discharges make up about 40 percent of all flashes and about 90 percent of the cloud-to-ground discharges lower negative charge (although these facts were not known to Wilson at the time), found the ratio of positive to negative field changes to be 1.56 for a total of about 900 discharges (cloud-to-ground and intracloud) observed at varying distances. At distances below 5 km, the ratio was near 3 (Wilson, 1920, Fig. 2). The implication is that cloud-to-ground flashes discharge negative charge to ground and that the thundercloud dipole is positive. Wilson found about the same mean value of moment change (he measured $\Delta E \ D^3$ and used the appropriate equivalent of Eq. (3.38) to calculate ΔM) for positive and negative field changes. His stated mean for the moment change is about 100 coul-km. (Note that $\Delta E \ D^3$ is smaller than the true moment change if D is not much greater than H, if the flash is not further than about 10 km away.) Wilson estimated the average charge involved in a flash to be between 10 and 50 coul.

Further evidence for Wilson's positive dipole picture was provided by Appleton et al. (1926). Working in England they found negative field changes to be almost twice as frequent as positive field changes at distances from the lightning channel greater than 50 km.

Schonland and Craib (1927), working in South Africa, measured cloud fields and field changes with an experimental setup identical to that described by Wilson (1920). In South Africa the great majority of lightning flashes takes place within the cloud, discharges from the bottom of the cloud to ground being much less frequent. One would expect therefore that distant storms would give predominantly negative field changes and near storms predominantly positive field changes. Out of 798 field changes from thunderclouds over 8 km away, 666 were negative changes. For thunderclouds observed at distances less than 6 km, there were 39 positive field changes and 9 negative field changes. In addition the steady fields were found to be positive when the thundercloud was distant and negative when it was near. A positive electrical dipole model for the cloud was indicated. Schonland and Craib found the mean value of $\Delta E \ D^3$ for 82 discharges at distances between 10 and 30 km to be 94 coul-km, the actual values ranging from 6.2 to 0.1 times the mean.

Although in the preceding discussion we have made a strong case for the existence of a positive electric dipole in the thundercloud, there was considerable disagreement in the early literature as to the polarity of the dipole and the sign of the predominant charge in the thundercloud base. A good account of this lively dialog is given by Jensen (1933).

By the early 1930s, however, it was becoming clear that Wilson's positive dipole model of the cloud was essentially correct. Wormell (1927, 1930) had measured the vertical electric currents from a point discharge as a function of distance from numerous thunderstorms. The direction of the vector electric field beneath a thundercloud was deduced from the observed direction of the current flow. In his 1930 paper Wormell reported current records which were supplemented in some cases by direct measurements of the electric field. Wormell concluded that his data strongly suggested that the great majority of both thunderclouds and showerclouds were bipolar clouds, the upper charge being positive.

The work of Schonland and Craib (1927) was extended by Schonland (1928) to include many more lightning discharges. In addition, Schonland made visual observations or photographic records of many strokes whose field changes were measured.

Jensen (1930, 1932a,b, 1933), working in the United States, obtained numerous correlated field-change measurements and lightning photographs. He obtained correlated data on 185 flashes which were closer than the field-change reversal distance. Most of the downward-branched cloud-to-ground discharges exhibited positive field changes, indicating a lowering of negative charge. Jensen reported, however, cloud-to-ground discharges showing negative field changes, apparently due to discharges from a positive portion of the cloud (see Sec. 3.7.6). Photographs of several of these discharges showed the multitudinous fine filamentary "streamers" characteristic of laboratory discharges from positive electrodes.

Schonland and Allibone (1931) reported that of 404 flashes for which field changes and correlated optical data were available, at least 95 percent originated from a negative cloud charge. Schonland and Allibone also experimented with a 1-million-volt impulse generator in order to show that branching could occur from either a positive or a negative rod electrode toward a grounded plane. This type of experiment showed that the direction of branching could not be used as a criterion for the type of charge from which the lightning originated. Simpson (1926) had previously argued that the downward branching observed in cloud-to-ground lightning could only occur from a positive cloud charge.

Halliday (1932) working in South Africa provided further data of the type obtained by Schonland and Craib (1927) and by Schonland (1928).

The argument that the lower part of the cloud was predominantly charged negative was further strengthened by the observation that lightning currents to transmission lines, in all the early measurements, were found to transport negative charge to ground. These early current measurements, made in the United States around 1930, will be discussed in the next chapter.

Since the field and field-change measurements of the 1910s, 1920s, and early 1930s, a number of ground-based and airborne measurements have been made of the steady electric field of the thundercloud. In general, these measurements support the earlier work. We will not concern ourselves with the details of the steady field measurements since they do not, in general, relate to the lightning discharge. For references to the steady field measurements made between the mid-1930s and the present the reader is referred to Gunn (1948), Wormell (1953), Fitzgerald (1965), and Vonnegut et al. (1966).

3.7 THE ELECTROSTATIC FIELD CHANGE DUE TO CLOUD-TO-GROUND LIGHTNING

3.7.1 Introduction

Wormell (1939) working in England employed the capillary electrometer to make extensive measurements of the moment changes due to lightning flashes. He found the most frequent value of the moment destroyed in both positive and negative field changes to be about 120 coul-km, the average being about 220 coul-km. The average is considerably larger than the most frequent value because of the occasional occurrence of discharges with very large moment changes (up to 1,200 coul-km). Pierce and Wormell (Pierce, 1955a) in the late 1940s in England used an oscilloscope to measure the moment changes due to individual strokes, leaders, and interstroke processes. On the average, both stepped-leader moment changes and return-stroke moment changes were found to be of the order of 30 coul-km. Measurements have been made in New Mexico by Brook et al. (1962) of the moment change and charge transferred by strokes, by continuing current, and by interstroke processes. The average moment change for a stroke was found to be about 22 coul-km, the average charge transferred per stroke about 2.6 coul. Flashes with only discrete strokes produced an average moment change of 151 coul-km (19.4 coul), while flashes with continuing current intervals (about half the flashes observed) produced an average moment change of 346 coul-km (33.9 coul). The average moment change due to a continuing current interval was 135 coul-km, the average charge lowered in the interval 12 coul. Wang (1963a,b) working in Singapore has found a mean moment change for cloud-to-ground flashes of 256 coul-km, the average charge transferred per flash being 31 coul.

In general, the structure of the observed electric field change due to the lightning discharge has been correlated with photographic observations of the lightning discharge. In Fig. 3.13 are shown the main features of the simultaneous photographic and electric field records of two lightning flashes as given by Kitagawa et al. (1962). (The electric-field meter and the electric-field-change meter have been discussed in Sec. 3.5.1.) The electric-field-change meter records events occurring on a relatively short time scale, but has a time constant such that the meter reading returns to zero with an exponential decay time of about 70 μsec. The most obvious feature of the records of Fig. 3.13 is the abrupt field

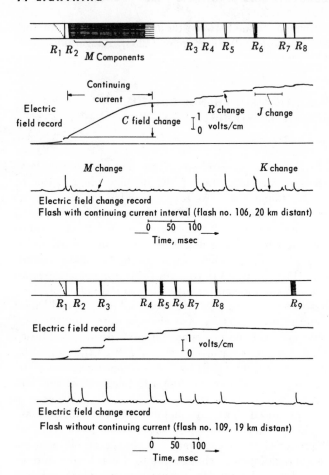

Fig. 3.13 *Examples of simultaneous photographic and electric field measurements. The records have been modified somewhat from the originals for illustrative purposes. (Adapted from Kitagawa et al. (1962).)*

change due to the return stroke, the R change. Also shown on Fig. 3.13 are the C change due to continuing current flowing to ground, the M change due to M components, the K change which occurs in the interstroke time interval, and the J change which has been attributed to the so-called junction process occurring in the cloud between strokes. Not shown are the F change, a slow positive field change which often follows the final stroke of a flash, and the detailed field changes due to the leader processes.   In the sections to follow we shall consider in detail the various electrostatic field changes and the lightning properties which have been inferred from those field changes.

3.7.2 The stepped leader

Electric field changes preceding the initial stroke in a flash are shown in Fig. 3.14 as a function of distance from the discharge.   The designations B, I, L for the field changes preceding the return-stroke change R are due

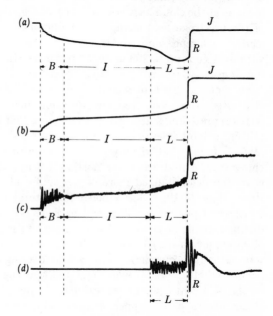

Fig. 3.14 *Diagrams of typical field changes of discharges to ground. (a) Electrostatic fields at 2 km. (b) Electrostatic fields at 5 km. (c) Electrostatic, intermediate, and radiation fields at 50 km. The relative amplitude of the R part has been reduced. (d) Radiation fields at 500 km in frequency band 200 Hz to 20 kHz. Durations: B from 2 to 10 msec; I from 0 to 400 msec; L from 4 to 30 msec. (Adapted from Clarence and Malan (1957) and Schonland (1956).)*

to Clarence and Malan (1957). It is not clear that these designations are meaningful in terms of physical processes, but they will be useful as definitions. The designation B stands for breakdown, I for intermediate, and L for leader. The B portion is characterized by a significant field change in a period of a few milliseconds, the I portion by a slow or irregular field change, and the L portion by a more rapid field change. The signs of the field changes vary with distance as shown in Fig. 3.14. Clarence and Malan (1957) suggest that the B stage is due to electrical breakdown between the p and N charge centers in the cloud, that the I stage is due to negative charging of the breakdown channel by subsidiary discharges, and that the L or leader stage occurs after the discharge channel at the cloud base is sufficiently charged. Thus only the L portion of the prestroke field change is attributed to the photographed stepped leader. Clarence and Malan (1957) measured prestroke field changes with durations up to 200 msec, 50 percent of the measured values exceeding 30 msec, 10 percent exceeding 120 msec. Kitagawa and Brook (1960) have reported prestroke field changes with durations between 10 and 200 msec, the most frequent duration being about 30 msec. Field changes preceding a first stroke as measured far from the stroke by Pierce (1955a) exhibit essentially the same features shown in Fig. 3.14c. Pierce (1955a) found that in only about 15 percent of the 332 prestroke field changes measured was the I portion distinguishable. In the remaining cases the field changed more or less uniformly over the total prestroke time. Pierce (1955) has labeled those total prestroke field changes in which the I portion was distinguishable as $L(\beta)$; those in which it was not distinguishable as $L(\alpha)$. The average duration for an $L(\alpha)$ change was 50 msec; for an $L(\beta)$ 175 msec (during 95 msec of which a change in field was actually occurring). The α and β designations were taken from Schonland et al. (1938) who suggested that the rapid B portion followed by the quiescent I portion were due to the initial fast and subsequent slow stages of a type β stepped leader. This opinion is no longer held by Schonland (1956). Schonland (1956) introduced the designation B to label all the prestroke field changes preceding the L portion. That is, what Schonland (1956) labels B, Clarence and Malan (1957) have subdivided into B and I. Kitagawa (1957a) working in Japan has also studied the prestroke field changes. His results are expressed in terms of the older Schonland notation.

It is appropriate now to look at the B field change and its suggested physical interpretation. Clarence and Malan (1957) state that in their studies the B field change is negative for distances D less than 2 km and is positive for D greater than 5 km. In 43 records obtained between 2 and 5 km, 21 of the B field changes were positive-starting and 22 were negative-starting. For distances between 2 and 5 km, the sign of the B change usually reversed during the course of the field change. Using the leader model described in Sec. 3.2, we can discuss Clarence and Malan's interpretation of their data: that the B field change is due either to a negative leader moving downward from the N to the p region or to a positive leader moving upward from the p to the N region, the two choices being equally probable. Consider first the negative field changes at distances less than 2 km. From Fig. 3.5 we see that these could be caused by a negative leader moving downward between, say, 3 and 1 km (x/H from 1 to 0.3) if H/D is greater than about 1.1. Thus for $D = 2$ km, H is greater than 2.2 km. From Fig. 3.8 we see that the negative field change could also be due to a positive leader moving upward if H/D is greater than about 0.7. Thus for D equal to 2 km, H is greater than 1.4 km. Consider now the positive field changes at distances greater than 5 km. From Fig. 3.5 we see that the positive field change could be due to a negative leader moving downward if H/D is less than about 0.7. Thus for D equal to 5 km, H is less than about 3.5 km. From Fig. 3.8 we see that the positive field change could also be due to a positive leader moving upward between, say, 1 and 3 km if H/D is less than about 0.3. Thus for D equal to 5 km, H is less than 1.5 km. Consider now the field change occurring between 2 and 5 km. From Fig. 3.5 we see that a negative leader moving downward will cause first a negative and then a positive field change while descending from 3 to 1 km if H/D is about 0.9. With H/D equal to 0.9 and H equal to 3 km, D is 3.3 km. From Fig. 3.8 we see that a positive leader moving upward will cause first a positive and then a negative field change while ascending upward from 1 to 3 km if H/D is about 0.5. With H/D equal to 0.5 and H equal to 1 km, D is 2 km. The heights calculated to be the approximate starting points of positive upward-moving leaders (1 km) or negative downward-moving leaders (3 km) are in reasonable agreement with the heights of the p region and the lower portion of the N region. The exact numbers obtained in the analysis depend on the model used to describe the charge movement. Clarence and Malan (1957) use the information in Fig. 3.2 (the model being a point charge moving upward or downward) to conclude that either a positive charge is moving upward from a minimum height of 1.4 km or a negative charge is moving downward from a maximum height of 3.6 km. According to Malan and Schonland (1951b) the bases of South African thunderclouds wherein resides the p charge are usually between 1.5 and 2.5 km above ground and the lower boundary of the negative charge center N is usually about 1 km above the cloud base. Malan (1955b) reports that clouds often emit light for periods of the order of 100 msec before the stepped leader emerges from the cloud base. This observation lends support to the hypothesis that the stepped leader is preceded by a breakdown process in the cloud.

Clarence and Malan (1957) report the measurement of low-amplitude electric field pulses that are of different character during each of the three defined prestroke periods. The measuring system used had a frequency response to 300 kHz and only storms closer than 80 km were observed. The B pulses are reported to have longer interpulse time intervals and larger amplitudes than the L pulses. The I stage is characterized by pulses of high frequency and very small amplitude. That the I stage is relatively quiet compared to the B stage was first noted by Schonland et al. (1938).

We consider now the L portion of the discharge. The model introduced in Sec. 3.2 for the cloud-to-ground negatively charged leader (whose predicted field change is shown in Fig. 3.5) is applicable to the stepped leader in only a qualitative way. This is so primarily because the stepped leader is branched resulting in a nonuniform charge distribution. Although the field change of Fig. 3.5 is plotted as a function of height above ground x, in practice field-change data are taken with linear time sweep and, in the absence of adequate correlated photographic data, must necessarily be converted to x by assuming a constant leader velocity. Since the average velocity toward ground is not constant, particularly for type-β stepped leaders, additional difficulties are encountered if a quantitative comparison of data with Fig. 3.5 is attempted. Nevertheless, the qualitative features of the model are applicable, and we should expect the L changes to be negative and hook-shaped when near and positive when far away. Examples of the near and far L field changes are shown in Fig. 3.14. Further we would expect to be able to differentiate on the basis of field change measurements between type α and type β leaders.

Schonland (1956) states that the duration of the L field change corresponds roughly with that of the photographed leader process. He draws diagrams of field changes due to type α and type β stepped leaders as would be observed at 3 and 15 km. At close distance the type α leader has a hooked field change much as is shown in Fig. 3.5. According to Schonland's diagram, the type β leader which moves rapidly near the cloud and relatively slowly near ground (Sec. 2.3.2) initially exhibits at close distances a more rapid negative field change than does the type α leader. This effect would be expected if the type β leader were rapidly lowering negative charge earthward in its initial stages.

The electrostatic field of the stepped leader measured at close distances has no appreciable electrostatic steps (Chapman, 1939; Malan and Schonland, 1947). Schonland (1953) reports the step electrostatic field changes to be less than one-tenth the continuous field changes in the intervals between steps. According to Schonland (1953) the brief luminous steps, therefore, do not coincide with the lowering of appreciable charge. Short time-duration steplike electrostatic field changes due to prestroke processes have been reported by Kitagawa (1957a) and by Kitagawa and

Brook (1960).   For the total prestroke period they find the most frequent interval between the short duration field changes to be 40 to 60 μsec and suggest that the field changes are to be associated with the photographically measured steps of the stepped leader.   The average time interval between field changes for the New Mexico storms studied was about 80 μsec.

When observations of the stepped leader are made at distances for which the lowered negative charge results in an initial positive field change, intermediate and radiation field components may become significant.   From Eq. (3.24) these field components are proportional to the current and rate-of-change of current, respectively.   According to Clarence and Malan (1957) the type α leader is characterized by a radiation field which consists of a uniform train of pulses whose amplitudes are often less than one-hundredth of that due to the return stroke; the type β leader is characterized by initial pulses which vary from slightly larger than the later β pulses to half as large as the return-stroke pulse.   After the large initial β pulses, successive β pulses resemble the type α pulses.   According to Hodges (1954) the type α radiation pulses have a modal value about one-fifteenth of the subsequent return-stroke pulses; the initial type β radiation pulses have a modal value of about 1/3.5 of the subsequent return-stroke pulses.   The range of values for measured type α leaders was one-tenth to one-fortieth; for type β, one-half to one-tenth.   Clarence and Malan (1957) recorded L radiation pulses with interpulse time intervals of from 30 to 100 μsec.   These values agree well with the pause time between photographed leader steps.

Schonland (1953) and Hodges (1954) have attempted to determine the current in the bright steps of stepped leaders by comparing the measured stepped-leader radiation-pulse amplitudes with the measured return-stroke radiation-pulse amplitudes.   They have assumed that both measured pulse amplitudes are given by the *same* constant factor multiplied by the rate of change of current.   Thus the ratio of the pulse amplitudes equals the ratio of the rates of change of current.   It is not clear that the assumption necessary for the calculation is valid.   Even if the radiated pulse heights were as assumed, the frequency content of the leader and the return-stroke pulses could be sufficiently different so that a wideband system would be needed to give accurate pulse-height ratios.   Nevertheless, the results obtained are reasonable.   Hodges (1954) using a rate of rise of return-stroke current of 9.25 ka/μsec finds for type α leaders a modal current rate of rise of 0.62 ka/μsec and for type β leaders 2.6 ka/μsec.   For a current lasting 1 μsec (the duration of step luminosity is 1 μsec or less, the resolution of the Boys camera being about 1 μsec), a peak step current of the order of 1 ka would be expected.   The charge lowered in the 1-μsec interval would be about 10^{-3} coul.   A fully developed leader 4 km in length with a total charge of 4 coul would have about 10^{-3} coul/m of length. Since a step length is of the order of 50 m, the charge which resides on a step length of a fully developed leader would appear to be at least an order of magnitude greater than the charge lowered in the bright step interval, in support of the view of Schonland (1953). Support for the view that the upper limit to the step current is of the order of 1 ka may be found in photographic data. A step current of, say, 10 ka with a duration of 1 μsec would probably be as bright on Boys camera photographs with 1-μsec resolution as is the return stroke. The step luminosity is found to be very weak compared with that of the return stroke. (Note that channel luminosity is not necessarily linearly proportional to channel current.) Further support for a step current of the order of 1 ka can be inferred from the magnetostatic field measurements of Norinder and Knudsen (1957). These data will be discussed in Sec. 3.9.

In order to determine how much charge resides on the stepped leader from leader-field measurements it is necessary to know what percentage of the prestroke field change is attributable to the stepped leader and to have an adequate physical model for the relative distribution of the stepped leader charge. Such information is not available. Pierce (1955a) reports that the total prestroke field change was, on the average, 120 percent larger than the succeeding return-stroke field change for prestroke field changes without the I section, and 240 percent larger for prestroke field changes with the I portion. Pierce (1955a) gives an average moment change for the total prestroke process of about 30 coul-km. Pierce (1955b) suggests that his measurements indicate that the stepped leader carries a charge of about 8 coul. Clarence and Malan (1957) do not present any quantitative data regarding field changes. Schonland (1956) suggests that the stepped leader carries at least 9 coul into the air, some of which is used in neutralizing part of the p-charge region. Brook et al. (1962) found that the minimum charge lowered by first strokes was 3 coul, the most frequent value of charge brought down by first strokes lying between 3 and 4 coul. For single-stroke flashes the average charge lowered was 4.6 coul. First strokes were found to lower charges up to about 20 coul. Care must be taken in inferring leader charge from these data since the magnitude of charge lowered by a stroke following a stepped leader may be greater or less than the charge on the stepped leader. It may be greater if current flows out of the cloud charge reservoir during the stroke; it may be less if the return stroke does not tap all the charge deposited on and around the stepped leader. A reasonable order-of-magnitude estimate for the stepped leader charge would be 5 coul. For a charge of this general magnitude to be lowered in tens of milliseconds would require a current flow of the order of magnitude of

100 amp. Williams and Brook (1963) measured the magnetic fields of two stepped leaders and estimated their average currents at 50 and 63 amp. Currents in upward-moving stepped leaders (Sec. 4.4) are of the order of 100 amp. Hence the available data on steady stepped-leader currents is in reasonable agreement with the data concerning the leader charge.

3.7.3 The dart leader

It has been determined from photographic measurements (see Sec. 2.3.3) that dart leaders usually travel to ground with relatively uniform velocity. In order for the leader field change shown in Fig. 3.5 (as a function of x or of t) to be an adequate representation of the dart-leader field change the dart-leader channel must be uniformly charged. Schonland et al. (1938) have presented evidence that this is the case by examining the ratio of the leader field change to the succeeding return-stroke field change. We see from Fig. 3.9, derived with the assumption of a uniformly charged leader, that for $H/D \ll 1$, the ratio of leader to return-stroke field change is unity. As we have noted on page 57, the effective negative charge center of a uniformly charged dart leader is located at $H/2$ when the leader touches ground. The return stroke lowers the effective charge center from $H/2$ to ground. Thus for $H/D \ll 1$, and a uniformly charged leader, we would expect the positive field change due to the leader (effective charge motion from H to $H/2$) to equal the positive field change due to the return stroke (effective charge motion from $H/2$ to 0). If the dart leader is not uniformly charged, the effective charge center will not be at $H/2$, (except for the special case of a charge distribution symmetric about $H/2$), and the ratio of leader to return-stroke field change will not be unity. For example, if the upper portion of the dart leader is heavily charged, the effective negative charge center when the leader touches ground may be at, say, $3H/4$. If the return stroke lowers this effective charge from $3H/4$ to ground, the ratio of leader to return-stroke field change observed at small H/D will be 1:3. Schonland et al. (1938) found that the most frequent value of the ratio was unity. They considered in their study 46 dart leaders and 26 stepped leaders of the α type (early Schonland notation: no distinct B, I, or L phases). They reported that 85 percent of their measured leaders had effective charge centers between $H/4$ and $3H/4$, the more extreme values being associated with the stepped leaders.·

As is indicated in Fig. 3.5, the dart-leader field change will be initially positive if the field change is observed at H/D less than about 0.7 and initially negative if observed at H/D greater than about 0.7. Succeeding dart leaders in multiple-stroke flashes are of increasing length H, tapping negative charge from higher and higher sections of the N region. Thus if field-change measurements are made with a fixed D (between 3 and 9 km), one might expect to observe positive field changes for the first few dart leaders in a multiple-stroke flash followed by negative field changes for the remaining dart leaders. Malan and Schonland (1951b) give selected examples of the data of Appleton et al. (1926) which show this effect. These data are reproduced in Fig. 3.15. The field change shown at the extreme left of Fig. 3.15 is due to the stepped leader. The field changes occurring in the time intervals between return stroke and subsequent dart leaders have been omitted from the diagrams. Flash 208 has a long stepped-leader field change with apparently clear B and I stages although the time constant of the equipment used was insufficient to keep the output signal from decaying during the I stage. The dart leader for which a negative field change is clearly observed is marked with an arrow. Dart leaders previous to these have positive field changes; those subsequent have initially negative, hook-shaped field changes.

The heights of dart leaders or of the resulting return stroke can be determined in a number of ways. Malan and Schonland (1951b) list

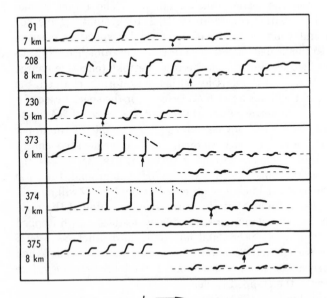

Fig. 3.15 *Electric field changes due to successive leader–return-stroke sequences in multiple-stroke flashes. (Adapted from Malan and Schonland (1951b).)*

five methods: (1) The time for the leader field change t_e can be compared with the photographed leader duration t_p and photographed length h to obtain the true leader height $H = ht_e/t_p$. The assumption is made that the dart leader velocity is uniform. Occasionally dart leaders slow down as they approach the ground. Brook et al. (1962) have in these cases determined several values of velocity along the dart-leader channel and extrapolated to obtain the dart-leader velocity in the cloud. Malan and Schonland (1951a) consider the correlated electrical-photographic technique separately for stepped leaders and for dart leaders. Application of the method to dart leaders is relatively straightforward. Application to stepped leaders is complicated by the lack of understanding of the field-change data for changes preceding the first stroke in a flash and by the nonuniform velocity of the stepped leader. (2) The height of the dart leader for which the field change goes from positive to negative can be determined if D is known: $H \cong 0.7D$. The distance to the stroke can be determined by measuring the interval between the light emission and the arrival of thunder, by using radar, or from the analysis of a photograph showing cloud base, stroke, and ground if the cloud-base height is known. Thus, from Fig. 3.15, flash 91, the leader height is about 5 km. (3) From Fig. 3.9 it can be seen that the ratio of leader to return-stroke field change is a relatively sensitive function of H/D if measurements are made close to the discharge. Thus if D is known, H can be determined from a measurement of the ratio. (4) A method based on the sign reversal of the field change due to the so-called J-streamer process. This technique will be discussed in Sec. 3.7.5. (5) The field change due to a leader or a return stroke can be considered to be a function of the three variables H, D, and ΔQ, the charge involved, if the discharge is assumed to be vertical. If simultaneous field-change measurements are made at two separate stations for which D is known, ΔQ and H may be calculated. If simultaneous measurements are made at three separate stations and if the discharge is vertical, H, D, and ΔQ may all be determined. In general, a complete specification of the field change due to a straight, nonvertical discharge requires seven parameters (the three coordinates of each endpoint and ΔQ); and field-change measurements must be made at seven separate stations to allow a determination of the seven parameters.

Malan and Schonland (1951a) using the first four of the five listed methods found an average increase in height of about 0.7 km per stroke, the height intervals tending to be smaller the greater the number of strokes in a flash. The maximum leader height was about 10 km above ground. Hacking (1954) making measurements at three stations and using method (5) verified that successive strokes originate in regions of increasing height, but in addition found that horizontal channel displacements of up to a few kilometers were often present. Hacking (1954) was unable to deduce quantitative values for H, D, and ΔQ from measurements made at three stations, presumably because the discharges were not vertical.

Brook et al. (1962) have determined H by method (1). They found that the most frequent (25 occurrences) height difference between successive strokes not separated by continuing-current intervals was 0.3 km. For a height difference of about 0.7 km, 14 occurrences were recorded. A diagram showing leader height as a function of stroke number is given in Fig. 3.16. The average number of strokes per flash for the data of Fig. 3.16 is seven (six if single-stroke flashes are included), somewhat higher than the average of three or four reported from South Africa.

The charge lowered by the dart leader can be determined from dart-leader field changes or, alternatively, can be inferred (see discussion in last paragraph of Sec. 3.7.2) from the determination of charge lowered by the return stroke following the dart leader. The former does not appear to have been done. Brook et al. (1962) using Eq. (3.37) with H determined by method (1) and D determined from photographic measurements report that the minimum charge brought down by strokes subsequent to the first is 0.21 coul, the most frequent value lying between 0.5 and 1 coul. Thus the dart leader would appear to carry less charge than

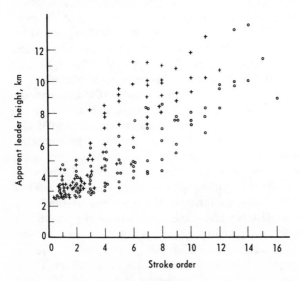

Fig. 3.16 *Apparent leader height vs. stroke order for flashes without continuing-current intervals (circles) and flashes with continuing-current intervals (plus signs). (Adapted from Brook et al. (1962).)*

does the stepped leader. If 1 coul were lowered in a millisecond or so by the dart leader, the resulting dart-leader current would be of the order of 1 ka.

3.7.4 The return stroke

The return-stroke field change R can be considered as composed of two parts. The first part, labeled R_b by Malan and Schonland (1951a) consists of a rapid positive field change (for strokes bringing negative charge to earth) in a time of the order of 100 μsec. This stage is shown in Fig. 3.11. Following the R_b stage the positive electric field change may continue to occur but at a slower rate for times up to several milliseconds. This latter stage has been labeled R_c by Malan and Schonland (1951a). The beginning of the R_c stage can be seen in Fig. 3.11. In some of the literature the label b is used for the R_b stage and c for the R_c stage. The b and c notation was introduced by Appleton and Chapman (1937).

Schonland et al. (1938) measured R_b field changes of 60 first strokes and found durations between 50 and 250 μsec, the most frequent value being 165 μsec. Since their photographically measured ground-to-cloud propagation time for 56 first return strokes was between 20 and 160 μsec, they associated the R_b phase with the return stroke propagation period. Schonland et al. (1938) reported R_c durations for 39 first strokes from 70 to 900 μsec. They related the R_c field change to the channel luminosity following the return-stroke propagation period. It would appear from their published oscillograph traces that the upper time limit on the R_c phase was determined by the time constant of the measuring equipment. Malan and Schonland (1951a) state that the R_c phase lasts as a rule from 1 to 3 msec. Malan and Schonland (1947) report that the R_c stage was absent in 60 percent of 199 strokes observed at close distances. The source of the R_c field change is probably (1) the lowering of available charge from the cloud through the stroke channel to ground and/or (2) the flow of negative charge residing in the air around the channel into the channel and subsequently to ground. The first suggestion is certainly reasonable and occurs, for example, in the case of long continuing currents. The latter suggestion is based on the stepped-leader theory due to Bruce (1941, 1944) to be discussed in detail in Chap. 7. In essence, Bruce pictured the leader channel as composed of a highly conducting, negatively charged arc core surrounded by a negatively charged corona region. Pierce and Wormell (1953) suggested that the return-stroke wavefront discharges and ties to near-ground potential only the arc core. Thus after the passage of the return stroke the core will be at positive potential with respect to the surrounding corona charge. Radial-charge movement into the channel will follow. Calculations of the corona current to be

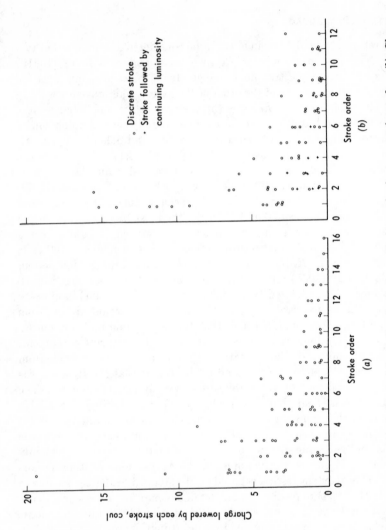

Fig. 3.17 (a) Charge lowered by strokes in flashes without continuing-current intervals. (b) Charge lowered by strokes in flashes with continuing-current intervals. (Data adapted from Brook et al. (1962).)

expected have been made by Pierce (1958) and by Rao and Bhattacharya (1966). Peak corona currents of the order of 1 ka and current durations of milliseconds are predicted. A further discussion of these ideas is given in Sec. 7.6.

Brook et al. (1962) have measured charge transfer due to strokes and due to continuing-current intervals. They list detailed data for 24 flashes including the charge transferred per stroke and the stroke height in addition to similar data for continuing-current intervals. In Fig. 3.17 are shown data concerning the charge lowered by strokes as a function of stroke order. Strokes following stepped leaders were found to always lower at least 3 coul. The largest calculated charge transfer for a first stroke was 19.4 coul. The minimum charge transferred by a stroke following a dart leader was 0.21 coul, due to the sixteenth stroke in flash 124. Many strokes subsequent to the first deliver less than 1 coul to ground.

Wang (1963b) working in Singapore has measured an average moment change of 51 coul-km for the most intense stroke of a flash. The corresponding average charge transferred was 5.7 coul. Wang (1963b) reports that the cloud-to-ground discharges studied consisted on the average of five or six component strokes, each stroke carrying a charge of about 2 to 5 coul to earth.

3.7.5 Processes occurring between strokes

We can distinguish between three general types of processes which may occur in the time interval between strokes: (1) Continuing current flow between cloud and ground resulting in a transfer of negative charge from cloud to ground, (2) the motion of charge within the cloud, and (3) the motion of charge surrounding the cloud, including charge motion between the cloud and upper atmosphere. Any hypothesis regarding the motion of charge will result in a predicted electric field variation measurable at the ground. In practice, measured field data are used to infer the existence of various charge motions.

We will list now the labels given to the observed interstroke field changes: The C field change is a slow, positive field change accompanied by channel luminosity. The indication is that the C change is due to continuing current flow. The J field change is a slow field change which is negative for close strokes and positive for distant strokes. The existence of a field-change reversal with distance indicates that the J change is due to charge motion within the cloud. The F field change is a slow, positive field change which may occur after the final stroke in a flash. It is apparently similar to the C change. The M field change is a short-duration change associated with a luminous M component and is probably indicative of a flow of negative charge to ground. The K field change is a short-duration field change probably due to abrupt charge motion within the cloud. In addition to the above-named field changes, Malan (1955a) reports a slow, negative interstroke field change observed in South Africa at distances of 25 to 100 km. This field change is reported by Malan (1965) to comprise 44 percent of all interstroke field changes measured in South Africa. We shall discuss now in detail the various interstroke field changes.

The C field change has been studied with correlated electrical and photographic observations by Brook et al. (1962) and Kitagawa et al. (1962). A C change is shown in Fig. 3.13. Brook et al. (1962) found an average moment change for the continuing current interval of 135 coul-km. The durations of the continuing current were measured to be between 40 and 500 msec, the average time being 150 msec. Charges lowered in continuing current intervals were calculated to be between 3.4 and 29.2 coul, the average being about 12 coul. The magnitude of the continuing currents was found to vary over a relatively narrow range from a minimum of 38 amp to a maximum of 130 amp, independent of time of current flow. Williams and Brook (1963) report continuing-current values derived from magnetometer measurements. For 14 long-continuing strokes they found an average current of 184 amp, an average charge transfer of 31 coul, and an average duration of 174 msec. Brook et al. (1962) and Kitagawa et al. (1962) report that about half the 200 flashes observed had a continuing-current interval and that about a quarter of all the interstroke time intervals were characterized by the C field change.

If field measurements are made far from a lightning discharge, both continuing current flow to ground and processes in the cloud raising positive charge or lowering negative charge will yield a positive electric field change. Brook et al. (1962) infer from their measurements that the field change due to the in-cloud processes, the J change, is too small (maximum moment change 2 coul-km) to be detected at distances beyond about 50 km. Thus they argue that the slow, positive interstroke field changes measured beyond 50 km are probably due to continuing current. At these distances it is not possible to obtain correlated photographs of the discharge channel to confirm this hypothesis. Pierce (1955a) made measurements far from the discharge channel and found that about a quarter of the interstroke field changes were slow, positive changes, the average moment change being 40 to 50 coul-km (Pierce, 1955b). In the remaining 75 percent of the interstroke intervals no field change could be detected. Pierce (1955a) attributes his measured interstroke field changes to J processes, but, as argued by Brook et al. (1962), they are probably due to continuing current flow. Malan (1955a) reports that in

South Africa a quarter of the slow field changes occurring between strokes are positive for flashes at a distance greater than 20 km, the rest being zero or negative. In a later publication Malan (1963) reports that 19 percent of the slow interstroke field changes measured in South Africa for distant flashes are positive and attributes those to continuing current. Of the remaining field changes, 37 percent are reported to be zero and 44 percent negative. Malan (1955, 1963) suggests that the source of the negative field change may be the rearrangement of space charge above the cloud. Negative interstroke field changes have not been reported by other investigators.

We discuss now the F field change. The F field change occurs after the final stroke in a flash. It is a slow, positive field change which would appear to be very similar to the C change. Malan (1954) suggests that the F change is due to the continuous discharge to ground of part of the negatively charged column higher than that reached by the final stroke. Malan (1954) reports that the F change occurs most frequently after flashes having fewer than four strokes. Half the 71 single-stroke flashes, 43 percent of the 181 two- and three-stroke flashes, and 21 percent of the 107 four-stroke flashes observed by Malan (1954) had F changes. Only 6 percent of the 110 flashes containing more than 6 strokes had F changes. Malan (1954) presents data which indicate that the average current flowing during the F change is equal to the average current flowing during the J processes of the preceding interstroke intervals. Pierce (1955a) found that 59 percent of single-stroke flashes were followed by slow, positive field changes (he termed them S). The corresponding percentages for flashes with two to four strokes and five or more strokes were 44 and 30, respectively. Brook et al. (1962) do not distinguish between the continuing current which flows between strokes and that which flows after the final stroke of a flash. However, of the 12 flashes discussed that do contain continuing current, 4 out of the 5 two- and three-stroke flashes have continuing current following the final stroke. The remaining seven flashes all consist of over seven strokes. Only two of the seven flashes are followed by an F change. In contrast with the data of Malan (1954) and Pierce (1955a), Kitagawa et al. (1962) found that continuing luminosity was rare following single-stroke flashes (Sec. 2.5.3).

We discuss now the J field change. The J field change can be defined as a slow field change which is negative for near strokes, positive for distant strokes, and is accompanied by no appreciable luminosity of the channel between cloud base and ground. That is, we define the J field change as being caused by charge motion in the cloud. In order to measure the J change it is probably necessary to be within about 20 km of the discharge.

Malan and Schonland (1951a) have measured J changes occurring

during 105 flashes to ground (388 strokes). Typical J changes are shown
in Fig. 3.18. For strokes within 5 km, Malan and Schonland (1951a)
report 80 negative field changes, 8 zero field changes, and 2 positive field
changes. For strokes between 12 and 30 km they report 7 negative field
changes, 29 zero field changes, and 64 positive field changes. Malan and
Schonland (1951a) state that some of the zero change in the 12- to 30-km
range "may well have been positive since the J field changes at a considerable distance were rather small for accurate measurement." For
strokes between 5 and 12 km Malan and Schonland (1951b) report 71
negative field changes, 64 zero field changes, and 63 positive field changes.
In addition to the J field change reversing with distance from the discharge, the J change may reverse within a single flash (see C in Fig. 3.18).
That is, within a single flash early interstroke intervals may have a
positive J change and later ones a negative J change. This type of
reversal occurs typically when observations are made within the 5- to
12-km range.

The data indicate that in the interstroke interval there is negative
charge moving downward or positive charge moving upward within the
cloud. Further, the height of the region across which the charge is
transferred increases with stroke order. The foregoing statements can be
confirmed from an examination of Fig. 3.5 and Fig. 3.8. It is appropriate
now to consider the physical interpretation of the J or junction process.
Schonland (1938) suggested that in the time between strokes, junction
discharges progress from previously untapped negative charge centers to
the top of the previous channel to ground, thereby connecting the top of
the channel with a new source of charge. Bruce and Golde (1941)
suggested that the junction process was more likely to proceed from the top
of the previous return stroke toward the new charge centers. Malan and
Schonland (1951b) concur with the latter point of view. As we have seen,

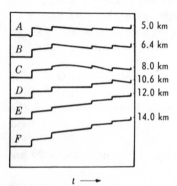

Fig. 3.18 *J-field changes observed at various
distances. (Adapted from Schon-
land (1956).)*

each succeeding return stroke is longer than the previous one. The lengthening is presumably due to the J process.

In order to attempt to determine whether the J process involves the upward movement of positive charge or the downward movement of negative charge, the detailed behavior of the J change must be examined. From Fig. 3.5 it can be seen that negative charge moving downward from a negative charge center at H yields an electric field change which, when observed at $D \sim H$, begins with a negative slope. If the height over which the J process occurs is sufficient, the slope of the field change will alter from negative to positive with increasing time. From Fig. 3.8 it can be seen that positive charge moving upward from a positive charge center at H yields an electric field change which, when observed at $D \sim 2H$, begins with a positive slope. If the height over which the J process occurs is sufficient, the slope of the field change will alter from positive to negative with increasing time. Malan and Schonland (1951a) report on eight cases in which the slope of the J field changed during a single inter-stroke interval. These eight J field changes were of long duration (0.18 sec average) and were presumably of great vertical length. Seven of the eight field changes were indicative of the motion of positive charge upward, the eighth of negative charge downward. A great deal more data are needed before definite conclusions can be drawn. Further, the source of the positive charge to be moved upward is not clear. Charges of the order of 1 coul would have to be available at the channel top in order to account for the J-process moment change, since a maximum value for the J-process moment change of 1.62 coul-km was found by Brook et al. (1962). Pierce (1955b) has suggested that the top of the discharge channel and corona regions around the top of the channel could be positively charged during the first few milliseconds following the return stroke while the channel was still electrically connected to ground.

As we have seen, theory allows a value of H/D to be estimated for which the slope of the J-process field change alters. If D is known, H can be determined. The calculation of H from J field-change data is method (4) of Sec. 3.7.3 (due to Malan and Schonland, 1951b) for determining the height of leaders and subsequent return strokes. Radar observations of the J process are considered in Sec. 3.10.

We discuss now the M field change and the K field change. Malan and Schonland (1947) reported correlated photographic and electric-field measurements from 37 cloud-to-ground flashes (199 strokes) most of which were at distances less than 6 km. Short-duration, hook-shaped field changes were found to occur during the R_c portion of the return-stroke field change while the discharge channel was still faintly luminous. These hooked-shaped field changes were correlated with increases in the

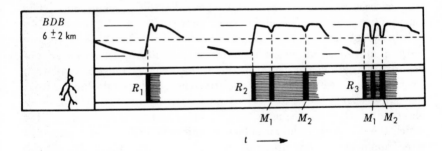

Fig. 3.19 *Electric field changes due to M components as observed near the lightning discharge correlated with channel luminosity. (Adapted from Malan and Schonland (1947).)*

channel luminosity, the photographically observed M components. In Fig. 3.19 examples of M component field changes are given. Malan and Schonland (1947) found that 60 percent of the return strokes examined did not have a R_c portion. These showed no M components. The hooked-shaped, M-component field changes observed at close distances began with a negative field change and were followed without any pause by a larger positive field change. The net field change was usually from one-fifth to one-hundredth that of the R_b change. The hook process was found to occupy between 200 and 800 μsec, although longer values were observed. Longer hook duration was found to be associated with longer time intervals between the return stroke and the occurrence of the M component. Kitagawa et al. (1962) have found that M changes occur also during periods of continuing luminosity (continuing current). They have correlated M changes, which for distant strokes appear as unidirectional pulses on the electric-field-change-meter records, with increases in channel luminosity. Examples of these data are shown in Fig. 3.13. Kitagawa et al. (1962) present data showing that during the first 15 msec after a return stroke the time interval between M components tends to increase rapidly with elapsed time; that between 15 and 40 msec this tendency decreases; and that after 40 msec the time interval between M component shows no dependence on the elapsed time. Kitagawa et al. (1962) have shown that the K change, the small, rapid electric field change which occurs in the intervals between and after the strokes of a multiple-stroke flash (Kitagawa, 1957b; Kitagawa and Brook, 1960), is to be associated with the photographed M component during periods of continuing channel luminosity. That is, the M field changes observed at a distance during continuing luminosity are K changes. Further, the most frequent time between all K changes and between M-component changes occurring during continuing luminosity was found to be about 6 msec.

Kitagawa et al. (1962) suggest that there is no essential difference between K changes associated with M components and K changes which occur during the nonluminous interstroke periods. Kitagawa et al. (1962) report that in the absence of continuing luminosity (and M components), K changes are sometimes associated either with luminosity in the cloud or with a luminous wave (similar to the dart leader) which propagates down the channel but fails to reach the ground. On the basis of their investigations, Kitagawa et al. (1962) suggest that the K change "is evidence of the movement of penetrative streamers into fresh regions of cloud, streamers whose occurrence must be determined wholly by conditions inside the cloud." Further, Kitagawa and Brook (1960) and Kitawaga (1965) state that the measured slow J field change can be interpreted as due to the instrumental time integration of a series of rapid K field changes of duration less than 1 msec and moment change between a few hundredths and 1 coul-km. In this view, the J change is the smoothed trace of the electric field record which actually consists of a number of very small K-change steps. Data regarding K changes in intracloud and in air discharges are given in Sec. 3.8.

We consider now the slow negative field changes reported by Malan (1965) to occur in South Africa in 44 percent of the interstroke periods for flashes at distances of 25 to 100 km. Malan (1965) presents a theory to explain the negative field changes. The physical basis for the theory is the increase in negative space charge above a cloud due to the increased field which exists there after negative charge has been lowered to ground by a stroke. Malan (1965) using certain assumptions calculates the ratio of the field change (which he calls U) due to the space-charge increase to the J field change. The U change is of opposite sign from the J change. If the magnitude of the two fields is the same at 20 km (the ratio set equal to unity), the U change will be at most a factor of 3 greater than the J change between 50 and 100 km. The space-charge field change would not appear to be of significant magnitude to account for the reported slow negative field changes. Further, it is not clear that the space charge can redistribute itself in the newly created electric field in a matter of 50 msec or so. Malan (1965) suggests that in some cases the fields may be large enough to initiate a glow discharge between the space charge and ionosphere, thus causing a significant decrease in the space-charge relaxation time.

It is possible that small steady currents flow to ground during the interstroke period and that these currents do not cause sufficient channel luminosity to darken photographic film at the distances that cameras are normally employed. Brook et al. (1962) have discussed this possibility. Ten amp flowing for 50 msec between strokes would result in a charge transfer of 0.5 coul and, assuming a height of 5 km, a moment change of

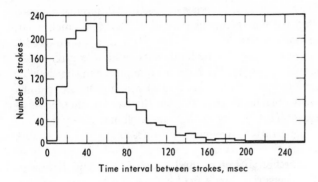

Fig. 3.20 *The time interval between strokes vs. frequency of occurrence as measured in South Africa. (Adapted from Schonland (1956).)*

2.5 coul-km. The "dark" current would yield a positive field change which would serve to decrease the negative J changes occurring at distances within the reversal zone.

In Fig. 3.20 data are given for the interstroke time interval as measured from electric-field-change records in South Africa. Data on the time interval between strokes are also presented in Sec. 2.5.3 and Fig. 2.10. Further information regarding the time interval is given in Sec. 7.7.

3.7.6 Discharges lowering positive charge to earth

While most cloud-to-ground lightning flashes result in the transfer of negative charge from the cloud to the ground, discharges do occur which lower positive charge. Positive charge resides in both the p and the P regions of the cloud, so that it is not unreasonable to expect "positive" discharges to occur.

Schonland and Allibone (1931) concluded from electric field measurements and visual observations made on more than 50 South African thunderstorms that net negative charge was lowered in at least 95 percent of the 404 flashes examined. By implication, positive flashes were also recorded. Halliday (1932) reports similar observations, finding 267 negative cloud-to-ground flashes and 16 positive, that is, about 6 percent positive. Jensen (1933) in the United States obtained correlated photographic and electric-field-change data for lightning flashes. He found 19 cloud-to-ground flashes accompanied by negative field changes, 93 cloud-to-ground flashes accompanied by positive field changes, and 21 cloud-to-ground flashes whose field changes were indeterminate.

Bruce and Golde (1941) have summarized the data available up to 1941 on the polarity of lightning flashes to earth. As pointed out by Bruce and Golde some of the data is in error in that flash polarity was determined from field change data without knowledge of whether the field changes were due to cloud discharge or to cloud-to-ground discharges. The percentage of flashes lowering a net positive charge as calculated from the data of 17 published papers varies from essentially 0 to about 30 percent.

Pierce (1955a,b) made electric field measurements with sufficient time resolution so that the R portions of flashes bringing positive charge to ground could be delineated. Making measurements far from the flashes he found 34 field changes which were negative and had negative R portions. Of these, six had more than one recorded R portion. By contrast, positive field changes with positive R portions were found in 373 cases. The negative field changes with negative R portions were in general similar to the positive field changes with positive R portions with the following exceptions: (1) The field changes preceding the initial negative R change were of significantly longer duration than those preceding the initial positive R change; (2) the moment change preceding the initial negative R change was almost twice that preceding the initial positive R change; and (3) there was usually only one R element in a positive discharge. Pierce (1955b) suggests that these effects indicate that the flash lowering positive charge is initiated in the P region. As further evidence of this supposition he states that flashes lowering positive charge most frequently occur in the later stages of a storm. In the later stages the P region should be less shielded from ground by the N region than in the early stages. Pierce (1955a,b) has recorded flashes in which the initial R change was negative and subsequent R changes positive, indicating that the first return stroke lowered positive charge while subsequent return strokes lowered negative charge. This type of field change was more common than that containing only negative R changes. The field change with initial negative R change and subsequent positive R change typically took place at the end of a storm.

From measurements of the magnetic fields accompanying 2,286 lightning strokes, Norinder (1956), working in Sweden, has inferred the charge sign lowered by those strokes. He reports that about 92 percent of the strokes lowered predominantly negative charge and about 3 percent predominantly positive charge, the remainder lowering both negative and positive charge. The magnetic field measurements of Norinder and co-workers will be further considered in Sec. 3.9 and in Chap. 4.

Additional data regarding flashes bringing positive charge to earth have been accumulated by direct measurement of lightning currents. These data will be discussed in Chap. 4, and portions are contained in Sec. 2.5.4. A photograph thought to be of a downward-moving, positively charged leader has been published by Berger and Vogelsanger (1966) (see Sec. 2.5.4).

3.8 THE ELECTROSTATIC FIELD CHANGE DUE TO CLOUD DISCHARGES

Although cloud-to-ground lightning has been the subject of considerable study, investigations into the nature of discharges within a cloud, between clouds, or between a cloud and air have not been so numerous. Further, there is considerable discrepancy between cloud-discharge parameters obtained by different workers. While lightning flashes to ground are characterized by rapid R-component field changes occurring every 50 msec or so and lasting for times of the order of 1 msec, intracloud discharges produce slow, relatively smooth field changes. Both cloud and cloud-to-ground discharges have about the same total time duration, generally a fraction of a second. The qualitative characteristics of the field changes due to cloud discharges were first discussed by Schonland et al. (1938). Pierce (1955a) has measured the time duration of 685 slow, negative field changes and 143 slow, positive field changes, all observed far from the discharges. Pierce (1955a) attributes the slow, negative field changes to "air or cloud discharges which either lower positive charge, or more probably, raise negative charge"; the slow, positive field changes to "flashes which do not reach the earth, and which probably involve the downward movement of negative charge." Pierce (1955a) finds a mean duration of 245 msec for the negative field changes and 145 msec for the positive field changes. Thus slow, positive field changes are of shorter duration and are apparently less frequent than are slow, negative field changes. Both the positive and the negative slow field changes were found to yield a typical moment change of about 100 coul-km.

If a cloud discharge occurs between two spherical charge centers which are not vertically oriented, seven parameters are needed to describe completely the resultant electric field change. These parameters are the three spatial coordinates of each charge and the magnitude of the charge involved. If the cloud discharge is vertical, only five parameters are needed. If a complete specification of the discharge parameters is not required, the field change due to a vertical cloud discharge can be expressed in terms of four parameters, as in Eq. (3.41). In Eq. (3.41), two spatial coordinates are contained in the distance D. If electric field change measurements are made at seven separate stations, the resultant data are sufficient to completely specify the seven discharge parameters. Such measurements were first reported by Workman et al. (1942), working in New Mexico. Workman et al. analyzed data for over 100 cloud discharges (three storms.) Charge transferred was found to vary from 0.3 to 100 coul. The average vertical separation of charges in an intracloud discharge was found to be 0.6 km, the positive charge being above the negative. The occurrence of the negative charge above the positive was also observed. Horizontal separations of charge were reported to vary between 1 and 10 km with an average value of about 3 km. Thus the intracloud discharges were found to be more nearly horizontal than vertical. Reynolds and Neill (1955) also working in New Mexico and using techniques developed by Workman et al. (1942) have presented data on 35 cloud discharges, 28 of which were found to have the positive charge center above the negative. The average height of the negative charge center was found to be about 5.5 km above ground level (about 8 km above sea level). Typical vertical separation of charge centers was about 0.6 km. Horizontal separations ranged from very small up to about 0.5 km. The charge destroyed in a cloud discharge was found to range from 1 to 40 coul, generally being about 10 to 20 coul. Typical moment changes were therefore of the order of 10 coul-km, about an order of magnitude less than reported for cloud discharges by Wilson (1920), Wormell (1939), and Pierce (1955a). Data in support of the observations of Reynolds and Neill (1955) have been obtained by Tamura et al. (1958) in Japan who report cloud-discharge moment changes (measured at eight sites) of between 6 and 72 coul-km with a most frequent value of about 20 coul-km. On the other hand, Hatakeyama (1958) also working in Japan reports average cloud-discharge moment changes of about 400 coul-km; and Wang (1963a,b) working in Singapore reports average cloud-discharge moment changes of about 200 coul-km.

Smith (1957), working in central Florida, simultaneously recorded at two stations 13.2 km apart the slow electric field changes produced by cloud discharges. Data were obtained for which the field changes at the two stations were of opposite signs, indicating that the two stations were on opposite sides of the field-change reversal distance. Smith (1957) analyzed electric-field-change data from 54 discharges, each of which produced a field change containing a maximum or minimum at one or both of the recording stations. These data plus a knowledge of which station was closer to a given discharge allowed a determination of the sign and direction of motion of the charge within the cloud. To make this determination diagrams similar to those given in Figs. 3.5, 3.6, 3.7, and 3.8 were employed. Smith (1957) used as discharge models a vertical positive dipole (positive point charge above negative) and a vertical negative dipole (negative point charge above positive) in which one point charge was allowed to move. He found that in 39 of the 54 observations the dipole was positive. In the 39 positive dipole cases, negative charge was raised 30 times and positive charge lowered 9 times. In the 15 negative dipole cases, positive charge was raised 9 times and negative lowered 6 times. Smith (1957) infers from his data that the discharge of positive dipoles takes place at greater altitude than does that of negative dipoles. Smith (1957) reports that about 80 percent of the slow field changes measured can be explained satisfactorily by the uniform vertical movement of a single charge (point charge model) within the cloud.

Kitagawa and Brook (1960) have obtained electric field records of about 1,400 cloud discharges. The data were recorded using the electric-field meter and the electric-field-change meter discussed in Sec. 3.5.1. Typical field-change data are shown in Fig. 3.21. Kitagawa and Brook (1960) have divided the cloud-discharge field change into 3 portions: (1) An initial portion, (2) a very active portion, and (3) a later or J (junction) type portion. (1) The initial portion is characterized by pulsations of relatively small amplitude with a mean pulse interval of about 680 μsec. The duration of the initial portion of the cloud discharge ranges from 50 to 300 msec. Thus significant differences are reported between the initial stages of cloud and of cloud-to-ground discharges: The time interval between pulses in the initial portion and the duration of the initial portion of the cloud discharge are reported to be significantly longer than the stepped-leader interpulse interval and time duration. In contrast to this data Schonland et al. (1938) and Schonland (1956) report that the pulsations superposed on the slow electric field change due to a cloud discharge exhibit the same interpulse time intervals as do the pulsations from the stepped leader. Kitagawa and Brook (1960) report that they find the initial field-change characteristics of cloud and of cloud-to-ground discharges so different that from the first 10 msec of the field records they can predict with over 95 percent certainty whether a discharge will reach ground or will remain within the cloud. (2) The very active portion of the cloud-discharge field change exhibits initially a relatively rapid field change on the electric-field meter and corresponding large pulses on the electric-field-

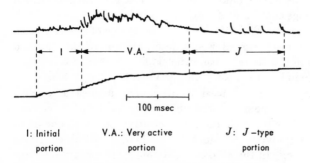

Fig. 3.21 *Diagram of the typical field change due to a cloud discharge. The upper and lower traces represent simultaneous data recorded by the electric-field-change meter and electric-field meter respectively. (Adapted from Kitagawa and Brook (1960).)*

change meter. (3) The later or J-type portion of the cloud-discharge field change is similar to the J portion of the ground discharge. K changes occur at intervals from 2 to 20 msec. The frequency distribution of K-change intervals for 671 ground discharges and for the J-type portion of 1,318 cloud discharges is plotted by Kitagawa and Brook (1960). The distributions for the two types of discharges are almost identical, strongly suggesting that the in-cloud discharge mechanism between strokes of a ground discharge and the later portion of a cloud discharge are essentially the same. The field change in the J-type portion is not as rapid as in the preceding very active stage. The J-type portion of a cloud discharge is very distinct in character from the initial and very active portions. The difference between the initial and the very active portions, however, is not always distinct, the transition from one to the other usually being more or less gradual.

Kitagawa and Brook (1960) report that out of about 1,400 cloud discharges studied, 50 percent contained all three of the above mentioned portions, 40 percent consisted of very active and J-type portions, and the remaining 10 percent lacked the J-type portion and consisted of either the initial or very active portion, or both.

Takagi (1961) and Ogawa and Brook (1964) have compared the electric field changes due to intracloud discharges to model field changes similar to those given in Figs. 3.5 to 3.8 and have found the observed variations of electric field in the initial part of the discharge to be consistent with a positive dipole model in which positive charge was lowered. Ogawa and Brook (1964) report that the field changes taking place during the initial and very active portions of the cloud-discharge field-change are probably to be associated with the downward propagation of a positively charged leader. Takagi (1961) suggests that the effective velocity of the downward-moving leader is about 10^4 m/sec. For an isolated storm Ogawa and Brook (1964) give data showing both the steady cloud-field and the cloud-discharge field changes (about 600) as a function of the distance to the storm. The field-change reversal distance differs from the steady-field reversal distance [see discussion preceding Eq. (3.42)]. Ogawa and Brook (1964) report that the final, or J-type, portion of the cloud discharge consists primarily of negative recoil streamers (see Sec. 2.3.7) which are associated with the K field changes. The recoil streamers are presumably initiated when the downward-moving, positive discharge contacts a region of concentrated negative charge. The recoil streamers, photographed by Ogawa and Brook (1964) in cloud-to-air discharges, originate at the tip of the advancing discharge and travel back along the channel. According to Ogawa and Brook (1964), the K changes occurring during the later portion of the cloud discharge have a time duration of between 1 and 3 msec and an average moment change of about 8 coul-km (a value somewhat higher than the K-change moment associated with cloud-to-ground lightning for which an upper limit of about 2 coul-km was measured). The most frequent number of K changes per discharge was 6. Usually no field change was noticeable in the intervals between the K changes, the net field change during the final part of the discharge being effectively the sum of the individual K changes. Ogawa and Brook (1964) report that K changes do not occur in the initial and very active portions of cloud discharges, that they are apparently not a part of the initial discharge-propagation process.

Takagi (1961) and Ogawa and Brook (1964) consider the negative recoil streamers (K changes) to be essentially negative return strokes. Ogawa and Brook (1964) suggest that the K-change channel is between 1 and 3 km in length and that since the K-change duration is of the order of 1 msec, the propagation velocity is about 2×10^6 m/sec. Ogawa and Brook (1964) further suggest from their measurements that the charge neutralized per K change is about 3.5 coul and that the average current associated with a K change is about 1 to 4 ka.

The values of recoil-streamer charge, current, channel length, propagation velocity, and frequency of occurrence reported by Takagi (1961) are all within a factor of 2 or 3 of the values obtained by Ogawa and Brook (1964). Takagi (1961) suggests that the intracloud discharge and the in-cloud interstroke processes of a cloud-to-ground flash are similar, except for the direction of discharge propagation. That is, Takagi (1961) considers the J process to be an upward-moving, positive discharge down which propagate negative recoil streamers. Takagi (1961) suggests that a recoil streamer of large intensity may become a dart leader propagating to ground down the previous return-stroke channel.

Photoelectric measurements of cloud luminosity correlated with electric-field-change data have been obtained by Takagi (1961). The complete luminous cloud discharge was found to have a duration of 0.2 to 0.5 sec. The slow field change which is attributed by Takagi (1961) to a downward-moving positive leader was accompanied by weak continuous luminosity. The continuous luminosity fluctuated considerably, but on the average rose to a peak value at a time from initiation equal to about one-third of the total luminous duration. Strong luminous pulses were found to coincide with the K field changes. The pulse luminosity rose to peak in about 0.5 msec and fell to half value in about the same time. Similar measurements of the luminosity of cloud discharges have been made by Malan (1955b) (see Sec. 2.3.7) and by Brook and Kitagawa (1960).

Takeuti (1965), working in Japan, reports on cloud-discharge field changes observed at three separate stations. In addition, he references previous publications (not all easily accessible) reporting field-change measurements of staff members of the Research Institute of Atmospherics, Nagoya University. Takeuti (1965) considered as discharge models the vertically oriented positive dipole, the vertically oriented negative dipole, and the horizontal dipole. Most cloud discharges were found to fit the first model. Charge transfer between an upper positive charge and a lower negative charge was reported to often exceed 100 coul. The discharge length was usually several kilometers. The overall occurence of positive descending charge (24 occurrences) was found to be about twice that of negative ascending charge. These statistics are not in good agreement with those of Smith (1957), but support the observations of Takagi (1961) and Ogawa and Brook (1964). Vertically oriented negative dipoles were observed in two storms. The discharges appeared to occur at a higher altitude and to transfer less charge, 5 to 40 coul over a distance of several kilometers, than did the positive-dipole discharges. Again, the data is in disagreement with the conclusions of Smith (1957). Quantitative data for three horizontal discharges were obtained. The horizontal discharges were found to occur at heights of 6 to 12 km and to exhibit lengths of 2 to 8 km. Two of the horizontal discharges were reported to transfer 5 to 10 coul, the third between 130 and 250 coul.

The velocity of propagation of a cloud discharge can be estimated from the path length (determined by field measurements) and the time duration of the field change. Takeuti (1965) reports that the mean propagation velocity was 1 to 2×10^4 m/sec, independent of discharge polarity and path length (2 to 8 km). Further, the discharges observed were reported to progress at relatively constant velocity throughout the path length.

3.9 THE MAGNETOSTATIC FIELDS OF LIGHTNING

The magnetostatic field (see Sec. 3.3) present during a lightning discharge is directly proportional to the discharge current. Thus measurements of the magnetostatic field can be used to infer discharge current. We will be concerned in this section with magnetic measurements made at distances of more than a few meters from the discharge channel. Magnetic measurements made in very close proximity to the channel will be considered in Chap. 4.

Ballistic magnetometers have been used by Hatakeyama (1936) and by Meese and Evans (1962) to measure the time integral of the lightning current, that is, the charge transferred. The ballistic magnetometer is essentially a bar magnet suspended on a wire. Angular momentum is imparted to the magnet by the magnetic field of the lightning discharge. After the magnetic field of the discharge has gone to zero, the magnet exhibits damped simple-harmonic motion about its equilibrium position.

If the magnetic field duration is short compared to the period of oscillation of the magnetometer, the initial angular momentum of the magnet will be proportional to the time integral of the applied magnetic field. Hence the maximum angle of displacement of the magnet can be related to the charge transferred. According to Meese and Evans (1962), if the magnetic field due to the lightning discharge has a duration which is not short compared to the period of oscillation of the magnetometer, the calculated value of charge transferred will be an underestimation of the actual value.

Meese and Evans (1962), using a magnetometer whose period was about 2 sec, determined the charge transferred for 16 flashes which occurred near Tucson, Arizona. They found values of charge transferred per flash between 23 and 1,065 coul. The average for all flashes was 256 coul. The average was 143 coul if the two largest values of charge transferred were ignored. It has been pointed out by E. T. Pierce that the charge transferred for a given discharge as measured by Meese and Evans is strongly correlated with the distance to that discharge; that is, larger charge transfer was found for strokes which were farther away. This observation casts some doubt on the validity of the measurements. Hatakeyama (1936), working in Japan, determined charge transfer for five lightning discharges of unknown character whose distances were known approximately. His data are analyzed assuming the five discharges were cloud-to-ground discharges, were intracloud discharges, and were horizontal discharges. If the discharges were cloud-to-ground, the charge transferred for four of the five discharges was between about 20 and 100 coul, the fifth being between 100 and 200 coul. If the discharges were vertical intracloud discharges, the charge transferred was calculated to be 1.5 to 2 times larger than in the cloud-to-ground case. If the discharges were horizontal, charge transferred for discharge lengths between 3 and 10 km was between 19 and 254 coul.

Magnetic measurements of lightning continuing currents and of two stepped leaders were reported by Williams and Brook (1963). Measurements were made using a fluxgate magnetometer with a 20 msec time response. An average continuing-current value of 184 amp was found for 14 continuing-current intervals of average time duration 174 msec. The average negative charge lowered in a continuing-current interval was found to be 31 coul. These current and charge values are a factor of 2 or 3 larger than the values determined by Brook et al. (1962) from electric-field-change measurements. Williams and Brook (1963) have measured stepped-leader currents of 50 and 63 amp for two stepped leaders in which the duration of the magnetic effects due to the leader was greater than the 20-msec response time of the magnetometer. Most of the records did not show magnetic effects due to stepped leaders. Williams and Brook

(1963) suggest that the lack of magnetic effects may indicate that the stepped leader current is generally less than 50 amp.

Norinder and his co-workers in Sweden have made extensive measurements of the magnetic fields associated with lightning. These measurements were made using loop antennas coupled to oscilloscopes. One of the primary goals of this research has been the determination of the time variation of the lightning current. We shall consider these data in the next chapter. Norinder and Knudsen (1961) report that photographic measurements show that the length of the lightning discharge between cloud base and ground in Sweden has a mean value of 1.4 km, somewhat smaller than values measured in temperate or tropical latitudes. Further, they report that of 1,135 lightning flashes observed, 79 percent were cloud-to-ground discharges, a larger percentage than reported in temperate or tropical latitudes. Norinder and Knudsen (1961) suggest that the high percentage of cloud-to-ground discharges in Sweden is to be associated with the low cloud base.

Norinder and Knudsen (1961) and Norinder and Dahle (1945) give data determined from magnetic field measurements regarding the number of strokes per flash in Sweden. In the earlier study (125 flashes) the most frequent number of strokes per flash was found to be 3 or 4; in the latter study (169 flashes) the most frequent number of strokes per flash was found to be two. In these two studies the relative magnitude of the magnetic flux density associated with the various strokes of a flash was determined. The magnetic flux density is, at least roughly, proportional to the current since measurements were made within 20 km of the discharges. Norinder and Knudsen (1961) report that for two-stroke flashes, the magnetic flux density from the first stroke was found to be about $2\frac{1}{2}$ times that of the second stroke. For three-stroke flashes, the magnetic flux density from the first stroke was found to be about $3\frac{1}{2}$ times that of the second and third strokes, these having about equal flux densities. For four-stroke flashes, the magnetic flux density of the first stroke was about $3\frac{1}{2}$ times that of the second, and about 14 times that of the third and fourth strokes. In the earlier study Norinder and Dahle (1945) reported similar findings for two- and three-stroke flashes, but found that in flashes with four to seven strokes, the second or third strokes had the highest magnetic flux density and presumably the highest current. Norinder and Vollmer (1957) report that the most frequent time interval between strokes in multiple-stroke flashes in Sweden is 30 msec.

Norinder and Knudsen (1957) have recorded the magnetic fields associated with the breakdown processes preceding 300 first strokes. In addition they have recorded the magnetic fields of 200 predischarges which were not followed by return strokes. The recording time was at most 10 msec prior to a return stroke, a time shorter than the usual duration of predischarge electric field changes (see Sec. 3.7.2). The few milliseconds just prior to a return stroke were found to be occupied by magnetic field pulses which occurred at a most frequent time interval of 50 to 100 μsec. These pulses Norinder and Knudsen (1957) have associated with the portion of the stepped leader occurring outside the cloud, although a time duration of a few milliseconds is somewhat short for the expected effects. Prior to this "final" predischarge phase is a period of a few milliseconds characterized by slower pulsations of the magnetic field. In this "initial" predischarge phase, pulses are reported to occur at time intervals generally over 100 μsec, although the most frequent time interval is given as about 100 μsec. The initial predischarge stage exhibits essentially the same characteristics whether or not it is followed by the "final" predischarge phase and the return stroke. Norinder and Knudsen (1957) found that in 40 percent of the cases studied the period between initial pulses showed no magnetic field, whereas in the remaining cases the period between the initial pulses was occupied by smaller pulses superposed on a relatively steady magnetic field. These smaller pulses had similarity to the pulses occurring just prior to the return stroke. Of the larger initial pulses measured, about 60 percent were predominantly negative. They were therefore probably magnetostatic pulses associated with the downward movement of negative charge. Many pulses showed some bipolar character. In these pulses the radiation component of the magnetic field was probably appreciable. Norinder and Knudsen (1957) have compared the peak values of the larger initial predischarge pulses with the peak magnetostatic fields due to return strokes. The predischarge pulse amplitudes were found to be most frequently about 15 percent of the return-stroke amplitudes. If the magnetic fields of the larger initial pulses are related to channel current with the same constant of proportionality as is the return-stroke field, and if both pulses and return-stroke waveshapes are adequately recorded, then the larger initial predischarge pulses are indicative of a current flow of a few thousand amperes.

3.10 THE RADIATION FIELDS OF LIGHTNING

We have considered briefly in Sec. 3.7.2 the radiation fields associated with the stepped leader. Stepped-leader radiation pulses are shown in Fig. 3.14d. The radiation field data of Fig. 3.14 were obtained by Clarence and Malan (1957) with a system whose upper-frequency cutoff was 20 kHz, and thus their data represent a somewhat distorted picture of the actual radiation fields present. In Fig. 3.22 is shown schematically the radiation fields of the lightning flash as a function of time and of

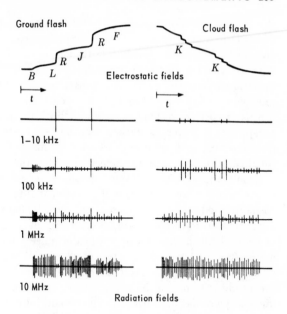

Fig. 3.22 *Electrostatic field changes and the corresponding radiation fields at different frequencies for typical cloud-to-ground flash and typical intracloud flash both at a distance of about 20 km. Amplitude scales for different frequencies are not the same. (Adapted from Malan (1958, 1963).)*

frequency. The data of Fig. 3.22 were obtained by Malan (1958, 1963) by placing tuned filters between the antenna and the recording oscilloscope. Electrostatic and radiation fields were displayed simultaneously on a dual-beam oscilloscope. It is important to remember that the radiation detected at a given frequency by the tuned antenna may not be due to a single discharge process, but that a series of radiation pulses may set the antenna system into forced oscillations even though the frequency of oscillation is not that characteristic of one of the radiation pulses. With this word of warning, we consider Malan's observations.

In the 1- to 10-kHz range the predominant radiation in cloud-to-ground flashes is emitted by the return stroke. In cloud discharges, the K changes are a source of low-frequency radiation. At 100 kHz the return stroke of flashes to ground and the K process in cloud discharges still provide the largest radiation fields, but appreciable radiation is emitted during the breakdown and interstroke processes of the ground flash and throughout the cloud flash. At 100 kHz the radiation field amplitudes from cloud and ground flashes are approximately equal. (At 1 kHz the ground-flash radiation-field amplitude is about 40 times that of the cloud flash; at 10 kHz about 10 times that of the cloud flash.) At 1 MHz the radiation pattern is similar to the pattern at 100 kHz except that the relative importance of the K and R fields is less. At 10 MHz the radiation is almost continuous throughout both ground and cloud discharges except that it decreases markedly in the 2- to 15-msec period following the return stroke in a ground discharge. This period of time corresponds roughly to the R_c phase of the return-stroke field change (see Sec. 3.7.4).

Brook and Kitagawa (1964) have measured radiation fields at frequencies of 420 and 850 MHz from lightning flashes 10 to 30 km distant. Also obtained were simultaneous electrostatic field measurements. They report that the stepped leader is a strong source of microwave radiation. In general, all the larger stepped-leader pulses on the electric-field-change meter (Sec. 3.5.1) records are accompanied by radiation pulses. The pulse duration is much less than the time between steps. The dart leader was found to be one of the strongest sources of microwave radiation. The microwave radiation during the dart leader phase was found to be almost always shorter in duration than the dart-leader electric field change. The dart-leader radiation was very often found to cease between 50 and 150 μsec before the return-stroke phase. The dart-leader radiation was generally continuous with occasional strong pulses of duration 50 to 80 μsec. The return-stroke radiation at microwave frequencies varied over wide limits. In about half the cases studied, microwave radiation was not coincident with the return stroke but appeared 60 to 100 μsec after the initiation of the return stroke. The quiet period of 2 to 15 msec following the return stroke reported by Malan (1958, 1963) for a frequency of 10 MHz was not especially noticeable at microwave frequencies. The entire period between return strokes was found to be relatively quiet except for the occurrence of K changes. The K-change process in both cloud and ground discharges was found to be a strong source of microwave radiation. The K changes, with a duration of 500 to 750 μsec as measured on the electric-field-change meter, usually produced several short radiation pulses as well as continuous emission of lower amplitude.

Brook and Kitagawa (1964) report that cloud discharges appear to be as strong or stronger microwave radiators than ground discharges. The initial portion of a cloud discharge has continuous microwave radiation as well as strong pulse emission. The very active part produces large and frequent radiation pulses. The J-type portion gives rise to K-change radiation pulses of the same magnitude as the K-change microwave radiation characteristic of the ground discharge.

Brook and Kitagawa (1964) suggest that the measured microwave radiation is due primarily to breakdown processes, and that, with the exception of the stepped leader, those processes occur in the cloud. This suggestion is supported by the observation that (1) microwave radiation is often absent during the time the return-stroke wavefront propagates from ground to cloud, indicating that the return stroke is not a significant source of microwave radiation, and (2) microwave radiation is almost always absent in the final 50 to 150 μsec of the dart leader phase indicating that the dart-leader tip or channel is not the principal source of the microwave radiation occurring during the dart-leader phase. On the other hand, microwave radiation which can be associated with the stepping process is present during the entire period preceding the first return stroke.

The temporal behavior of the electric component of the radiation field due to a return stroke is shown in Fig. 3.11. If the temporal behavior is measured or is derived theoretically, a Fourier analysis of the temporal behavior will yield an amplitude spectra for those frequencies present in the pulse. The theoretical approach usually taken is (1) to calculate an effective electric dipole moment M for the return stroke, (2) to derive the electric field intensity from M using the last term on the right of Eq. (3.24), and (3) to Fourier-analyze the time-dependent electric field intensity to obtain a frequency spectrum. Analyses of this type are given by Hill (1966) and by Dennis and Pierce (1964). The electric dipole moment M is calculated by assuming a functional form for the lightning charge variation as a function of time and of channel height. On the basis of the different waveshapes of current (see Sec. 4.3) a distinction can be made in theory between radiation fields (frequency spectra) of first strokes and of subsequent strokes. Some experimental data regarding the differences between first and subsequent stroke spectra are given by Hart (1967) and by Bradley (1965). In general, the experimental and theoretical return-stroke frequency spectra are in reasonably good agreement. The peak amplitude for the frequency spectrum of the return stroke generally occurs between 3 and 7 kHz. The amplitude of the frequency spectrum of a typical return stroke is relatively small above 100 kHz.

Arnold and Pierce (1964) have calculated that a series of stepped-leader pulses or a series of K-change pulses will have a frequency spectra with peak amplitudes in the vlf region (3 to 30 kHz). Arnold and Pierce (1964) find peak amplitudes of the frequency spectra for stepped-leader radiation, K-change radiation, and return-stroke radiation at 20, 8, and 5 kHz, respectively, with corresponding relative peak amplitudes of 1:2:10. Arnold and Pierce (1964) compare their calculations with the experiments of Steptoe (1958).

Review papers containing references to measurements of the electromagnetic radiation emitted by lightning and by thunderstorms in the frequency range from hertz to hundreds of megahertz have been published by Horner (1964), Taylor (1965), Kimpara (1965), and Pierce (1967). The following general features are evident. At vlf a few discrete radiation pulses are emitted, and these are to be associated with macroscopic features (return strokes, K changes, etc.) of the lightning discharge. As the frequency increases above the vlf band the number of pulses per discharge increases but the peak amplitude of the pulses decreases. From about 10 kHz to over 100 MHz the peak electric field intensity varies approximately inversely with frequency.

In Table 3.1, values of peak electric field intensity are given as a function of frequency. The values were obtained by Pierce (1967) by averaging the results of various investigators. The results of any one study may differ from the average by more than an order of magnitude. Horner (1964) and Kimpara (1965) have presented the frequency-spectrum measurements of various investigators in graphical form.

Horner (1964) has reviewed the literature concerning the reflection of electromagnetic signals (radar) from the lightning discharge. Of particular interest are the data of Hewitt (1957). Hewitt found that echoes from successive in-cloud interstroke discharge processes came from greater and greater vertical extents of the cloud, in support of the idea of the K and/or J processes. Further, the increase in altitude of the interstroke discharges was continuous during the time between strokes, and the occurrence of a new stroke only enhanced some or all the echoes previously present. Echoes were received from vertical heights up to 10 km. Echoes were most intense and most prolonged in the lower part of the cloud. The lower height limit from which the echoes were received remained essentially constant throughout a given flash. The

TABLE 3.1

Peak electric field intensity of vertically polarized radiation from lightning. Intensity normalized to a distance of 10 km and a detector bandwidth of 1 Hz. To obtain the field intensity for a detector bandwidth B in Hz multiply the values given by $B^{1/2}$ for frequencies above about 50 kHz and by B for frequencies below about 50 kHz. Over the frequency range covered by this table, the electric field intensity varies approximately inversely with frequency. There may, however, be wide deviations from the inverse-frequency relation in limited spectral regions. Data represent the average of the results of various investigators. Results of any one study may differ from the average by more than one order of magnitude. Adapted from Pierce (1967).

Frequency, (MHz)	10^{-2}	10^{-1}	1	10	10^2	10^3
Field, μV/m	2×10^4	2×10^3	2×10^2	10	2	3×10^{-1}

horizontal extent of the discharge activity in the lower regions (4 to 7 km in height) of the cloud was frequently of the order of 1 to 2 km. The echoing properties of the interstroke discharges were found to decrease during the latter part of the interstroke period, and it was found that a further stroke did not occur until this decrease had taken place. Hewitt states that echoes from air discharges and those from in-cloud interstroke processes are remarkably similar.

Ligda (1956) has reported the radar observation of very long, horizontal lightning discharges. These discharges showed considerable branching. The longest discharge observed was reported to be over 150 km in length.

References

1. Appleton, E. V., and F. W. Chapman: On the Nature of Atmospherics-IV, *Proc. Roy. Soc. (London)*, **A158**:1–22 (1937).

2. Appleton, E. V., R. A. Watson-Watt, and J. F. Herd: Investigations on Lightning Discharges and on the Electric Fields of Thunderstorms, *Proc. Roy. Soc. (London)*, **A221**:73–115 (1920).

3. Arnold, H. R., and E. T. Pierce: Leader and Junction Processes in the Lightning Discharge as a Source of VLF Atmospherics, *J. Res. NBS/USNC-URSI*, **68D** *(Radio Science)*: 771–776 (1964).

4. Berger, K., and E. Vogelsanger: Photographische Blitzuntersuchungen der Jahre 1955–1965 auf dem Monte San Salvatore, *Bull. SEV*, **57**:2–22 (1966).

5. Bradley, P. A.: The VLF Energy Spectra of First and Subsequent Return Strokes of Multiple Lightning Discharges to Ground, *J. Atmospheric Terrest. Phys.*, **27**:1045–1053 (1965).

6. Brook, M., and N. Kitagawa: Electric-field Changes and the Design of Lightning-flash Counters, *J. Geophys. Res.*, **65**:1927–1931 (1960).

7. Brook, M., and N. Kitagawa: Radiation from Lightning Discharges in the Frequency Range 400 to 1000 Mc/s, *J. Geophys. Res.*, **69**:2431–2434 (1964).

8. Brook, M., N. Kitagawa, and E. J. Workman: Quantitative Study of Strokes and Continuing Currents in Lightning Discharges to Ground, *J. Geophys. Res.*, **67**:649–659 (1962).

9. Bruce, C. E. R.: The Lightning and Spark Discharges, *Nature*, **147**:805-806 (1941).

10. Bruce, C. E. R.: The Initiation of Long Electrical Discharges, *Proc. Roy. Soc. (London)*, **A183**:228–242 (1944).

11. Bruce, C. E. R., and R. H. Golde: The Lightning Discharge, *J. Inst. of Elec. Eng. (London)*, **88** (pt. 2):487–524 (1941).

12. Chapman, F. W.: Atmospheric Disturbances Due to Thundercloud Discharges, pt. 1, *Proc. Phys. Soc. (London)*, **51**:876–894 (1939).

13. Chapman, S.: Corona-point Discharge in Wind and Application to Thunderclouds, in L. G. Smith (ed.), "Recent Advances in Atmospheric Electricity," pp. 277–288, Pergamon Press, New York, 1958.

14. Clarence, N. D., and D. J. Malan: Preliminary Discharge Processes in Lightning Flashes to Ground, *Quart. J. Roy. Meteorol. Soc.*, **83**:161–172 (1957).

15. Dennis, A. S., and E. T. Pierce: The Return Stroke of the Lightning Flash to Earth as a Source of VLF Atmospherics, *J. Res. NBS/USNC-URSI*, **68D** (*Radio Science*): 777–794 (1964).

16. Ette, A. I. I.: Laboratory Studies of Point-discharge from Multiple Points in Irregular Configuration, *J. Atmospheric Terrest. Phys.*, **28**:983–999 (1966).

17. Fitzgerald, D. R.: Measurement Techniques in Clouds, in S. C. Coroniti, (ed.), "Problems of Atmospheric and Space Electricity," pp. 199–214, American Elsevier Publishing Co., New York, 1965.

18. Gunn, R.: Electric Field Intensity Inside of Natural Clouds, *J. Appl. Phys.*, **19**:481–484 (1948).

19. Hacking, C. A.: Observations on the Negatively-charged Column in Thunderclouds, *J. Geophys. Res.*, **59**:449–453 (1954).

20. Halliday, E. C.: The Polarity of Thunderclouds, *Proc. Roy. Soc. (London)*, **A138**:205–229 (1932).

21. Hart, J. E.: VLF Radiation from Multiple Stroke Lightning, *J. Atmospheric Terrest. Phys.*, **29**:1011–1014 (1967).

22. Hatakeyama, H.: An Investigation of Lightning Discharge with the Magnetograph, *Geophys. Mag.*, **10**:309–319 (1936).

23. Hatakeyama, H.: The Distribution of the Sudden Change of Electric Field on the Earth's Surface Due to Lightning Discharge, in L. G. Smith (ed.), "Recent Advances in Atmospheric Electricity," pp. 289–298, Pergamon Press, New York, 1958.

24. Hewitt, F. J.: Radar Echoes from Inter-stroke Processes in Lightning, *Proc. Phys. Soc. (London)*, **B70**:961–979 (1957).

25. Hill, R. D.: Electromagnetic Radiation from the Return Stroke of a Lightning Discharge, *J. Geophys. Res.*, **71**:1963–1967 (1966).

26. Hodges, D. B.: A Comparison of the Rates of Change of Current in the Step and Return Processes of Lightning Flashes, *Proc. Phys. Soc. (London)*, **B67**:582–584 (1954).

27. Holzer, R. E., and D. S. Saxon: Distribution of Electrical Conduction Currents in the Vicinity of Thunderstorms, *J. Geophys. Res.*, **57**:207–216 (1952).

28. Horner, F.: Radio Noise from Thunderstorms, in J. A. Saxton (ed.), "Advances in Radio Research," vol. 2, pp. 121–204, Academic Press, Inc., New York, 1964.

29. Jensen, J. C.: Further Studies on the Electrical Charges of Thunderstorms, *Monthly Weather Rev.*, **58**:115–116 (1930).

30. Jensen, J. C.: The Relation of Branching of Lightning Discharges to Changes in the Electrical Field of Thunderstorms, *Phys. Rev.*, **40**:1013–1014 (1932a).

31. Jensen, J. C.: The Relation of Lightning Discharges to Changes in the Electric Field of Thunderstorms, *Trans. Am. Geophys. Union*, **13**:190–191 (1932b).

32. Jensen, J. C.: The Branching of Lightning and the Polarity of Thunderclouds, *J. Franklin Inst.*, **216**:707–747 (1933).

33. Jones, D. S.: "The Theory of Electromagnetism," pp. 152–154, The Macmillan Company, New York, 1964.

34. Kasemir, H. W.: The Thundercloud, in S. C. Coroniti (ed.), "Problems of Atmospheric and Space Electricity," pp. 215–235, American Elsevier Publishing Company, New York, 1965.

35. Kimpara, A.: Electromagnetic Energy Radiated from Lightning, in S. C. Coroniti

(ed.), "Problems of Atmospheric and Space Electricity," pp. 352–365, American Elsevier Publishing Company, New York, 1965.

36. Kitagawa, N.: On the Electric Field-change due to the Leader Processes and Some of Their Discharge Mechanism, *Papers Meteorol. Geophys.* (*Tokyo*), **7**:400–414 (1957a).

37. Kitagawa, N.: On the Mechanism of Cloud Flash and Junction or Final Process in Flash to Ground, *Papers Meteorol. Geophys.* (*Tokyo*), **7**:415–424 (1957b).

38. Kitagawa, N.: Types of Lightning, in S. C. Coroniti (ed.), "Problems of Atmospheric and Space Electricity," pp. 337–348, American Elsevier Publishing Company, New York, 1965.

39. Kitagawa, N., and M. Brook: A Comparison of Intracloud and Cloud-to-ground Lightning Discharges, *J. Geophys. Res.*, **65**:1189–1201 (1960).

40. Kitagawa, N., M. Brook, and E. J. Workman: Continuing Currents in Cloud-to-ground Lightning Discharges, *J. Geophys. Res.*, **67**:637–647 (1962).

41. Kitagawa, N., and M. Kobayashi: Distribution of Negative Charge in the Cloud Taking Part in a Flash to Ground, *Papers Meteorol. Geophys.* (*Tokyo*), **9**:99–105 (1958).

42. Ligda, M.: The Radar Observation of Lightning, *J. Atmospheric Terrest. Phys.*, **9**:329–346 (1956).

43. Malan, D. J.: Les décharges dans l'air et la charge inférieure positive d'un nuage orageux, *Ann. Geophys.*, **8**:385–401 (1952).

44. Malan, D. J.: Les décharges orageuses intermittentes et continu de la colonne de charge négative, *Ann. Geophys.*, **10**:271–281 (1954).

45. Malan, D. J.: La distribution verticale de la charge négative orageuse, *Ann. Geophys.*, **11**:420–426 (1955a).

46. Malan, D. J.: Les décharges lumineuses dans les nuages orageux, *Ann. Geophys.*, **11**:427–434 (1955b).

47. Malan, D. J.: Radiation from Lightning Discharges and Its Relation to the Discharge Process, in L. G. Smith (ed.), "Recent Advances in Atmospheric Electricity," pp. 557–563, Pergamon Press, New York, 1958.

48. Malan, D. J.: "Physics of Lightning," The English Universities Press Ltd., London, 1963.

49. Malan, D. J.: The Theory of Lightning, in S. C. Coroniti (ed.), "Problems of Atmospheric and Space Electricity," pp. 323–331, American Elsevier Publishing Company, New York, 1965.

50. Malan, D. J., and B. F. J. Schonland: Progressive Lightning, pt. 7, Directly Correlated Photographic and Electrical Studies of Lightning from Near Thunderstorms, *Proc. Roy. Soc.* (*London*), **A191**:485–503 (1947).

51. Malan, D. J., and B. F. J. Schonland: An Electrostatic Fluxmeter of Short Response-time for Use in Studies of Transient Field-changes, *Proc. Phys. Soc.* (*London*), **B63**:402–408 (1950).

52. Malan, D. J., and B. F. J. Schonland: The Electrical Processes in the Intervals between the Strokes of a Lightning Discharge, *Proc. Roy. Soc.* (*London*), **A206**:145–163 (1951a).

53. Malan, D. J., and B. F. J. Schonland: The Distribution of Electricity in Thunderclouds, *Proc. Roy. Soc.* (*London*), **A209**:158–177 (1951b).

54. Meese, A. D., and W. H. Evans: Charge Transfer in the Lightning Stroke as Determined by the Magnetograph, *J. Franklin Inst.*, **273**:375–382 (1962).

55. Norinder, H.: Magnetic Field Variations from Lightning Strokes in Vicinity of Thunderstorms, *Arkiv Geofysik*, **2**:423–451 (1956).

56. Norinder, H., and O. Dahle: Measurements by Frame Aerials of Current Variations in Lightning Discharges, *Arkiv Mat. Astron. Fysik*, **32A**:1–70 (1945).

57. Norinder, H., and E. Knudsen: Pre-discharges in Relation to Subsequent Lightning Strokes, *Arkiv Geofysik*, **2**:551–571 (1957).

58. Norinder, H., and E. Knudsen: Some Features of Thunderstorm Activity, *Arkiv Geofysik*, **3**:367–374 (1961).

59. Norinder, H., and B. Vollmer: Variation Forms and Time Sequence of Multiple Lightning Strokes, *Arkiv Geofysik*, **2**:515–531 (1957).

60. Ogawa, T., and M. Brook: The Mechanism of the Intracloud Lightning Discharge, *J. Geophys. Res.*, **69**:5141–5150 (1964).

61. Pierce, E. T.: Electrostatic Field-changes Due to Lightning Discharges, *Quart. J. Roy. Meteorol. Soc.*. **81**:211–228 (1955a).

62. Pierce, E. T.: The Development of Lightning Discharges, *Quart. J. Roy. Meteorol. Soc.*, **81**:229–240 (1955b).

63. Pierce, E. T.: Some Topics in Atmospheric Electricity, in L. G. Smith (ed.), "Recent Advances in Atmospheric Electricity," pp. 5–16, Pergamon Press, New York, 1958.

64. Pierce, E. T.: Atmospherics: Their Characteristics at the Source and Propagation, in "Progress in Radio Science 1963–1966," pt. 1, pp. 987–1039, International Scientific Radio Union, Berkeley, Calif., 1967.

65. Pierce, E. T., and T. W. Wormell: Field Changes Due to Lightning Discharges, in H. R. Byers (ed.), "Thunderstorm Electricity," pp. 251–266, University of Chicago Press, Chicago, 1953.

66. Rao, M., and H. Bhattacharya: Lateral Corona Currents from the Return Stroke Channel and the Slow Field Change after the Return Stroke in a Lightning Discharge, *J. Geophys. Res.*, **71**:2811–2814 (1966).

67. Reynolds, S. E., and W. H. Neill: The Distribution and Discharge of Thunderstorm Charge-centers, *J. Meteorol.*, **12**:1–12 (1955).

68. Schonland, B. F. J.: The Polarity of Thunderclouds, *Proc. Roy. Soc. (London)*, **A118**:233–251 (1928).

69. Schonland, B. F. J.: Progressive Lightning, pt. 4, The Discharge Mechanism, *Proc. Roy. Soc. (London)*, **A164**:132–150 (1938).

70. Schonland, B. F. J.: The Pilot Streamer in Lightning and the Long Spark, *Proc. Roy. Soc. (London)*, **A220**:25–38 (1953).

71. Schonland, B. F. J.: The Lightning Discharge, "Handbuch der Physik," vol. 22, pp. 576–628, Springer-Verlag OHG, Berlin, 1956.

72. Schonland, B. F. J., and T. E. Allibone: Branching of Lightning, *Nature*, **128**:794–795 (1931).

73. Schonland, B. F. J., and J. Craib: The Electric Fields of South African Thunderstorms, *Proc. Roy. Soc. (London)*, **A114**:229–243 (1927).

74. Schonland, B. F. J., D. B. Hodges, and H. Collens: Progressive Lightning, pt. 5, A Comparison of Photographic and Electrical Studies of the Discharge Process, *Proc. Roy. Soc. (London)*, **A166**:56–75 (1938).

75. Simpson, G. C.: On Lightning, *Proc. Roy. Soc. (London)*, **A111**:56–67 (1926).

76. Simpson, G. C., and G. D. Robinson: The Distribution of Electricity in Thunderclouds, pt. 2, *Proc. Roy. Soc. (London)*, **A177**:281–329 (1941).

77. Simpson, G. C., and F. J. Scrase: The Distribution of Electricity in Thunderclouds, *Proc. Roy. Soc. (London)*, **A161**:309–353 (1937).

78. Smith, L. G.: Intracloud Lightning Discharges, *Quart. J. Roy. Meterol. Soc.*, **83**:103–111 (1957).

79. Steptoe, B. J.: "Some Observations on the Spectrum and Propagation of Atmospherics," Ph.D. thesis, University of London, England, 1958.

80. Takagi, M.: The Mechanism of Discharges in a Thundercloud, *Proc. Res. Inst. Atmospherics, Nagoya Univ.*, **8B**:1–106 (1961).

81. Takeuti, T.: Studies on Thunderstorm Electricity, (1) Cloud Discharge, *J. Geomagnetism and Geoelectricity*, **17**:59–68 (1965).

82. Tamura, Y., T. Ogawa, and A. Okawati: The Electrical Structure of Thunderstorms, *J. Geomagnetism and Geoelectricity*, **10**:20–27 (1958).

83. Taylor, W. L.: Lightning Characteristics as Derived from Spherics, in S. C. Coroniti, (ed.), "Problems of Atmospheric and Space Electricity," pp. 388–404, American Elsevier Publishing Company, New York, 1965.

84. Vonnegut, B., C. B. Moore, R. P. Espinola, and H. H. Blau, Jr.: Electrical Potential Gradients above Thunderstorms, *J. Atmospheric Sci.*, **23**:764–770 (1966).

85. Wang, C. P.: Lightning Discharges in the Tropics, (1) Whole Discharges, *J. Geophys. Res.*, **68**:1943–1949 (1963a).

86. Wang, C. P.: Lightning Discharges in the Tropics, (2) Component Ground Strokes and Cloud Dart Streamer Discharges, *J. Geophys. Res.*, **68**:1951–1958 (1963b).

87. Williams, D. P., and M. Brook: Magnetic Measurement of Thunderstorm Currents, pt. 1, Continuing Currents in Lightning, *J. Geophys., Res.*, **68**:3243–3247 (1963).

88. Williams, J. C.: Some Properties of the Lower Positive Charge in Thunderclouds, in L. G. Smith (ed.), "Recent Advances in Atmospheric Electricity," pp. 425–429, Pergamon Press, New York, 1958.

89. Wilson, C. T. R.: On Some Determinations of the Sign and Magnitude of Electric Discharges in Lightning Flashes, *Proc. Roy. Soc. (London)*, **A92**:555–574 (1916).

90. Wilson, C. T. R.: Investigations on Lightning Discharges and on the Electric Field of Thunderstorms, *Phil. Trans. Roy. Soc. (London)*, **A221**:73–115 (1920).

91. Workman, E. J., R. E. Holzer, and G. T. Pelsor: The Electrical Structure of Thunderstorms, *Tech. Notes Natl. Advisory Comm. Aeron. (Washington)*, no. 864 (47 pp.), 1942.

92. Wormell, T. W.: Currents Carried by Point-discharges beneath Thunderclouds and Showers, *Proc. Roy. Soc. (London)*, **A115**:443–455 (1927).

93. Wormell, T. W.: Vertical Electric Currents below Thunderstorms and Showers, *Proc. Roy. Soc. (London)*, **A127**:567–590 (1930).

94. Wormell, T. W.: The Effects of Thunderstorms and Lightning Discharges on the Earth's Electric Field, *Phil. Trans. Roy. Soc. (London)*, **A238**:249–303 (1939).

95. Wormell, T. W.: Atmospheric Electricity; Some Recent Trends and Problems, *Quart. J. Roy. Meterol. Soc.*, **79**:3–38 (1953).

4

Current Measurements

4.1 INTRODUCTION

The first measurements of lightning current were made by Pockels. From his laboratory experiments Pockels (1897, 1898) found that the residual magnetism induced in a piece of nepheline basalt by a unidirectional magnetic field depended neither on the duration nor the time variation of the field, but only on its maximum value. Since the residual magnetism could be related to the maximum magnetic field to which the basalt had been exposed, it could by theory be related to the current which gave rise to the magnetic field. Pockels (1897) reported that measurements on basalt specimens obtained from the vicinity of trees damaged by lightning gave peak current values of 6.4, 6.6, and 10 ka. The errors in the measurements were probably such as to yield underestimates of the actual peak currents. Pockels (1900) placed pieces of basalt a few centimeters from a lightning rod on the observation tower on Mount Cimone in the Apennines. For one lightning flash Pockels (1900) found a peak current of about 20 ka. For a series of four lightning flashes, Pockels (1900) found a peak current of about 11 ka.

In 1929 about 300 *lightning-stroke recorders* were placed on transmission-line towers in the United States. The lightning-stroke recorder operates by producing a Lichtenberg figure whose size is crudely proportional to the peak lightning current in the tower. This instrument will be discussed in Sec. 4.2.1. The first measurements of lightning current using the lightning-stroke recorder were made in 1929 and were reported by Sporn and Lloyd (1930) and by Smeloff and Price (1930). Sporn and Lloyd (1930) measured peak currents of 175 and 100 ka for two strokes which transported negative charge to ground. Smeloff and Price (1930) reported peak currents of 60 and 100 ka for two strokes which transported negative charge to ground. During the years 1929 and 1930 over 50 crude measurements of peak lightning current were made using the lightning-stroke recorder. (Bell and Price, 1931; Sporn and Lloyd, 1931; Gross and Cox, 1931). All measured currents represented a transport of negative charge to ground and most indicated a peak lightning current near or over 100 ka. The peak current values deduced were probably somewhat of an overestimation (see Sec. 4.2.1).

In 1932, Foust and Kuehni (1932) revived the *magnetic link* for measuring peak lightning current. The magnetic link serves essentially the same purpose as the pieces of basalt used by Pockels. Whereas basalt occurs in nature, the magnetic link is fabricated in the laboratory and thus can be tailored to the desired application. Considerable data regarding peak lightning current have been collected by attaching magnetic links to transmission-line towers (see Sec. 4.3). Magnetic links have also been used as elements in instruments for measuring the temporal characteristics of lightning currents (Sec. 4.2.2).

Measurements of the lightning-current waveshape can be made by allowing the lightning current to flow across a resistance of known value. The resultant voltage vs. time characteristic can then be displayed on a cathode-ray oscilloscope. Detailed studies of this type have been carried out near Lugano, Switzerland (see Sec. 2.5.4), and at the Empire State Building in New York City (see Sec. 2.4.2). The measurement techniques will be discussed in Sec. 4.2.3, and the results of these studies will be discussed in Sec. 4.3 and 4.4.

Indirect measurements of the lightning-current waveshape have been made by Norinder and his co-workers in Sweden. In these studies, to be discussed in Sec. 4.2.4 and 4.3, a loop antenna and oscilloscope were used to record the magnetic flux density a few kilometers from the lightning discharge, and theory was used to relate the measurements to values of lightning current. The theory is made complex by the fact that the return-stroke wavefront which propagates from ground to cloud causes different currents to flow in different sections of the channel at different times. For example, the current measured at the ground may be maximum before appreciable current has started to flow a few hundred meters above the ground, that is, before the return stroke has ascended a few hundred meters above the ground.

Continuing currents and the magnetic fields associated with continuing currents can be relatively constant on a time scale measured in tens of milliseconds. Distant measurements of these magnetic fields using magnetometers with low time resolution will, with appropriate theory, allow the determination of values for the continuing current. Current measurements of this type have been made by Williams and Brook (1963) and have been considered in Secs. 3.5.4, 3.7.5, and 3.9. The electric fields associated with continuing currents increase monotonically as negative charge is slowly lowered to ground. The measurements of these slow field changes will, with appropriate theory, allow the determination of values for the continuing current. Current measurements of this type have been made by Brook et al. (1962) and have been considered in Sec. 3.7.5.

Almost all measurements of the time variation of lightning current, with the exception of the loop-antenna measurements of Norinder, have been made near ground level and represent the current flowing at the base of the lightning channel. It is certainly not apparent that the current in the channel at an arbitrary height above the ground should be identical with or even very similar to the current measured at ground level. For example, the initial current measured at the ground is determined by the breakdown process between the leader head and the ground. Upward-traveling discharges, which may bridge the gap between the ground and the downward-moving leader head, will register as current at ground. On the other hand, the initial current at a given position in the channel above ground (excluding any leader current which may be flowing) will be determined primarily by the form of the return-stroke wavefront as it passes that position. Thus, it should be kept in mind that when one discusses lightning current, it is usually the current measured at the ground which is being discussed, and that this current is not necessarily the current which is flowing or will flow at positions in the channel other than where the measurements are made.

4.2 MEASUREMENT TECHNIQUES

4.2.1 Instruments using Lichtenberg figures

In 1778, Lichtenberg (1778) reported an experiment in which a capacitor was discharged across a spark gap in which was placed an insulating plate covered with powder. One electrode of the spark gap was in contact with the insulating plate. It was found that the discharge caused the powder to be arranged in unusual patterns (Lichtenberg figures), and that the patterns had different characteristics for different polarities of the contact electrode. Trouvelot (1888) and Brown (1888) found that if a photographic plate (emulsion side in contact with an electrode) were substituted for Lichtenberg's insulating plate, figures similar to those found by Lichtenberg would be in evidence on the developed photograph.

In 1924, Peters (1924) introduced the *klydonograph*, a device to measure voltage by means of the Lichtenberg figures produced by the voltage. The klydonograph was the first practical instrument available for the measurement of lightning voltages on transmission lines. The klydonograph consists of a rounded electrode (connected to the source voltage to be measured) which rests on the emulsion side of a photographic film which in turn rests on the smooth surface of an insulating material backed by a metal plate electrode. The photographic film is often moved mechanically in order to be able to make several measurements on one piece of film. It is found experimentally that a Lichtenberg figure is produced even though the time duration of the impressed voltage is only a fraction of a microsecond; that the radius of the Lichtenberg figure is a function of the maximum value of the impressed voltage; that the shape and configuration of the Lichtenberg figure are functions of the wave shape of the impressed voltage; and that Lichtenberg figures produced by negative voltages are smaller and of a different character than those produced by positive voltages. McEachron (1926) has presented data regarding the characteristics of photographic Lichtenberg figures for different waveshapes of the applied voltage. Two klydonograph recording elements connected in parallel so that for a given applied voltage both positive and negative Lichtenberg figures are recorded is known as a *surge-voltage recorder*. The use of the surge-voltage recorder insures that a measurement of the magnitude of either positive or negative applied voltages can be made using the larger positive Lichtenberg figure. Under usual conditions, the accuracy of the surge-voltage recorder in measuring peak voltage is about ± 25 percent.

A Lichtenberg camera can also be employed as a current-measuring device. If the peak voltage across a known impedance is determined from the radius of a Lichtenberg figure, the corresponding current can be found by dividing the voltage by the impedance. This is the basis of operation of the lightning-stroke recorder (Sec. 4.1) which was generally attached across a section of a leg of a transmission-line tower. In the lightning-stroke recorder the electrodes rest on the outside of an insulator-covered film pack which can be easily changed in the field. Because no electrode makes direct contact with the film, the Lichtenberg figures produced cannot be as adequately related to voltage as can the Lichtenberg figures of the klydonograph or surge-voltage recorder. Further, in order to calculate the transmission-line-tower current from a voltage measurement, the impedance of the section of the tower leg in which the lightning current is flowing and across which the recorder is attached must be known. Since the impedance depends on the current waveshape and since neither waveshape nor impedance is known with any accuracy, the current measurements obtained using the lightning-stroke recorder should be considered rather inaccurate (Section 4.1).

A recent application of the klydonograph to the measurement of lightning current has been proposed by Griscom (1960). The *kine-klydonograph* produces time-resolved Lichtenberg figures by using electric time-delay networks to impress different portions of the voltage wave on different parts of the film. The voltage measured is produced by current flow across a noninductive resistor. The kineklydonograph is reported to have a time resolution of about 0.05 μsec. Results obtained using the kineklydonograph are given by Griscomb et al. (1965) and will be discussed in Sec. 4.5.

4.2.2 Instruments using magnetizable materials

The magnetic flux density measured near a long, straight current-carrying conductor is given by Eq. (3.20a). The flux density is directly proportional to current and inversely proportional to the radial distance from the conductor. The basalt used by Pockels (1897, 1898, 1900) and the magnetic link introduced by Foust and Kuehni (1932) become permanently magnetized when placed in a magnetic field. The degree of residual magnetization increases with the strength of the applied magnetic field and is proportional to the peak value of a time-varying magnetic flux density so long as the flux density does not change sign and saturation does not take place. Foust and Kuehni (1932) showed experimentally that essentially the same residual magnetism was induced in a link exposed to a pulsed current of microsecond duration as was induced in the same link by exposing it to direct current whose magnitude equaled the peak current of the pulse. Similar results were obtained by Pockels (1898).

The magnetic link may be constructed by combining a number of strips of cobalt steel or other magnetizable materials or may be made by sintering of magnetic material in powder form. Usually two links are placed at different distances from the current under study allowing a relatively wide range of peak currents to be measured and providing a consistency check on the data obtained. The residual magnetism in a magnetic link can be measured using various commercial and homemade magnetometers. Foust and Kuehni (1932) developed such a device which operated by providing a measure of the force of attraction or repulsion between the link and a permanent magnet. Foust and Keuhni (1932) called the combination of magnetic links and the device to measure residual magnetism in the links a *surge-crest ammeter*.

It is important to keep in mind that the magnetic link measures *peak* current. Since a lightning flash usually consists of a number of strokes, the residual magnetism of the link after being exposed to a flash is that characteristic of the stroke carrying the highest peak current. Thus peak-current statistics obtained using magnetic links do not apply to all strokes but only to the larger strokes. As we shall see in Sec. 4.3, it has been found that the first stroke in a flash usually has the highest peak current. Evidence of this fact has also been reported by Norinder and Knudsen (1961) as discussed in Sec. 3.9.

Three instruments which use magnetic links in the measurement of lightning properties have been described by Wagner and McCann (1940). These instruments are the *fulchronograph*, the *magnetic surge front recorder*, and the *magnetic surge integrator*. The fulchronograph (from the Latin, *fulmen* meaning *lightning*, the Greek, *chronos* meaning *time*, and the Greek, *graphein* meaning *to write*) consists of a slotted aluminum wheel along whose outer periphery magnetic links are placed. The wheel rotates allowing each link to pass between coils through which the current to be measured is allowed to flow. The operation of the fulchronograph is essentially the operation of a magnetic link except that the element of time is introduced by the rotation of the wheel. The maximum time resolution is about 50 μsec and at this time resolution the total recording time is about 20 msec. A low-speed fulchronograph is also described by Wagner and McCann (1940). The low-speed fulchronograph has a time resolution of about 2 msec and a recording time of about 1 sec. A single fulchronograph can record current in the range 100 amp to 50 ka. The wide current range is made possible by the use of two different magnetic circuits to couple the magnetic flux density to the links. The links on one side of the aluminum wheel are used to cover the current range 1 to 50 ka; the links on the other side the current range 100 amp to 6 ka.

The magnetic surge front recorder is an instrument used to measure the effective rate of rise of lightning current. The recorder consists of three circuits each containing a resistance and an inductance in series. Magnetic links are located near each of three inductances. The three circuits are connected in parallel across an inductance carrying the lightning current to be measured. The peak current in each of the parallel RL circuits is recorded by the links. With a knowledge of the peak currents and the values of resistance and inductance in each of the three circuits, and with the use of circuit theory, one can calculate the effective rate of rise of the lightning current. The recorder is used in practice to determine the slope on the current vs. time curve of a straight line drawn through the points having 10 and 90 percent of the peak current value.

The magnetic surge integrator is an instrument for recording the time integral of the lightning current, that is, the charge transported. It consists of a noninductive resistor which carries the lightning current and an inductor which is connected across the resistor. One or two magnetic links are placed near the inductor in order to measure the peak current in the inductor. The instantaneous current in the inductor is proportional to the time integral of the voltage across the resistor (also across the inductor). This voltage is directly proportional to the lightning current. Thus, the final maximum current in the inductor is proportional to the total charge which passed through the resistor. The time over which accurate reading may be obtained is of the order of 10 msec. The parameter which limits the time response to about 10 msec is the resistance associated with the inductor.

A device using magnetic links for the measurement of the time-duration of the lightning current has been described by Hyltén-Cavallius and Strömberg (1959). Four magnetic links are mounted on an insulating rod so that each is the same distance from the lightning current. One of the links is used in the conventional way to record peak current. The other three are surrounded with metal coils (inductors) which attenuate the flux density reaching them. The RL time constant of each of the three inductors is different. From a measurement of the residual magnetism in each link and a knowledge of the RL time constants of the coils, the approximate time for current to decay to half peak value can be determined.

4.2.3 The noninductive shunt and oscilloscope

In principle, if lightning current is allowed to pass through a resistance of known value, the resultant voltage waveshape can be displayed on an oscilloscope and the lightning current can therefore be determined with good accuracy. In practice, the task of obtaining an accurate current waveshape is complicated by the presence of undesirable circuit capacitance and inductance, by the lack of adequate common electric grounds, by traveling-wave phenomena (that is, by the inadequacy of lumped-parameter circuit theory to describe the measured voltages), and by the presence of strong electric and magnetic fields which induce spurious signals in the measuring circuits and electronics.

The main difficulties involved in measuring lightning currents using resistive (so-called, noninductive) shunts and oscilloscopes have been discussed by Berger (1955a) and by Berger and Vogelsanger (1965). Accurate measurements of the current rise to peak are especially difficult to obtain for strokes whose front durations are of the order of 1 μsec or less (most subsequent strokes in multiple-stroke flashes). This is so mainly because of the strong electric and magnetic fields associated with the early stages of the stroke. In the absence of electromagnetic pickup, Berger and Vogelsanger (1965) report that the lower limit on their measurement of front duration is a few tenths of a microsecond due to the effects of the inductance and capacitance of the antenna and the tower structure to which the shunt is attached. In the Lugano measurements, two shunts were placed in series in each of the two measuring towers. Each of the two series shunts (0.05 and 0.8 ohms) was used for a different current range. Several oscilloscopes were used to record the resultant voltages. The oscilloscope beams were not allowed to free-run when triggered, but rather were swept back and forth across the oscilloscope faces at predetermined speeds so that measurements could be obtained for relatively long times.

The shunts and oscilloscopes used prior to 1941 in the Empire State Building study are described by Hagenguth (1940). The modifications in equipment made after 1941 are described by Hagenguth and Anderson (1952). Originally a cathode-ray oscilloscope was used for recording fast-current components and a crater-lamp oscillograph was used for recording the slow continuous-current variation. Both of the recording devices were connected to a nonlinear shunt composed of thyrite and a linear resistance in parallel. With this shunt the low-speed oscillograph could record current from 50 amp to 24 ka while the high-speed oscilloscope could record currents from 1 to 200 ka. The equipment used in the studies taking place after 1941 consisted of various cathode-ray oscilloscopes attached to a constant resistance, noninductive shunt of 0.01 ohm. In order to adjust to the wide range of current values, thyrite suppression networks were used at the oscilloscopes. This arrangement had the advantage over the original setup of allowing the use of a less inductive shunt so that wave-front measurements could be made more accurately.

4.2.4 The loop antenna and oscilloscope

In Sec. 3.5.4 we have considered some of the aspects of the measurement of magnetic flux density using a loop antenna. A good description of the difficulties involved in performing such an experiment is given by Norinder and Dahle (1945). One of the primary problems encountered is that of shielding the loop antenna from electric fields. It was found that the presence of a simple tubular shield induced distortions in the measured waveforms. This problem was solved by using 14 separate tubular shields, each enclosing a part of the loop antenna and each grounded. In this way the waveform distortion was reduced to tolerable levels, but unfortunately the electrostatic shielding capability was diminished because of the spaces between adjacent shielding tubes. It was therefore necessary to construct a special shielding roof above the loop antenna. The shielding roof consisted of a number of copper wires stretched in the direction of the axis of the loop and at a distance above the loop. The combination of shielding roof and separate tubular shields provided adequate electrostatic shielding without appreciable signal distortion.

The RC integrating circuit described in Sec. 3.5.4 provides a voltage output which is directly proportional to the magnetic flux density. If circuit parameters are chosen so as to yield the integrating effect, the voltage across the capacitor is necessarily a small fraction of the voltage induced in the loop, which may also be relatively small, and hence an amplifier may be needed between the output of the integrating circuit and the input of the oscilloscope. The amplifier described by Norinder and Dahle (1945) had an upper-frequency cutoff of about 200 kHz and that described by Norinder (1956) had an upper-frequency cutoff of about 100 kHz. It follows that magnetic field variations occurring on a time scale faster than 5 to 10 μsec could not be measured with accuracy. In particular, the lower limit to the magnetic field (or current) rise-time measurements possible with these amplifiers will be of the order of a few microseconds.

The conversion of magnetic field data to current data is not straight-forward, as we have discussed in Sec. 4.1. In order to effect the conversion it is necessary to assume values for the temporal variation of the lightning current as a function of position along the lightning channel. The zeroth-order approximation to this physical situation involves the assumption that the same current flows along the whole length of the channel at the same time. (This assumption may well be a good one after the return stroke has traversed the channel.) With this assumption and the stipulation that the magnetic fields be measured within about 10 km of the lightning channel, the calculation of current from measured magnetic flux density follows directly from Eq. (3.19). Norinder and Dahle (1945) and Papet-Lépine (1961) have considered various higher-order approximations to the physical situation of return-stroke propagation and resultant current flow. They conclude that, to quote from Papet-Lépine, "the sum of the effects caused by propagation, damping, and progress will at last give a form of variation which is that which corresponds to the form given by the law of Biot and Savart." That is, they conclude that if Eq. (3.19) is used to convert measured flux density to current, the current variation so obtained will be a reasonable representation of the current existing at the base of the lightning channel. The results of the current determinations of Norinder and co-workers would appear to support this conclusion. The peak currents reported by Norinder and Dahle (1945) (see Sec. 4.3 and Fig. 4.3) are in reasonably good agreement with the results of other investigators. The current rates of rise reported by Norinder and Dahle (1945) (see Sec. 4.3 and Fig. 4.4) are a factor of 2 or 3 slower than those of other investigators. Part of this discrepancy may be due to the low upper-frequency cutoff of the amplifier used in the magnetic field measurements.

It is worth noting that in some instances Norinder (1956) has measured lightning currents with very slow current rates of rise and that these slow currents (times to peak current of the order of 100 μsec) have been mis-interpreted by Malan (1963) as characteristic of the data obtained by magnetic field measurements. The origin of these slow magnetic field changes is not apparent.

4.2.5 Miscellaneous measurement techniques

Before the development of the klydonograph, peak voltages on trans-mission lines were measured by a relatively crude device consisting of a number of parallel spark gaps with different spacings. A knowledge of the largest gap broken down, which was indicated by the markings on thin pieces of paper placed within the gaps, plus laboratory data on the gap breakdown voltage as a function of gap size allowed a rough deter-mination of the peak lightning voltage.

A similar device has been used by Hyltén-Cavallius and Strömberg (1959) to measure the rate of rise of lightning current. The device, called a *steepness indicator*, is placed in close proximity to the lightning current. The steepness indicator consists of five loop antennas, each antenna being connected to a separate spark gap. Between the electrodes of each spark gap is a small amount of explosive material which detonates when the voltage induced in the antenna by the rate of rise of magnetic flux density (and of current) exceeds the breakdown voltage of the gap. Each antenna is composed of a different number of coiled loops so that a differ-ent voltage is impressed across each spark gap for a given rate of rise of current. From an observation of which gaps have been fired the rate of rise of current can be deduced.

4.3 CURRENT IN CLOUD-TO-GROUND LIGHTNING DISCHARGES

In this section we consider the current characteristics of cloud-to-ground discharges initiated by downward-moving leaders. Although the bulk of these data has been obtained from measurements of lightning discharges to towers or to high buildings, it is often reasonable to assume that discharges to tall structures are not too different from discharges to normal ground. This is the case if both discharges are initiated by down-ward-moving leaders and if the downward-moving leaders to the tall structures are not appreciably affected by the presence of the structures over the greater portion of the leader path; that is, if the structures do not appreciably distort the electric field far above the tops of the structures. We shall discuss currents in discharges initiated by upward-moving leaders in Sec. 4.4.

Data regarding lightning properties measured by various investigators are presented in Figs. 4.1 to 4.7. It should be noted that in most experiments there are practical limits to the range of values which can be measured (e.g., an upper limit on the rate of rise of current measured with shunt and oscilloscope). In some cases the practical limits may enclose all reasonable values of the parameter being measured. On the other hand, when different limits are employed by various investigators to measure a given parameter, it is to be expected that the data obtained by the various investigators will not necessarily be in good agreement. Often, the limits applicable to a given measurement are not stated. Further, many investigators draw their frequency-distribution curves so that although 100 percent of the events measured exceed the zero value, 100 percent do not exceed any small value above zero. The implication is that the parameter under consideration has been measured with no lower measuring limit, whereas in actual fact this is often not the case. This is not the case, for example, in the measurement of the time to peak current by Hagenguth and Anderson (1952) (Fig. 4.5). Where possible in the figures original data have been replotted to show the existence of the lower measuring limit.

In order to determine whether a measured current peak is due to a return stroke or to an M component, it is necessary to obtain correlated photographic evidence of the presence or absence of a leader preceding the high channel luminosity. Such photographic evidence is not always available. In the presentation of data in this section, we necessarily assume that all measured properties of current peaks are properties of return strokes.

In Fig. 4.1 are shown typical current vs. time curves obtained by Berger and Vogelsanger (1965) (see also the review paper by Berger, 1967) for that class of strokes with peak current greater than or equal to 10 ka. In Fig. 4.1 each stroke current is drawn twice, once on a fast time scale, t_1, and once on a slow time scale, t_2. Examples of currents initiated by downward-moving, negatively charged stepped leaders are shown in Fig. 4.1a; by downward-moving, negatively charged dart leaders in Fig. 4.1b; and by downward-moving, positively charged stepped leaders in Fig. 4.1c. The current waveform for a first stroke lowering negative charge begins as a relatively constant or slowly increasing function and may remain so for milliseconds before evolving into a current front with a rate of rise of current of 10 to 20 ka/μsec. However, the current for a subsequent stroke lowering negative charge exhibits no current flow before the fast current front. The current rise time on subsequent strokes is usually too fast to be accurately measured. It is almost always less than a microsecond and may be less than a few tenths of a microsecond. Berger and Vogelsanger (1965) report that strokes initiated by downward-moving, positively charged leaders make up about 15 percent of all strokes initi-

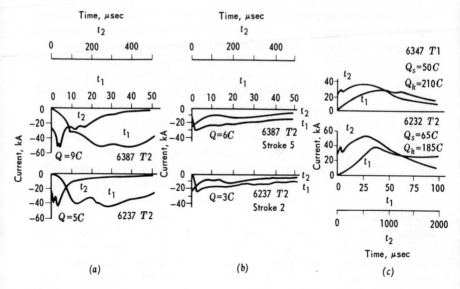

Fig. 4.1 *Current vs. time for representative lightning strokes in the class of strokes with peak currents greater than or equal to 10 ka; measurements by Berger and Vogelsanger (1965). For each stroke, data are presented on two time scales: t_1, fast; t_2, slow. (a) Two first strokes, initiated by downward-moving, negatively charged stepped leaders. Charge Q is the time integral of current to 1 msec. (b) Two subsequent strokes, initiated by downward-moving, negatively charged dart leaders. Charge Q is the time integral of current to 1 msec. (c) Two first strokes, initiated by downward-moving, positively charged stepped leaders. The charge Q_s is the time integral of current to 2 msec; Q_k, the time integral of current after 2 msec.*

ated by downward-moving leaders. In general the positive currents exhibit a longer time to peak current, a longer time to current half-value, and a greater charge transfer than do the negative currents. A peak positive current of 180 ka has been reported by Berger and Vogelsanger (1965). Positive discharges were found by Berger and Vogelsanger (1965) to lower an average charge of 87 coul.

Statistical data accumulated by Berger and Vogelsanger (1965) on strokes bringing negative charge to ground are given in Fig. 4.2. The data presented are only for peak currents greater than or equal to 10 ka. According to Berger (1955b), peak-current values below 10 ka exist in about 75 percent of all strokes and flashes; peak values below 2 ka exist in about 35 percent of all strokes and in about 50 percent of all flashes. It is apparent from Fig. 4.2 that first strokes in multiple-stroke flashes have greater peak current, larger charge transfer, and slower current rate of rise than do subsequent strokes. The maximum negative peak current reported by Berger and Vogelsanger (1965) was 105 ka for a first stroke. Subsequent strokes which are initiated by downward-moving, negatively charged leaders but which occur in flashes initiated by upward-moving leaders exhibit relatively small peak current and charge transfer.

Statistical data regarding peak lightning currents are given in Fig. 4.3. The median peak current given by Berger and Vogelsanger (1965), about 30 ka, is substantially higher than that of other investigators, possibly because of the exclusion of currents below 10 ka from the plotted data. Berger and Vogelsanger (1965) report that for 85 strokes initiated by downward-moving, negatively charged leaders and having peak currents above 2 ka, the median peak current was about 25 ka. The data of McCann (1944) show that if only the highest-current stroke in a flash is considered, the median current, about 20 ka, is considerably larger than the median of all strokes, 5.5 ka. From McCann's fulchronograph and oscilloscope records it would appear that the highest stroke current is usually associated with the first stroke. These findings are in substantial agreement with those of Berger and Vogelsanger (1965) given in Fig. 4.2 which show median first-stroke currents to be about twice the value of median subsequent-stroke currents, and with those of Norinder and Knudsen (1961) given in Sec. 3.9. The maximum current recorded by McCann (1944) was 160 ka.

Lewis and Foust (1945) present the most extensive data (2,721 events)

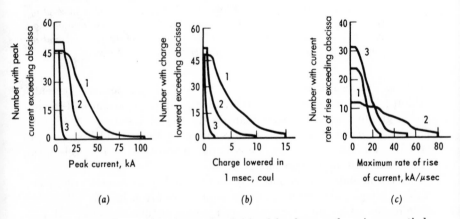

Fig. 4.2 *Properties of lightning strokes initiated by downward-moving, negatively charged leaders. (1) First strokes initiated by downward-moving stepped leaders; (2) subsequent strokes initiated by downward-moving dart leaders; (3) negative strokes within flashes originating with upward-moving leaders. Only strokes with peak currents equal to or greater than 10 ka are considered. (Data adapted from Berger and Vogelsanger (1965).)*

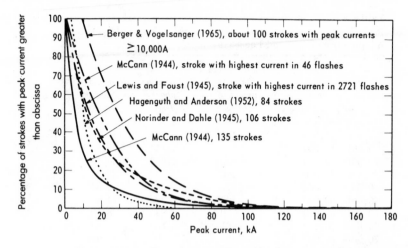

Fig. 4.3 *Frequency distribution of peak current for lightning strokes initiated by downward-moving leaders. Strokes lowering negative charge and strokes lowering positive charge are both included.*

available concerning peak lightning currents. The data are summarized in Table 4.1 and plotted in Fig. 4.3. The current measurements were made by placing magnetic links on various transmission-line towers and thus represent the maximum current per *flash*. It was assumed that the lightning current divided equally down the legs of the transmission-line towers. Thus, if a measurement was made on one tower leg, the total current per tower was calculated by multiplying the measured value by the number of legs. Further, the peak currents recorded on various towers for a given flash were added to give the total peak current per flash. The errors are probably such as to overestimate the actual peak lightning current. The median value of peak current reported was about 15 ka. The largest current was 218 ka. Six percent of the currents were above 60 ka. About 82 percent of the flashes to various transmission-line towers lowered negative charge. The highest percentage of flashes lowering positive charge, 33.2 percent, occurred on the 100 kv line of the Public Service Company of Colorado. The line ranged in altitude from about 2 to 4 km. The lowest percentage of flashes lowering positive charge, 3.3 percent, occurred on the Wallenpaupack-Siegfried 220 kv line in Pennsylvania. This line ranged in altitude from about 0.2 to 0.7 km.

Data on peak lightning current measured at the Empire State Building are given by Hagenguth and Anderson (1952) and are reproduced in Fig. 4.3. All but 2 of the 84 current peaks measured with a shunt and oscilloscope were associated with the lowering of negative charge. The

TABLE 4.1
Peak lightning currents as measured using magnetic links.
Adapted from Lewis and Foust (1945).

Range of current, amp	No. of flashes with peak current in range	No. at or above level	Percentage at or above level
1,000– 5,000	567	2,721	100
5,001– 10,000	611	2,154	79.2
10,001– 20,000	640	1,543	56.7
20,001– 30,000	296	903	33.2
30,001– 40,000	227	607	22.3
40,001– 50,000	140	380	14.0
50,001– 60,000	80	240	8.82
60,001– 70,000	61	160	5.88
70,001– 80,000	22	99	3.64
80,001– 90,000	21	77	2.83
90,001–100,000	11	56	2.06
100,001–110,000	11	45	1.65
110,001–120,000	9	34	1.25
120,001–130,000	9	25	0.918
130,001–140,000	7	16	0.588
140,001–150,000	2	9	0.331
150,001–160,000	3	7	0.257
160,001–170,000	0	4	0.137
170,001–180,000	1	4	0.147
180,001–190,000	0	3	0.110
190,001–200,000	1	3	0.110
200,001–210,000	0	2	0.073
212,000	1	2	0.073
218,000	1	1	0.037
Total	2,721		
Maximum negative	218,000		
Maximum positive	212,000		

median peak current was 10 ka. The maximum current recorded was 58 ka and was associated with the lowering of positive charge. The average error possible in the measurement of peak current was estimated by Hagenguth and Anderson (1952) to be of the order of ±20 percent.

Norinder and Dahle (1945) have measured peak current using a loop antenna and oscilloscope. These data are shown in Fig. 4.3. About 35 percent of the measured peak currents were between 10 and 20 ka. The largest measured current was 130 ka.

Statistical data on the rate of rise of stroke current for strokes initiated by downward-moving, negatively charged leaders are given in Fig. 4.4.

The data of Berger and Vogelsanger (1965) represent the maximum rate of rise during the increase in current.   These values are larger, as they should be, than the corresponding time for current to increase from 10 to 90 percent of peak value as presented by Hagenguth and Anderson (1952) and McCann (1944).   The rise times measured by Norinder and Dahle (1945) are limited by the frequency response of the amplifier used and may be affected by the techniques used to obtain current waveshapes from the measured data (see Sec. 4.2.4).   Current rate-of-rise data for first strokes and for subsequent strokes as determined by Berger and Vogelsanger (1965) are given in Fig. 4.2c.   Supporting evidence for the view that first strokes have longer rise times than subsequent strokes is presented by Norinder, Knudsen, and Vollmer (1958) from data obtained using a loop antenna and oscilloscope.

Statistical data on the time to peak current for strokes initiated by downward-moving, negatively charged leaders are given in Fig. 4.5.   The measurements of McCann (1944) were made with shunt and oscilloscope and with a magnetic surge front recorder.   The measurements of Hagenguth and Anderson (1952) were made with shunt and oscilloscope. Hyltén-Cavallius and Strömberg (1959) have determined the time to

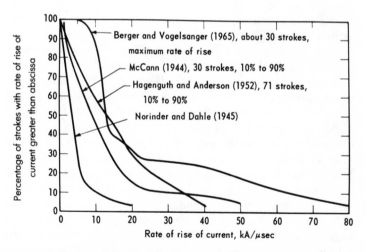

Fig. 4.4 *Frequency distribution for rate of rise of current for lightning strokes initiated by downward-moving, negatively charged leaders. Data of McCann (1944) and Hagenguth and Anderson (1952) represent slope of line through points on current wavefront at 10 percent and 90 percent of peak current; data of Berger and Vogelsanger (1965) represent maximum rate of rise of current on wavefront.*

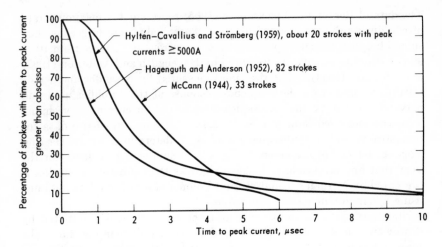

Fig. 4.5 *Frequency distribution of time to peak current for lightning strokes initiated by downward-moving, negatively charged leaders.*

peak current by dividing measured current rate of rise (see Sec. 4.2.5) into the peak current as measured by magnetic links. This technique should lead to the calculation of a shorter time to peak current than is actually the case. It should be noted that the measured time to peak current obtained with a shunt and oscilloscope depends somewhat on the sensitivity of the measuring apparatus to small currents which may precede the rapid current increase. Berger and Vogelsanger (1965) report that preceding a first stroke these small currents may flow for milliseconds.

Statistical data on the time for the stroke current to decrease to half peak value are given in Fig. 4.6. The maximum value reported by McCann (1944) was 90 μsec. In the case of the Empire State Building study (Hagenguth and Anderson, 1952) the maximum duration was longer than the oscilloscope sweep time and was estimated to be 120 μsec. The maximum time to half-value measured by Hyltén-Cavallius and Strömberg (1959) was 200 μsec although considerable error in this value is possible due to the technique of measurement (see Sec. 4.2.2). Berger and Vogelsanger (1965) show oscillographs of stroke currents with times to half-value of 180 and 250 μsec for negative strokes and 1,000 and 1,500 μsec for positive strokes.

Statistical data on the charge lowered per lightning flash for flashes bringing negative charge to earth are given in Fig. 4.7. The data presented include flashes initiated by downward-moving leaders and flashes initiated by upward-moving leaders. Berger and Vogelsanger (1965) report that flashes lowering negative charge initiated by downward-moving leaders lower an average of 11 coul while those initiated by upward-moving leaders lower an average of 22 coul. The maximum negative charge lowered in a flash was 220 coul. The flash was initiated by an upward-moving leader. Hagenguth and Anderson (1952) report that the maximum negative charge lowered in a flash to the Empire State Building was 164 coul. McCann (1944) reports a maximum charge transfer of 100 coul.

We consider now the charge lowered by a lightning *stroke*. Although a current peak may have a magnitude as high as 100 ka, the time to current half-value is too short to allow the transfer of more than a few coulombs of charge during the effective time duration of the current peak. Most current peaks involve a charge transfer of less than 1 coul. Hagenguth and Anderson (1952) report that 50 percent of 83 current peaks measured had a charge greater than 0.15 coul, 6 percent exceeded 1.7 coul, and the maximum charge recorded was 4.9 coul produced by the 58 ka current peak of positive polarity. Similar data can be derived from the current traces presented by Berger and Vogelsanger (1965), examples of which are shown in Fig. 4.1. It follows from the discussion in the previous paragraph that most of the charge lowered during a lightning flash is not associated with the current peaks. The bulk of the charge is apparently lowered by relatively low amplitude currents in the milliseconds following the stroke-current peaks and by continuing currents during the tens to hundreds of milliseconds between strokes. Some statistics concerning

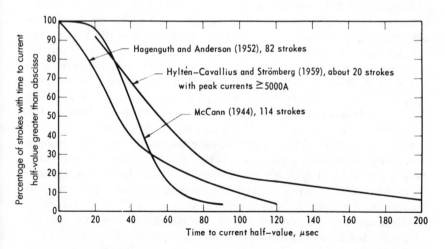

Fig. 4.6 *Frequency distribution of time for lightning stroke current to decrease to half peak value for strokes initiated by downward-moving, negatively charged leaders.*

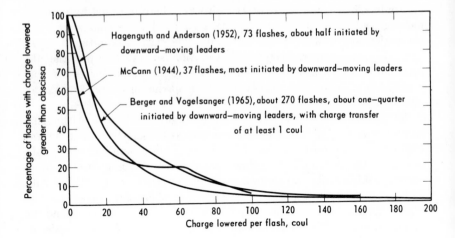

Fig. 4.7 *Frequency distribution of charge lowered per lightning flash for flashes bringing negative charge to earth. Discharges initiated by downward-moving leaders and discharges initiated by upward-moving leaders are both included.*

the flow of low-amplitude currents after the passage of the return stroke are given in Sec. 3.7.4. The frequency of occurrence of continuing currents has been discussed in Sec. 3.7.5.

Statistical data regarding the charge lowered by first and by subsequent strokes in 1 msec as determined by Berger and Vogelsanger (1965) are shown in Fig. 4.2. Only strokes with peak currents over 10 ka have been considered. The median charge for first strokes is about 4.5 coul, for subsequent strokes about 1.5 coul. From electric field measurements Brook et al. (1962) (see Secs. 3.7.1 and 3.7.5) have found that the most frequent value of charge brought down by first strokes lies between 3 and 4 coul; for subsequent strokes the most frequent value lies between 0.5 and 1 coul. It appears, then, that following the stroke-current peak, a current of the order of 1 ka must flow for a time of the order of 1 msec in order to provide for the observed charge transfer. Such currents have been reported by Hagenguth and Anderson (1952) who have designated them intermediate currents. The intermediate currents are probably to be associated with the R_c electric field change (see Sec. 3.7.4). According to Hagenguth and Anderson (1952) the intermediate current can be approximated by an exponentially decaying function whose initial value is of the order of 2.5 ka or less. The decay time to half-value of the exponential function was found to be over 1 msec for 50 percent of the intermediate current intervals and over 10 msec for 10 percent of the intervals.

We consider now the charge transferred by continuing currents in flashes initiated by downward-moving, negatively charged leaders. Berger and Vogelsanger (1965) report that in half the flashes containing continuing-current intervals, the charge lowered by continuing current was over 25 coul. The continuing currents were generally of the order of 100 to 300 amp. The maximum charge lowered by a continuing current was about 80 coul. Electric field measurements by Brook et al. (1962) have yielded values for the charge lowered in a continuing current interval of between 3.4 and 29.2 coul, the average being about 12 coul. The continuing-current amplitudes varied between 38 and 130 amp. Magnetic field measurements by Williams and Brook (1963) have yielded an average value for charge transfer by a continuing-current interval of 31 coul and an average value for continuing current amplitude of 184 amp.

The current waveform of a lightning stroke as measured at the ground can be approximately represented by the following analytical expression

$$I = I_0(e^{-\alpha t} - e^{-\beta t}) + I_1 e^{-\gamma t} \tag{4.1}$$

The parameters I_0, α, and β can be chosen to give approximations to the current rise time, peak current, and time to current half-value. The parameters I_1 and γ can be chosen to give approximations to the intermediate current. If necessary, an additional term may be added to Eq. (4.1) to take account of continuing current. Dennis and Pierce (1964) suggest the use of the following parameters for first strokes lowering negative charge: $\alpha = 2.0 \times 10^4 \ \text{sec}^{-1}$, $\beta = 2.0 \times 10^5 \ \text{sec}^{-1}$, $I_0 = 30$ ka; and for subsequent strokes: $\alpha = 1.4 \times 10^4 \ \text{sec}^{-1}$, $\beta = 6.0 \times 10^6 \ \text{sec}^{-1}$, $I_0 = 10$ ka. Reasonable parameters for the intermediate current term are $\gamma = 1.0 \times 10^3 \ \text{sec}^{-1}$, $I_1 = 2.5$ ka.

4.4 DISCHARGES INITIATED BY UPWARD-MOVING LEADERS

Upward-moving stepped leaders and the resultant lightning discharges were first observed at the Empire State Building in New York City (see Sec. 2.4.2). Considerable additional data concerning such discharges have been accumulated by Berger and his co-workers in Lugano, Switzerland (see Sec. 2.5.4).

The majority of the lightning discharges to the Empire State Building is initiated by an upward-moving stepped leader whose point of origin is the top of the building. The leader current merges smoothly into a continuous-current flow between cloud and building. There is no return stroke. In about half these discharges, the continuous current is interrupted by subsequent return-stroke current peaks initiated by downward-moving dart leaders. Hagenguth and Anderson (1952) report that the average maximum continuous-current flow is about 250 amp from an upward-initiated discharge and that the maximum continuous-current amplitude is of the order of 1,450 amp. The duration of the continuous current is of the order of tenths of seconds. It is of some interest to note that according to McEachron (1939), upward-initiated discharges without successive current peaks may produce no thunder.

About 75 percent of the discharges observed to the measuring towers near Lugano, Switzerland, are initiated by upward-moving stepped leaders. Current traces for representative discharges are given by Berger (1955b), by Berger and Vogelsanger (1965), and by Berger (1967). The current curves are characterized by current amplitudes between twenty and several hundred amperes and times to maximum current of from hundredths to tenths of seconds. Current curves were relatively smooth as the leader "stepped" upward toward the cloud. No general current waveform for the continuous-current discharge could be given. As in the Empire State Building study, the total continuous-current duration was of the order of tenths of a second. Continuous currents were often followed by current pauses during which time dart-leader–return-stroke combinations occurred. Berger and Vogelsanger (1965) report that the current peaks occurring during continuous-current flow are relatively weak compared to the peaks which occur during periods of no current flow.

Berger and Vogelsanger (1965) have identified upward-moving leaders of both charge polarities. About 85 percent of the upward initiated discharges lower negative charge; that is, the leaders are positively charged. A few of the upward-initiated discharges have currents of alternating polarities. Usually these begin with a current which indicates the raising of positive charge. These currents usually change polarity only once. The average charge lowered by upward-initiated discharges lowering negative charge (positively charged leaders) was found to be 22 coul. The average charge lowered by upward-initiated discharges lowering positive charge (negatively charged leaders) was found to be 64 coul. The maximum negative charge lowered was 220 coul; the maximum positive charge 310 coul.

4.5 OTHER CURRENT MEASUREMENTS

Measurements of lightning currents to an instrumented airplane have been reported by Fitzgerald (1967) and by Petterson and Wood (1968). The plane was flown within thunderclouds so that it would be struck by lightning. Values of peak current were found to be generally a few thousand amperes, and rise times to current peak were generally measured in milliseconds.

Observations of lightning currents in the flying cables of barrage balloons have been made by Davis and Standring (1947). The balloons were flown at an altitude of about 600 m. Measurements were made using magnetic links and a shunt and oscilloscope. Both continuous-current discharges and discharges containing discrete strokes were observed. In general, the observations were similar to those discussed in the previous two sections.

In a series of papers, Griscomb et al. (1965) describe the results of lightning-current measurements made using a kineklydonograph. The measurements were made of lightning strikes to masts erected above transmission-tower tops, to transmission-line ground wires, and to a tall building in Pittsburgh, Pennsylvania. Seventeen flashes lowering negative charge were recorded. Peak currents ranged from 11 to 50 ka. One positive discharge was reported. Its peak current was 80 ka. Griscomb et al. (1965) reported the measurement of a maximum current rate of rise of 230 ka/μsec during a 0.05 μsec interval. This would appear to be the most rapid rate of rise of current reported in the literature.

Williams and Brook (1963) have measured the magnetic fields produced by stepped leaders and in two cases were able to estimate the leader currents. Values of 50 and 63 amp were calculated. Williams and Brook (1963) suggest that the absence of detectable magnetic fields on most of their stepped-leader records indicates that stepped-leader currents are generally less than 50 amp.

References

1. Bell, E., and A. L. Price: Lightning Investigation on the 220-kV System of the Pennsylvania Power and Light Company (1930), *Trans. AIEE*, **50**:1101–1110 (1931).

2. Berger, K.: Die Messeinrichtungen für die Blitzforschung auf dem Monte San Salvatore, *Bull. SEV*, **46**:193–201 (1955a).

3. Berger, K.: Resultate der Blitzmessungen der Jahre 1947–1954 auf dem Monte San Salvatore, *Bull. SEV*, **46**:405–424 (1955b).

4. Berger, K.: Novel Observations on Lightning Discharges: Results of Research on Mount San Salvatore, *J. Franklin Inst.*, **283**:478–525 (1967).

5. Berger, K., and E. Vogelsanger: Messungen und Resultate der Blitzforschung der Jahre 1955–1963 auf dem Monte San Salvatore, *Bull. SEV*, **56**:2–22 (1965).

6. Brook, M., N. Kitagawa, and E. J. Workman: Quantitative Study of Strokes and Continuing Currents in Lightning Discharges to Ground, *J. Geophys. Res.*, **67**:649–659 (1962).

7. Brown, J.: On Figures Produced by Electric Action on Photographic Dry Plates, *Phil. Mag.*, **26**:502–505 (1888).

8. Davis, R., and W. G. Standring: Discharge Currents Associated with Kite Balloons, *Proc. Roy. Soc. (London)*, **A191**:304–322 (1947).

9. Dennis, A. S., and E. T. Pierce: The Return Stroke of a Lightning Flash to

Earth as a Source of VLF Atmospherics, *J. Res. NBS/USNC-URSI*, **68D** (*Radio Science*):777–794 (1964).

10. Fitzgerald, D. R.: Research Aircraft Lightning Strike Experience in Thunderstorms, *14th Gen. Symp. Union Géodésique Géophys. Intern.*, St. Gallen and Luzerne, Switz., Sept. 28 to Oct. 6, 1967.

11. Foust, C. M., and H. P. Kuehni: The Surge-crest Ammeter, *Gen. Elec. Rev.*, **35**:644–648 (1932).

12. Griscomb, S. B.: The Kine-klydonograph—A Transient Waveform Recorder, *Trans. AIEE (PAS)*, **79**:603–610 (1960).

13. Griscomb, S. B., R. W. Caswell, R. E. Graham, H. R. McNutt, R. H. Schlomann, and J. K. Thornton: Five-Year Field Investigation of Lightning Effects on Transmission Lines, *Trans. IEEE (PAS)*, **84**:257–280 (1965).

14. Gross, I. W., and J. H. Cox: Lightning Investigation on the Appalachian Electric Power Company's Transmission System, *Trans. AIEE*, **50**:1118–1131 (1931)

15. Hagenguth, J. H.: Lightning Recording Instruments, *Gen. Elec. Rev.*, **43**:105–201, 248–255 (1940).

16. Hagenguth, J. H., and J. G. Anderson: Lightning to the Empire State Building, pt. 3, *Trans. AIEE*, **71**(*pt.3*):641–649 (1952).

17. Hyltén-Cavallius, N., and Å. Strömberg: Field Measurements of Lightning Currents, *Elteknik*, **2**:109–113 (1959).

18. Lewis, W. W., and C. M. Foust: Lightning Investigation on Transmission Lines, pt. 7, *Trans. AIEE*, **64**:107–115 (1945).

19. Lichtenberg, G. C.: Novo methodo naturam ac motum fluidi electrici investigandi, *Societatis Regiae Scientiarum Gottingensis*, **T8**:168–180 (1778).

20. Malan, D. J.: "Physics of Lightning," The English University Press Ltd., London, 1963.

21. McCann, G. D.: The Measurement of Lightning Currents in Direct Strokes, *Trans. AIEE*, **63**:1157–1164 (1944).

22. McEachron, K. B.: Measurements of Transients by the Lichtenberg Figures, *Trans. AIEE*, **45**:712–717 (1926).

23. McEachron, K. B.: Lightning to the Empire State Building, *J. Franklin Inst.*, **227**:149–217 (1939).

24. Norinder, H.: Magnetic Field Variations from Lightning Strokes in Vicinity of Thunderstorms, *Arkiv Geofysik*, **2**:423–451 (1956).

25. Norinder, H., and O. Dahle: Measurements by Frame Aerials of Current Variations in Lightning Discharges, *Arkiv Mat. Astron. Fysik*, **32A**:1–70 (1945).

26. Norinder, H., and E. Knudsen: Some Features of Thunderstorm Activity, *Arkiv Geofysik*, **3**:367–374 (1961).

27. Norinder, H., E. Knudsen, and B. Vollmer, Multiple Strokes in Lightning Channels, in L. G. Smith, (ed.), "Recent Advances in Atmospheric Electricity," pp. 525–542, Pergamon Press, New York, 1958.

28. Papet-Lépine, J.: Electromagnetic Radiation and Physical Structure of Lightning Discharges, *Arkiv Geofysik*, **3**:391–400 (1961).

29. Peters, J. F.: The Klydonograph, *Elec. World*, **83**:769–773 (1924).

30. Petterson, B. J., and W. R. Wood: Measurements of Lightning Strikes to Air-

CURRENT MEASUREMENTS **137**

craft, Report No. SC-M-67-549, Sandia Laboratory, Albuquerque, New Mexico, January, 1968.

31. Pockels, F.: Über das magnetische Verhalten einiger basaltischer Gesteine, *Ann. Physik Chem.*, **63**:195–201 (1897).

32. Pockels, F.: Bestimmung maximaler Entladungs-strom-stärken aus ihrer magnetisirenden Wirkung, *Ann. Physik Chem.*, **65**:458–475 (1898).

33. Pockels, F.: Über die Blitzentladungen erreichte Stromstärke, *Physik. Z.*, **2**: 306–307 (1900).

34. Smeloff, N. N., and A. L. Price: Lightning Investigation on 220-kV System of the Pennsylvania Power and Light Company (1928 and 1929), *J. AIEE*, **49**: 771–775 (1930).

35. Sporn, P., and W. L. Lloyd, Jr.: Lightning Investigation on 132-kV System of the Ohio Power Company, *J. AIEE*, **49**:259–262 (1930).

36. Sporn, P., and W. L. Lloyd, Jr.: 1930 Lightning Investigations on the Transmission System of the American Gas and Electric Company, *Trans. AIEE*, **49**:1111–1117 (1931).

37. Trouvelot, E. T.: Sur la forme des décharges electriques sur les plaques photographique, *Lumière Elec.*, **30**:269–273 (1888).

38. Wagner, C. F., and G. D. McCann: New Instruments for Recording Lightning Currents, *Trans. AIEE*, **59**:1061–1068 (1940).

39. Williams, D. P., and M. Brook: Magnetic Measurements of Thunderstorm Currents, (1) Continuing Currents in Lightning, *J. Geophys. Res.*, **68**:3243–3247 (1963)

5

Lightning Spectroscopy

5.1 EARLY HISTORY

*The principal features of the spectrum are a more or less bright continuous
spectrum crossed by numerous bright lines, so numerous indeed as to perplex
one as to their identity.   This perplexity is increased by the constantly changing appearance due to a variable illuminating-power.   This variable character of the appearance is unquestionably the peculiar feature of the spectrum!
It is not that the whole spectrum varies in brightness in the same degree, but
that the relative intensities are variable, not only among the various lines, but
between these and the continuous spectrum.*

J. Herschel: On the Lightning Spectrum, *Proc. Roy. Soc. (London)*, **15**:61–62
(1868).

*All observers of lightning-spectra agree in having seen the line-spectrum of
nitrogen; but most of them have seen, in addition to this, sometimes a continuous spectrum, sometimes a band spectrum, the chemical origin of which is
unknown.*

A. Schuster: On Spectra of Lightning, *Proc. Phys. Soc. (London)*, **3**:46–52
(1880).

Lightning spectroscopy has been practiced for over a century.   Thus,
as a diagnostic technique for the study of lightning, its origins predate
the use of photography and of electric field, magnetic field, and current
measurements.

In spectroscopy, the radiation (light) emitted by a source is decomposed into its various wavelengths. The salient features of the spectrum are then analyzed. The spectrum for the ultraviolet, visible, and infrared wavelengths is typically produced by allowing the light under study to pass through a dispersing prism or to pass through or be reflected off a diffraction grating (transmission or reflection grating, respectively). Lightning spectrometers fall into two general categories, slit and slitless. The slit spectrometer produces wavelength dispersion of the image of a thin slit which is placed in front of the dispersing element and which is illuminated by the lightning under study. The slitless spectrometer produces wavelength dispersion of the image of the lightning channel itself by allowing the light from the channel to fall directly on the dispersing element. The resulting spectrum is a series of monochromatic images (spectral lines) of the lightning channel superimposed on a continuum. We shall have more to say about lightning spectrometers in Sec. 5.2.

Prior to about 1900, most lightning spectroscopy was concerned with the *visual* determination of the general characteristics of the lightning spectrum and of the identification by wavelength of the brightest spectral features. That is, the recording device for the lightning spectrum was the human eye. Most of the significant spectral work on lightning prior to 1880 is reviewed by Schuster (1880). The researchers working in the nineteenth century were able to identify the strong hydrogen and nitrogen lines in the lightning spectrum and to locate their wavelengths with surprisingly good accuracy.

The first lightning spectrum recorded on a photographic plate is apparently due to Meyer (1894). Meyer's results were unspectacular, but they did serve to show that slitless lightning spectra could be obtained by placing a dispersing element (in Meyer's case a diffraction grating) in front of an ordinary camera. Pickering (1901) published the first good photographic lightning spectrum. The lightning spectra reported by Pickering (1901) were obtained by J. H. Freese who used an objective prism placed in front of a telescope. Pickering (1901) noted that there was variation from flash to flash between the salient features of the lightning spectrum. Fox (1903) compared his slitless lightning spectra with the slit spectrum of a laboratory spark in air and with the lightning spectra of several previous investigators. He made the observation that the relative intensities of line pairs may change between the upper and lower ends of the lightning channel. Larsen (1905) published two slitless lightning spectra and compared these to the spark spectrum. Steadworthy (1914) obtained slitless lightning spectra which were analyzed by J. B. Cannon (Steadworthy, 1914; Cannon, 1918). Perhaps the best of the early work on lightning spectra was that of Slipher (1917), who was the first to report photographically recorded slit spectra of lightning. The slit spectrometer, as we shall see in Sec. 5.2, allows a more accurate determination of wavelength than does the slitless spectrometer. Slipher (1917) identified with quite good accuracy the important spectral lines of lightning between 3830 and 5000 Å and identified these lines as being due to either oxygen or nitrogen. Further, he compared his measurements of the wavelengths of spectral lines with those of the laboratory spark spectrum and with those of Pickering (1901) and Fox (1903). Cannon's (1918) analysis of wavelengths from Steadworthy's (1914) spectra are in very good agreement with the data of Slipher (1917).

The early era of lightning spectroscopy can be considered to end with the work of Dufay (1926). Using a slit spectrometer, Dufay obtained spectral data from 2860 to 6550 Å. Dufay's spectra represent the first ultraviolet lightning data. Dufay was able to identify a number of molecular bands in the lightning spectrum. While Dufay was performing his experiments, modern atomic physics was in the process of formulation. Bohr's model of the atom, which had been successful in explaining certain patterns exhibited by some spectral lines, was being replaced in the 1920s by the quantum theory of the atom, by quantum mechanics. Modern atomic physics would provide the tools necessary for the further development of lightning spectroscopy.

5.2 EXPERIMENTAL TECHNIQUES

5.2.1 The slit spectrometer

A schematic diagram of a slit spectrometer is shown in Fig. 5.1. Light from the source under study is focused or falls diffusely on the entrance slit. The thin entrance slit can then be considered the source of light to be analyzed. The light from the entrance slit is rendered parallel (collimated) by the collimating mirror. The diffraction (reflection) grating then splits the light into its spectrum and the focusing mirror focuses the spectrum on a photographic plate or other detection device (see Sec. 5.2.4). In all lightning experiments performed to date, light from the lightning discharge has been allowed to fall diffusely on the entrance slit. The entrance slit will accept all light that is in its "field of view," and thus the resultant spectrum may be due to light emission from many different spatial positions. For example, the spectra of both corona discharges in the cloud and the main discharge channel may appear superimposed in the same photographic plate. Further, the diffuse light from a single lightning discharge which reaches the slit spectrometer is of insufficient intensity to expose the photographic film. It is, therefore, necessary to expose the spectrometer to a number of lightning discharges to attain sufficient exposure. It is not practical to control what types

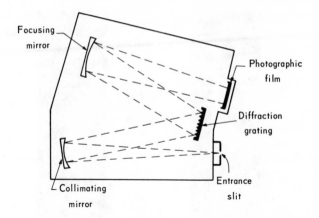

Fig. 5.1 *Schematic diagram of a slit spectrometer. Light from source illuminates entrance slit. Light rays entering entrance slit are rendered parallel by collimating mirror, diffracted in wavelength by diffraction grating, and focused on the photographic film by focusing mirror.*

of discharges (cloud, ground, air, corona) the spectrometer records during the integration time.

There are two primary advantages in using a slit spectrometer rather than a slitless spectrometer: (1) Wavelengths of spectral features can be accurately determined on a slit spectrum by superimposing on the spectrum another spectrum whose line identifications are well known. The position of the slit with respect to the optics and grating determines the wavelength position on the film. In slitless spectroscopy, the position of the source, the discharge channel, determines the wavelength position on the film. (2) Very good wavelength resolution can be obtained so that closely spaced spectral lines can be separated and the shapes of spectral lines can be examined. In part, the wavelength resolution is determined by the slit width; narrowing the slit width will generally improve the resolution. In slitless spectroscopy, the effective slit width is the width of the radiating channel. For close strokes this effective width may be relatively large so that the finite diameter of the discharge channel places a lower limit on the wavelength resolution.

An example of the lightning spectra obtained by Wallace (1964a), using a slit spectrometer, is shown in Fig. 5.2. The principal emission features from Wallace's spectra are listed in Table 5.1. In Fig. 5.2, in Table 5.1, and in other lightning spectral data to be presented, the following notation

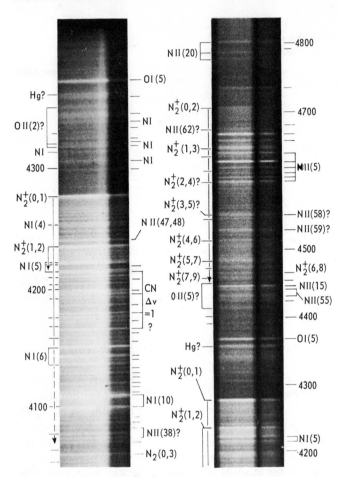

Fig. 5.2   *A spectrum of lightning in the wavelength region 4050 to 4839 Å obtained by Wallace (1964a) using a slit spectrometer.   (Courtesy, University of Chicago Press.)*

is used for atomic spectra: the roman numeral after the element abbreviation designates the ionization stage; that is, "I" is a neutral atom, "II" is a singly ionized atom.   The arabic number following in parentheses is the multiplet number assigned by Moore (1945).   For a given element and ionization state, the larger the multiplet number the higher the energy of the lower energy level to which the radiative transition occurs. For a discussion of the notation used in describing molecular and molecular-ion spectra, the reader is referred to Herzberg (1950).

TABLE 5.1

Principal emission features in the lightning spectrum. For those identifications that are considered established beyond all doubt, only the known laboratory wavelengths are given. For less certain identifications, the measured wavelengths are given and are underlined. Unidentified features in the region 3888 to 4220 Å are listed by Wallace (1960). Adapted from Wallace (1964b).

λ	Identification			λ	Identification
3159.3	N₂(1,0)			71.4	CN(1,1)
3292.4				83.4	CN(0,0)
97.6				94.6	N₂(3,6)
3329.4				3914.4	N₂⁺(0,0)
60.0	NH(0,0): Q head			3919.0	NII(17)
63.4				41.7	43.0 N₂(2,5)?
67.2				47.3	OI(3)
71.4	N₂(0,0)			3955.4	{55.9 NII(6)? / 54.4 OII(6)?}
	NH(0,0) .	O₂(0,14)		73.0	73.3 OII(6)?*
85.1	84.8 P₇	85.0 P₁₅, R₁₉		83.0	82.7 OII(6)?*
88.6	88.5 P₈	88.4 P₁₇, R₂₁		95.0	NII(12)
91.9	92.1 P₉	92.2 P₁₉, R₂₃		98.4	N₂(1,4)
95.8	95.7 P₁₀	96.5 P₂₁, R₂₅		4026.1	NII(40)
3400.0	99.3 P₁₁	01.1 P₂₃, R₂₇		35.1	NII(39)
09.8	09.9 P₁₄?	11.4 P₂₇, R₃₁?		41.3	NII(39)
16.3	16.8 P₁₆?	17.2 P₂₉, R₃₃?		43.5	NII(39)
27.4	27.0 P₁₉?			59.4	N₂(0,3)
37.2	NII(13)			72.7	73.0 NII(38)?*
3532.				82.1	82.3 NII(38)?*
33.2	N₂⁺(5,4)			99.9	NI(10)
36.7	N₂(1,2)			4110.0	NI(10)
3538.0	N₂⁺(4,3)			37.4	37.6 NI(6)
43.5				52.2	51.5 NI(6)
48.2	N₂⁺(3,2)			67.1	67.8 CN(3,4)?
53.3				80.8	81.0 CN(2,3)?
64.4	N₂⁺(2,1)			97.0	97.2 CN(1,2)?
72.				4216.0	{16.0 CN(0,1)? / 15.9 NI(5)}
76.9	N₂(0,1)			22.8	23.0 NI(5)
82.5	N₂⁺(1,0)			24.6	24.7 NI(5)
3710.6	N₂(2,4)			36.5	N₂⁺(1,2)
55.5	N₂(1,3)			41.8	NII(47,48)
3804.9	N₂(0,2)			54.7	NI(4)
30.	30.4 NI(11)?			78.1	N₂⁺(0,1)
51.0	CN(4,4)			98.4	
54.7	CN(3,3)			4305.0	05.5 NI(unc.)
61.9	CN(2,2)				

TABLE 5.1 (Continued)

λ	Identification	λ	Identification
13.5	13.1 NI(unc.)	70.7	
17.3	{17.7 NI(unc.) / 17.1 OII(2)?}	77.9	77.9 NII(62)?
19.5	19.6 OII(2)	84.2	
23.8		4709.3	$N_2^+(0,2)$
26.0		37.7	
32.5		79.7	NII(20)
36.3	36.5 NI(unc.)	88.2	NII(20)
42.0		92.3	
4348.8	49.4 OII(2)?	4803.3	NII(20)
56.1		61.5	$61.3H_\beta$
58.5	58.3Hg?	4914.9	NI(9)
68.3	OI(5)	35.0	NI(9)
91.7		69.6	
4403.4		77.1	
14.5	14.9 OII(5)?	94.3	94.4 NII(64)?
26.2		5001.9	01.5 NII(19)
32.7	NII(55)	05.0	{05.1 NII(19) / 05.1 NII(64)?}
42.0	NII(55)	45.3	45.1 NII(4)
47.0	NII(15)	5169.6	
51.6	52.4 OII(5)?	80.9	
56.6		86.4	
59.6	59.3 $N_2^+(7,9)$	99.1	
66.0	66.6 $N_2^+(6,8)$, (8,10)	5201.7	
75.3		81.2	NI(14)
85.1	85.6 $N_2^+(5,7)$	93.4	
4515.9	$N_2^+(4,6)$	5309.3	
30.4	30.4 NII(59)?	28.7	NI(13)
48.2		56.8	NI(13)
52.5	{52.5 NII(59)? / 53.8 $N_2^+(3,5)$?}	66.7	
4601.5	NII(5)	87.7	
07.2	NII(5)	5400.9	
13.9	NII(5)	10.9	10.8 OI(51,52)?
17.2		37.5	36.8 OI(11)?
21.4	NII(5)	42.8	
30.5	NII(5)	50.4	
38.5		60.7	Hg: contamination
43.1	NII(5)	5563.	{60.4 NI(25)? / 64.4 NI(25)?}
51.0	51.9 $N_2^+(1,3)$	77.4	[OI]: aurora
60.9			

TABLE 5.1 (Continued)

λ	Identification	λ	Identification
5617.9	16.5 NI(24)?	40.7†	
21.6	23.2 NI(24)?	45.1†	
24.3	25.4 NI(24)?	50.5†	
51.7		62.8	H_α
66.6	NII(3)	77.2†	
76.0	NII(3)	89.3	
79.6	NII(3)	99.0	Ne: contamination
86.2	NII(3)	6607.5	
5769.6	Hg: contamination	10.6	NII(31)
90.7	Hg: contamination	23.8	22.5 NI(20)?
6000.2	99.5 NI(16), deg. long λ?	29.5	30.5 NII(41)?
08.1	08.5 NI(16), deg. long λ?	39.4	37.0 NI(20)?
96.2	Ne: contamination	47.0	45.0 NI(20), deg. long λ
6143.1	Ne: contamination	55.0	53.4 NI(20), deg. long λ
58.1	57. OI(10)	78.3	Ne: contamination
6217.3	Ne: contamination	67 0.2	
66.5	Ne: contamination	17.0	Ne: contamination
6300.2	[OI]: aurora	23.9	23.1 NI(31), deg. long λ
05.9		6948.4	45.2 NI(29)?
34.4	Ne: contamination	66.6	65.4 ArI(1)?
83.0	Ne: contamination	81.5	
6402.0	Ne: contamination	7003.1	02.2 OI(21)?
22.0		32.6	Ne: contamination
27.0		7156.8	OI(38)
30.4		7384.5	84.0 ArI‡
42.	41.7 NI(23)?	7405.2	
56.4	{57.9 NI(22)? / 55.0 OI(9)?}	23.6	NI(3)
65.1		42.3	NI(3)
68.6	68.3 NI(22)?	68.3	NI(3)
6471.6		76.8	76.5 OI(55)
82.8	82.7 NI(21), deg. long λ	7503.7	ArI(8)
85.2	84.9 NI(21), deg. long λ	15.5	ArI†
98.9	99.5 NI(21), deg. long λ	53.7	
6506.5	Ne: contamination	88.6	
10.0†		7723.8	ArI(1)
21.4†		74.	OI(1)
25.6†		7886.7	86.3 OI(64)?
28.3†		7902.0§	
32.9	Ne: contamination	14.4§	
37.4†		49.	OI(35)

TABLE 5.1 (Continued)

λ	Identification	λ	Identification
93.6	Contamination	46.4	OI(4)
8006.2	ArI(3)	8521.4	ArI(8)
14.8	ArI(1)	67.4	NI(8)
8103.7	ArI(3)	94.0	NI(8)
15.3	ArI(1)	8629.2	NI(8)
27.5		55.9	NI(8)
37.6		80.2	NI(1)
67.1		83.4	NI(1)
72.6		86.1	NI(1)
86.	NI(2)	8703.2	NI(1)
8200.3	NI(2)	11.7	NI(1)
16.	NI(2)	18.8	NI(1)
33.7	33.0 OI(34)?	28.9	NI(1)
42.3	NI(2)	47.4	NI(1)
64.5	ArI(8)	8820.5	OI(37)
8343.2	Contamination?	9047.2	48. $3s'^2D-3p'^2F$ NI¶
8408.2	ArI(8)	61.3	60.6 NI(15)
24.7	ArI(3)	9264.7	63. OI(8)
		9392.8	92.8 NI(7)

* The features at 3973.0 and 3983.0 were incorrectly identified by Wallace (1960) with NII(38).
† Some or all of these features may be gaps between water vapor absorptions.
‡ Kayser (1939).
§ These are doublets with about a 4-Å separation.
¶ According to Eriksson's (1958) analysis.

Among the most prominent features of the lightning spectrum are the spectral lines due to neutral nitrogen (NI), neutral oxygen (OI), singly ionized nitrogen (NII), and singly ionized oxygen (OII). The upper energy levels of NI from which transitions yielding visible and infrared radiation occur lie primarily between 11 and 13 ev above the NI ground state. The ionization potential of NI is 14.53 ev. The upper energy levels of NII from which transitions yielding measurable ultraviolet, visible, and infrared lines occur lie primarily between 20 and 30 ev above the NII ground state. Neutral oxygen, OI, emits visible and infrared spectral lines primarily due to transitions from upper energy levels between 10 and 16 ev above the OI ground state. The normal ionization potential of OI is 13.61 ev. Singly ionized oxygen, OII, emits measureable ultraviolet, visible, and infrared spectral lines primarily due to transitions from upper energy levels between 25 and 37 ev above the OII ground state.

5.2.2 The slitless spectrometer

A schematic diagram of a slitless spectrometer capable of resolving a lightning flash into its component strokes is shown in Fig. 5.3. In slitless spectroscopy, the lightning stroke itself acts as the thin source of illumination (the effective slit). Light from the channel is dispersed in wavelength, and images of the channel are recorded on film at those wavelengths at which the channel radiates appreciably. A near-infrared lightning spectrum obtained with a slitless spectrometer is shown in Fig. 5.4. With the spectrometer of Fig. 5.3, time resolution is obtained by moving the recording film. The drum speed is about 1 rps. The film moves imperceptibly during the time of a stroke (whose period of high luminosity lasts hundreds of microseconds) but moves a considerable distance during the time between strokes (tens of milliseconds). The slitless spectrometers used by Salanave have a 1-cm channel-isolator slot which, with the optics employed, yields a spectral image on the film of about 100 m of the lightning channel for a channel 6 km distant. Spectra obtained have an inverse dispersion of about 25 Å/mm in first order, and the wavelength resolution is of the order of a few angstroms. The prism shown in Fig. 5.3 is not necessary for wavelength dispersion but rather is used for convenience in that it makes possible the "straight-through"

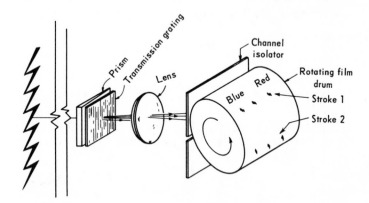

Fig. 5.3 *Schematic diagram of a slitless spectrometer capable of resolving a lightning flash into its component strokes. The direction change of the incident light produced by the prism is set to be approximately that necessary to yield a first-order grating spectrum that is parallel to the direction of the incident light. Thus the spectrometer can be aimed directly at the lightning discharge under observation. (Adapted from Orville (1966a).)*

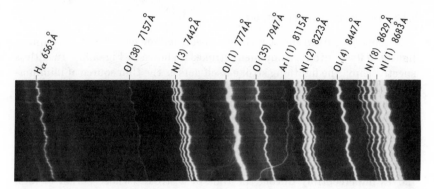

Fig. 5.4 *A stroke-integrated near-infrared spectrum of lightning obtained by Salanave (1966) using the slitless spectrometer shown schematically in Fig. 5.3. (Courtesy, Institute of Atmospheric Physics, University of Arizona, 1966.)*

viewing of the lightning spectrum. That is, the prism deflects the incident and essentially parallel light rays emitted by the lightning discharge so that the first-order spectrum diffracted from the transmission grating is directed parallel to the incident light. The spectrometer can thus be easily aimed at the source.

The slitless spectrometer has the following advantages over the slit spectrometer: (1) The slitless spectrometer can be used to view a single

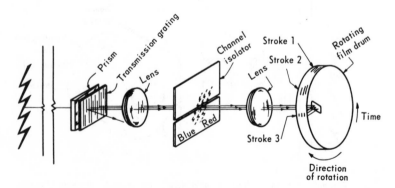

Fig. 5.5 *Schematic diagram of a slitless spectrometer capable of microsecond time resolution of the spectrum of a 10-m or smaller section of the lightning channel. Spectrometer in principle is similar to that shown in Fig. 5.3 except channel isolator slot is an order of magnitude or more smaller and drum speed is two orders of magnitude greater. (Adapted from Orville (1966a).)*

lightning channel since there is far less light loss than with the slit instrument; (2) it is possible with a slitless spectrometer to study the spectral properties of the channel as a function of position along the channel.   The slitless spectrometer suffers from the following disadvantages: (1) It is not possible to make very accurate wavelength identification with a slitless instrument; (2) for close strokes the wavelength resolution that can be obtained with a slitless spectrometer is limited by the luminous channel diameter.

A schematic diagram of a slitless spectrometer capable of time resolving on a microsecond time scale the spectral emissions from a 10-m section of lightning channel is given in Fig. 5.5.   Some time-resolved spectra of lightning are shown in Fig. 5.6a and b.   The spectrometer in principle is similar to that shown in Fig. 5.3, except that the channel-isolator slot is an order of magnitude or more smaller and the drum speed is two orders of magnitude greater.   The time resolution is determined by the time required to sweep the image of the slot width on the film.   For example, a 1-mm slot-width image and a writing rate of 0.2 mm/μsec produce a time resolution of 5 μsec.   Orville (1966a, 1968) has obtained time-resolved spectra by coupling a Bausch and Lomb replica grating and

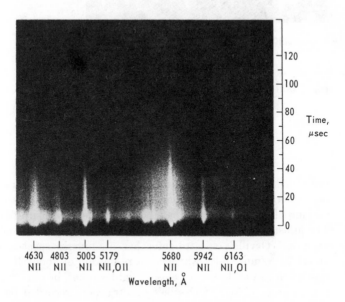

Fig. 5.6a   *A time-resolved visible spectrum of a lightning stroke obtained by Orville (1968) using the slitless spectrometer shown schematically in Fig. 5.5.*

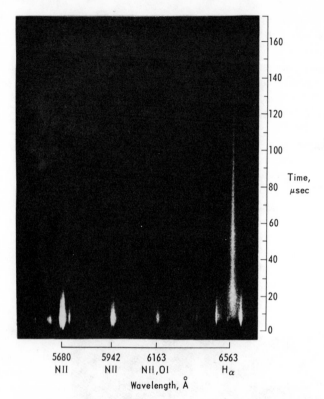

Fig. 5.6*b*   *A time-resolved spectrum of a lightning stroke showing the* H_α *region. Spectrum obtained by Orville (1968) using the slitless spectrometer shown schematically in Fig. 5.5.*

necessary optics to a Beckman and Whitley Model 318 high-speed streaking camera. The Beckman and Whitley camera is an electrically driven moving-film streak camera capable of writing speeds up to 0.3 mm/μsec. The filmstrip fits on the inner surface of a rotating drum and is rotated in vacuum. Orville's spectrometer has an inverse spectral dispersion of about 70 Å/mm and a wavelength resolution of about 10 Å. The time resolution can be 1 μsec for close, bright strokes, but in practice has been 2 to 5 μsec. Most of Orville's spectra were recorded on Kodak 2475 Recording film, a high-speed film with an ASA rating of 4000.

As an alternative method of recording time-resolved lightning spectra, Orville (1966a,b) has used a camera with fixed film and rotating mirror. While the spinning-mirror type of camera can provide higher time-resolution than can the rotating film-drum type, the spinning mirror must necessarily be small and thus considerable available light is wasted.

5.2.3 Quantitative spectroscopic measurements from photographic film

In Sec. 5.3 we shall consider the theory which makes possible the determination of lightning temperatures, particle densities, and pressures from the measurement of spectral line intensities and profiles. Before the theory can be applied, it is necessary to obtain the true line intensities and profiles from the measured film density. The response of a photographic emulsion (resultant film density) at a particular wavelength depends on the intensity of the source at the particular wavelength, the sensitivity of the emulsion at the same wavelength, the time of exposure, the humidity and temperature under which the emulsion is exposed, and the time, type, and temperature of the development it undergoes. The resultant film density is a nonlinear function of exposure, exposure being the product of source intensity and time. The fact that intensity and time, when varied to produce the same exposure, do not always produce the same density is known as reciprocity failure. Further, film sensitivity is a nonlinear function of wavelength. It follows from the above that the photographic film must be calibrated for its nonlinear responses if meaningful data are to be obtained.

The measurement of relative line intensities from Salanave's stroke-integrated spectra has been considered by Prueitt (1963). Prueitt placed in Salanave's spectrometer filmstrips from the same roll used to record lightning spectra and exposed the calibration strips to an Edgerton, Germeschausen, and Grier type FX-1 xenon flash tube via a step slit. The duration of the light from the flash tube (200 μsec) was approximately that of a lightning stroke so that errors due to reciprocity failure were minimized. The step slit allowed known relative exposure values to be impressed on the film so that characteristic curves of exposure vs. film density could be plotted for each wavelength. An exposure-vs.-wavelength curve for the xenon continuum supplied by the manufacturer was used for the wavelength calibration. In order to obtain relative line-intensity measurements from lines widely separated in wavelength, it was also necessary to correct for atmospheric absorption and for vignetting in the spectrograph. Additional details regarding the calibration process and sources of reference for additional reading are given by Prueitt (1963).

Orville (1966a, 1968) has described the calibration process used in determining relative spectral intensities from his time-resolved spectra. The procedure is similar to that of Prueitt (1963), except that in order to avoid reciprocity failure a xenon source with a time duration of luminosity of 3 μsec was used.

5.2.4 Spectroscopy utilizing photoelectric detectors

Photographic film is valuable for recording spectra in that film provides a permanent and essentially continuous record of spectra as a function of wavelength and time. The disadvantages of film are its nonlinear response and the associated calibration problems, its relatively poor sensitivity, and its relatively narrow intensity range. Where photographic film can at best record intensity variations over three orders of magnitude, photoelectric devices are capable of responding to intensity variations of seven or more orders of magnitude. Further, photoelectric devices are much more sensitive detectors of light than is film and photoelectric devices can provide an output current which is linearly proportional to the intensity being measured over most of the dynamic range. The main disadvantage of using photoelectric techniques to obtain such information as the characteristics of the lightning spectrum as a function of wavelength and time is that more than one photoelectric device (possibly many, depending on the wavelength resolution required) must be used to cover a given wavelength range, thus necessitating the simultaneous recording of a number of pulsed output signals. We consider now several photoelectric systems that have been used or proposed for the study of the lightning spectrum.

The slit spectrometer shown in Fig. 5.1 can be adapted for photoelectric recording of lightning spectra by replacing the photographic plate with a number of photocells or photomultipliers. To record the intensity of one spectral line, two photodetectors are needed, one to measure the intensity at the wavelength of interest, the measurement being of line plus continuum at that wavelength, and one to measure the continuum in the vicinity of the line. Then, the continuum intensity can be subtracted from the intensity of line plus continuum to yield spectral line intensity. Generally, individual spectral features of interest are isolated at the output of the spectrometer by exit slits, the photodetectors being placed behind the exit slits. The photodetector outputs are displayed on oscilloscopes and the oscilloscope records photographed. To take full advantage of the system described above, a portion of the discharge channel should be focused on the entrance slit.

A particular wavelength band can be effectively separated out of a beam of light without the use of a prism or diffraction grating. This can be done in one of two ways: (1) By use of an interference or other filter in front of a photodetector or (2) by use of a photodetecting surface which is only sensitive to a particular bandwidth. The former technique can be used to examine wavelength regions with a width as small as several angstroms, whereas the latter is generally reserved for wavelength regions several thousand angstroms wide. Krider (1965a,b) has used the former technique to study the time sequence of emission of spectral lines due to neutral and ionized elements and of continuum radiation from lightning. Krider (1966) and Krider et al. (1968) have used the latter technique to obtain an estimate of the absolute power radiated by lightning in the wavelength range from about 4000 to about 11,000 Å. The measurements of absolute power were made with a calibrated photodiode, the so-called "lite mike," manufactured by Edgerton, Germeshausen, and Grier.

Finally, it is probable that electronic image-converter and image-intensifier cameras will be used in the near future to record lightning spectra with good light gain and high time resolution. The image-converter or image-intensifier camera would be used in conjunction with either a slit or a slitless spectrometer. These cameras are capable of providing high-speed framing or streak photographs of the spectra displayed at their inputs. Spectroscopic systems with image-converter or intensifier recording have been used with considerable success in the laboratory.

5.3 THEORY

In this section we examine some of the techniques available for the determination of the particle densities and temperatures* within the lightning channel from an analysis of the lightning spectrum. The first step in analyzing a lightning spectrum must be a demonstration that the lightning channel under study is either optically thin or optically thick, or intermediate, to the particular wavelengths of interest. A channel is said to be optically thin to particular wavelengths if photons of those wavelengths can pass through the channel without appreciably interacting with the particles composing the channel. A channel is said to be optically thick to particular wavelengths if photons of those wavelengths are absorbed and reemitted many times in traversing the channel. An optically thick channel of constant temperature emits blackbody radiation characteristic of that temperature. We shall consider the details of the opacity determination at the end of this section and in Sec. 5.5.2.

If the measured radiation emitted by either optically thin or optically thick lightning channels is to be used to determine channel properties, it is usually necessary, from a practical point of view, to assume that local thermodynamic equilibrium (LTE) exists within the lightning channel as a function of position and time. We now define thermodynamic equilibrium and local thermodynamic equilibrium. A classical gas is said to be

---

* In this chapter, and in other parts of the book, the word *temperature* is often used in a loose sense to indicate a measure of the particle average energies since local thermodynamic equilibrium may not exist under all conditions within the discharge channel.

in thermodynamic equilibrium at temperature T if all energy states, continuous and discrete, are populated according to Boltzmann statistics. In particular, the kinetic energy distribution of each group of particles is described by a Maxwell-Boltzmann distribution function, the population of the discrete atomic energy levels is described by a Boltzmann distribution, and the relation between the populations of discrete atomic energy levels and continuum levels is described by a Saha equation. A gas is said to be in LTE as a function of position and time if each small volume of the gas at local temperature $T(\mathbf{r},t)$ satisfies the conditions for thermodynamic equilibrium. If the assumption of the existence of LTE is not made, one must resort to detailed calculations involving excitation and ionization cross section, recombination rates, transition probabilities, etc., many of which are not known, in order to reduce the measured radiation to particle densities and energies. Unfortunately, proof of the existence of LTE requires a detailed calculation involving the same unknown atomic parameters. We shall provide some justification for the assumption of LTE at the end of this section. Fortunately, whether LTE exists or does not exist, electron densities in the lightning channel can be determined from a measurement of the Stark broadening of certain spectral lines.

In LTE the atomic energy levels within an ionization state are populated according to Boltzmann statistics:

$$N_n = \frac{Ng_n}{B(T)} \exp\left(-\epsilon_n/kT\right) \tag{5.1}$$

where N_n is the number density of atoms in energy level n, N is the total number density of atoms, ϵ_n is the excitation potential of the nth level, k is the Boltzmann constant, T is the absolute temperature, g_n is the statistical weight of the nth level, and $B(T)$ is the partition function, given by

$$B(T) = \sum_j g_j \exp\left(-\epsilon_j/kT\right) \tag{5.2}$$

The following relation, the Saha equation, is valid for a system of particles in LTE:

$$n_e = \frac{N^i}{N^{i+1}} \frac{2}{h^3} (2\pi mkT)^{3/2} \frac{B^{i+1}}{B^i} \exp\left(-X/kT\right) \tag{5.3}$$

where n_e is the electron density, the superscripts give the state of ionization (N^0 is the number density of neutral atoms of a given type, N^1 is the number density of singly ionized atoms formed by ionization of those neutrals, etc.), X is the ionization potential from the ith to the $(i+1)$th ionization state, h is Planck's constant, and m is the electron mass.

The measured radiated power in an emission line (sometimes called the line intensity) from an optically thin gas per unit volume of gas at uniform temperature and density due to transitions from level n to level r is

$$I_{nr} = CN_n A_{nr} h \nu_{nr} \qquad (5.4)$$

where A_{nr} is the Einstein transition probability, ν_{nr} is the frequency of the emitted photon, and C is a geometric factor. If we use Eq. (5.1), that is, assume that LTE exists, the radiated power may be written

$$I_{nr} = \frac{CN g_n A_{nr} h \nu_{nr}}{B(T)} \exp\left(-\epsilon_n / kT\right) \qquad (5.5)$$

Equation (5.5) is usually used with a known temperature and measured intensity to determine particle density. The ratio of the radiated power due to transitions from level n to level r to the radiated power due to transitions from level m to level p is given by

$$\frac{I_{nr}}{I_{mp}} = \frac{g_n A_{nr} \nu_{nr}}{g_m A_{mp} \nu_{mp}} \exp\left[\frac{-(\epsilon_n - \epsilon_m)}{kT}\right] \qquad (5.6)$$

which can be solved for the temperature, as follows:

$$T = \frac{\epsilon_m - \epsilon_n}{k \ln\left(I_{nr} g_m A_{mp} \nu_{mp} / I_{mp} g_n A_{nr} \nu_{nr}\right)} \qquad (5.7)$$

Thus the measured power ratio (sometimes called the intensity ratio) of two spectral lines emitted by the same type of atom from an optically thin gas along with tabulated atomic parameters is sufficient to determine the temperature of that gas. In practice, to obtain an accurate temperature determination using Eq. (5.7), $(\epsilon_m - \epsilon_n)$ should be chosen to be greater than kT. For Eqs. (5.6) and (5.7) to be useful it is not necessary to require LTE to exist, but only to require that the energy levels involved in the pertinent transitions be occupied according to Boltzmann statistics. Should only the latter be true, the temperature under consideration will be the electron temperature, since it is electron excitation and deexcitation collisions that are responsible for the maintenance of the Boltzmann distribution.

In the case of a gas that is optically thick at several wavelengths and in LTE, a blackbody temperature can be determined for that gas by comparing the measured intensities at wavelengths for which the gas is optically thick with the Planck radiation law

$$I(T) = \frac{2h\nu^3}{c^2} \frac{1}{e^{h\nu/kT} - 1} \qquad (5.8)$$

Consider now atoms of type C denoted by subscript C and atoms of type D denoted by subscript D. The ratio of the power in a spectral line of atoms D to the power in a line of atoms C radiated from an optically thin gas is, using Eq. (5.5),

$$\frac{I_D}{I_C} = \frac{N_D g_{Dj} B_C A_D \nu_D}{N_C g_{Ci} B_D A_C \nu_C} \exp\left(\frac{\epsilon_{Ci} - \epsilon_{Dj}}{kT}\right) \tag{5.9}$$

If, for instance, atoms D are neutral atoms and atoms C are singly ionized ions formed by ionization of D, the elimination of N_D/N_C in combining Eq. (5.9) with the Saha equation, Eq. (5.3), yields an expression for electron density as a function of temperature and line power ratio. Once the temperature has been determined using Eq. (5.7) and the power ratio of the line radiation from a neutral atom (for example, NI) to the line radiation from a singly ionized atom (for example, NII) has been measured, the electron density can be calculated. As an alternative to this approach, one can use the tables of thermodynamic properties of air computed, for example, by Gilmore (1955, 1967) or by Hilsenrath and Klein (1965) and various measured power ratios—for instance, the power in a neutral oxygen (OI) spectral line to the power in a singly ionized nitrogen (NII) line—to determine electron density and additional properties (percentage ionization, pressure, various particle densities, stored energy) of the high temperature air. Use of the thermodynamic tables is equivalent to use of the Saha equation since the thermodynamic data are the result of the solution of a number of coupled Saha equations, an equation of charge conservation, and an equation of percentage composition.

Strictly speaking, the Saha equation is only valid if LTE is maintained. In practice, the Saha equation is often used in situations where the existence of LTE is questionable. There is, as we shall discuss, some question whether LTE exists within the lightning channel. It is difficult to determine what errors are incurred in using the Saha equation when deviations from LTE do exist.

Fortunately, electron density can be determined without resort to the assumption of LTE. The width of a spectral line which is broadened predominantly by the Stark effect is primarily dependent upon charged-particle number densities and only slightly dependent on particle energies. Stark shifts, widths, and profiles of many spectral lines are given by Griem (1964). Hydrogen is present in the lightning stroke by virtue of the decomposition of water vapor, and the H_α line of the Balmer series is considerably Stark broadened so that electron density may be determined from a measurement of its width. Figure 5.7 shows the full width of H_α at half-intensity as a function of electron density. The theoretical Stark profiles for the Balmer series of hydrogen have been well substantiated by experiment (Wiese, 1965).

We consider now the question of channel opacity. The opacity deter-

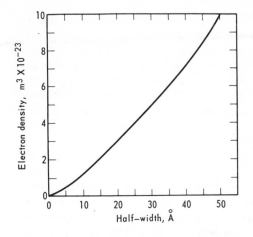

Fig. 5.7 *A theoretical plot of the width of* H_α *at half-intensity vs. electron density for 20,000°K. (Adapted from Uman and Orville (1964).)*

mination is made as follows: According to Eq. (5.6), if two spectral lines originate in an optically thin gas from the same or nearly the same upper atomic energy level, their intensity ratio will be independent of temperature (the exponential factor will be essentially unity) and will only depend on atomic constants. Thus theory can predict the intensity ratio of certain line pairs if the channel is optically thin. If the measurements support the theory, then the evidence is strong that the channel is optically thin to those wavelength regions in which the opacity determination is made. We shall consider specific examples of the opacity determination in Sec. 5.5.2. In the opacity determination we make use of Eq. (5.6), which is an equilibrium relationship. However, Eq. (5.6) with the exponential term set equal to unity may well be valid for spectral lines originating from closely spaced upper energy levels even in the absence of LTE. This will be the case if (1) the upper energy levels under consideration alone are Boltzmann distributed (due presumably to collision processes which transfer atoms between these close-lying energy levels) or (2) the collision cross sections for excitation of the atom under study into the upper energy in question are proportional to the statistical weights of the energy levels.

We consider now the question of whether LTE exists in the lightning channel as a function of position and time. The mechanism for maintaining a Boltzmann distribution of atomic energy levels in the lightning plasma is the balance between electron collisional excitation and collisional deexcitation of the atomic energy states in the absence of appreciable radiative transitions (Griem, 1964). In general, for a fixed electron density and electron temperature, there will be, for a given atom, an energy level above which a Boltzmann distribution is maintained and below which it is not.   The Boltzmann distribution can be expected to exist above a given energy level if above that level the collisional rate processes (electron collisional excitation and deexcitation) dominate the radiative processes (radiative decay and recombination).   Both the radiative and the collisional rates for a given transition are proportional to the emission oscillator strength of the transition.   The emission oscillator strength is a measure of the Einstein transition probability and is defined as follows:

$$f = \frac{mc^3 \epsilon_0}{2\pi e^2 \nu^2} A \tag{5.10}$$

where e is the electron charge, c the velocity of light, and ϵ_0 the free-space permittivity.   The emission oscillator strength cannot exceed unity. Cross sections for collisional excitation are roughly inversely proportional to the energy difference between the levels involved in the transition and, since energy levels usually become more closely spaced with increasing energy, the collisional excitation rates in general increase with energy. The largest radiative-decay rates are generally from excited states into the ground state or states near the ground state.   The radiative-decay rates are proportional to the oscillator strength and to the square of the energy difference between the upper and lower states; that is, the rates are given by the Einstein transition probability A.   Since the energy difference between the ground state and the first few excited states is, in general, large and since the energy difference between the first few excited states and higher excited states is usually small, the radiative-decay rates, in general, cannot increase much with increasing energy level. The radiative-decay rates may in fact decrease with increasing energy level.

To determine for a given atom, in the presence of a given electron density and temperature, the energy level above which a Boltzmann distribution is maintained requires a detailed calculation involving transition probabilities, collision cross sections, etc.   Calculations of this type have been performed by Griem (1964) and by Sampson (1966).   It would appear that for the Boltzmann distribution to extend to the ground states of air atoms for temperatures above about 20,000°K, an electron density in excess of 10^{24} m^{-3} is required.   As we shall see in Sec. 5.5.2, electron densities in the early stages of the return-stroke channel are near 10^{24} m^{-3}, but the density in the latter stages of the lightning stroke is near 10^{23} m^{-3}.

The visible line radiation from the return-stroke channel is mostly due to NII transitions between energy levels at or above the lowest $n = 3$ level.   It is from this NII radiation that the lightning temperature has been determined.   It can be shown (according to the criterion given by

Griem, 1964) that at electron densities of the order of 10^{23} m^{-3} for temperatures from 13,000 to over 35,000°K a Boltzmann distribution is maintained at and above the lowest $n = 3$ energy level in NII. Further, it can be shown (again, according to the criterion given by Griem, 1964) that for these temperatures the Boltzmann distribution will be maintained on a time scale measured in 0.1 μsec as will the equilibrium between electron and heavy-air-particle average energies. Since it is electron collisions which are responsible for the maintenance of the Boltzmann distribution, the temperature determined [using Eq. (5.7)] from line radiation due to NII transitions above the lowest $n = 3$ level is, strictly speaking, the electron temperature. (The temperature so determined is also known as the excitation temperature.) However, in view of the fact that for temperatures of interest above 13,000°K all particles in the discharge channel will, on a time scale measured in 0.1 μsec, have essentially the same average kinetic energies, it is reasonable to speak of a "channel temperature" even in the absence of "complete" LTE.

For a detailed discussion of the conditions under which LTE can be expected and the criteria for its existence, the reader is referred to the work of Griem (1964) and Sampson (1966).

It is not practical at present to attempt to measure radiative emission as a function of channel radius. Hence all lightning spectra thus far obtained yield at best the total radiation at a given wavelength emanating from a short length of channel regardless of where within the channel that radiation originated. The lightning spectra can most simply be analyzed if we assume that, at a given time, physical conditions are constant throughout a cross section of the channel. The actual temperature profile of the channel is determined primarily by the means available for transporting heat out of the channel. If, for example, thermal conduction is the dominant mechanism, there must be a temperature gradient throughout the channel; i.e., the temperature decreases with increasing radius. If radiation that escapes the channel is the dominant energy-loss mechanism, the temperature profile of the channel will, as assumed, be relatively flat. It is to be expected that the temperature profile will change with time.

5.4 LIGHTNING SPECTROSCOPY: 1940–1960

In the years 1940 to 1960 about 15 papers were published in which lightning spectroscopic data were reported. All spectra obtained were time-integrated over at least the time of one lightning flash. Spectra of single flashes were recorded with slitless spectrometers. (In order to obtain a spectrum using a slit spectrometer it is necessary, because of the relatively low light-gathering ability of the instrument, to time-integrate over a number of flashes.) The main emphasis of the 1940 to 1960 spectroscopic studies was on the identification of the lines and bands in the lightning spectrum and on the association of these spectral features with particular radiative transitions of the molecules, atoms, and ions in the lightning discharge. During these studies spectral features of lightning were identified in the wavelength range from about 2800 to about 9100 Å. We discuss the results of these spectroscopic studies in this section.

In 1941 and in a later paper, Israel and Wurm (1941, 1947) presented the results of a lightning study made with a slitless spectrometer. Israel and Wurm listed the observed spectral lines according to their multiplet designations and gave the upper energy levels (excitation potentials) for the observed lines. They noted that the lightning spectrum was composed primarily of spectral lines of neutral and singly ionized nitrogen and oxygen atoms superimposed on a continuum and that no spectral lines from doubly ionized elements were observable. In addition, Israel and Wurm identified some Balmer series hydrogen lines and molecular band spectra due to N_2^+ (first negative system around 3914 Å) and N_2 (second positive system around 3370 Å). Israel and Wurm observed that the intensity of spectral lines of high excitation potential decreased from the ground upward relative to the intensity of lines of lower excitation potential. They also reported that spectral lines due to calcium were present on that part of the spectrum corresponding to the ground contact of the lightning discharge.

In 1943, Nicolet (1943) classified the pre-1940 lightning spectral data, particularly the data of Slipher (1917), into multiplets and provided lists of spectral lines by element and by ionization stage.

In a series of five papers published between 1947 and 1949, M. Dufay, J. Dufay, and M. Tcheng published the results of a lightning spectral study which covered the wavelength region from 2910 to 6570 Å. In the study, both slit and slitless spectrometers were employed. Dufay (1947) reported on the violet and ultraviolet spectrum of lightning; Dufay and Tcheng (1949a) reported on slit spectra obtained in the wavelength range 3830 to 5670 Å; and Dufay and Dufay (1949) published and analyzed slitless lightning spectra. The results of the slit spectroscopy reported in the papers by Dufay (1947) and by Dufay and Tcheng (1949a) are repeated in the more detailed two-part publication by Dufay and Tcheng (1949b) and Dufay (1949).

Dufay and Dufay (1949) showed from their slitless spectra that the intensity of the lines of ionized nitrogen and oxygen decreased from the ground toward the cloud, these results confirming the observations of Israel and Wurm (1941, 1947). Further, they attempted to determine from the Stark broadening of H_β of the Balmer series of hydrogen the degree of ionization in the lightning channel. They calculated that the fraction of ionized atoms and molecules was of the order of 5×10^{-4}, a result apparently much in error (see Sec. 5.5.2 for the results of recent Stark broadening measurements). Dufay and Tcheng (1949a,b) have presented data obtained from the study of 10 slit spectrograms. They present detailed lists of spectral lines, their wavelengths, multiplet designations, and excitation potentials. They identified the N_2^+ negative bands, the N_2^+ positive system, and, tentatively, several other band systems including an absorption band due to water vapor. Dufay and Tcheng (1949a,b) noted a marked difference in apparent excitation from (slit) spectrum to spectrum. They have classified their spectra as due to strong, medium, weak, or very weak excitation. The strong-excitation spectrum is characterized by strong ion lines and by weak neutral lines and weak or absent bands; the very weak excitation spectrum is characterized by an absence of ion lines and the presence of band spectra and certain neutral lines. It should be kept in mind that the spectrum obtained with a slit spectrometer represents the integrated effect of the spectra of several lightning discharges and of the spectra from all portions of the discharge including that portion within the cloud. Dufay (1947, 1949) has studied the violet and ultraviolet lightning spectrum. Only one spectrum was obtained. The limit on the shortest observable wavelengths was set by the strong ozone absorption near 2900 Å. Dufay (1947, 1949) reported that the ultraviolet spectrum was characterized by a strong continuum on which were superimposed spectral lines and molecular emission bands. The single spectrum studied was judged to be in the category of the weak-excitation spectrum since the bands were strong and plentiful and the observable lines were not numerous or intense. A detailed list of the observed line and band spectral features was given. Dufay (1947, 1949) reported the identification of 23 fine-structure peaks in the N_2^+ band system and of 25 fine-structure peaks in the N_2 second positive system. Band systems of OH, NH, and NO were tentatively identified. Dufay (1947, 1949) calculated from the variation of the ozone absorption with wavelength the amount of ozone that existed between the spectrometer and the effective source of radiation. We shall have more to say about ozone absorption in Sec. 5.5.1.

Between 1950 and 1952, three papers were published concerning the infrared spectrum of lightning. Jose (1950) reported that the infrared slit spectrum from 7400 to 8800 Å was due exclusively to spectral lines from neutral oxygen and nitrogen, and he provided a table of those lines. Petrie and Small (1951) secured a slit spectrum of lightning in the wavelength range 7100 to 9100 Å. In addition to identifying the spectral lines from neutral oxygen and nitrogen, Petrie and Small identified 13 lines due to neutral argon. Knuckles and Swensson (1952) studied the wavelength range 6159 to 7157 Å and identified two previously unobserved and several previously misidentified spectral lines due to neutral oxygen and nitrogen.

Vassy (1954) compared the time-resolved spectrum of a long laboratory spark in air with the time-integrated lightning spectra obtained by previous investigators. Vassy concluded that the lightning spectrum most resembles that portion of the spectrum of the laboratory spark which occurred after the so-called "principal discharge," the short period (1 μsec) of high excitation and, apparently, high current. In the laboratory spark the so-called "post-luminence" period had a duration of about 20 μsec. On the basis of these observations, Vassy suggested that the features of the time-integrated lightning spectra are due primarily to the long time duration of the "post-discharge" and hence that the characteristic levels of excitation derived from time-integrated lightning spectra are not indicative of the short time duration "principal discharge" of lightning. In Sec. 5.5.2 we shall consider the determination of return-stroke properties both from time-integrated and from time-resolved (2 to 5 μsec) lightning spectra.

Wallace (1960) reported the analysis of a near-ultraviolet spectrum of lightning. The wavelength covered was 3670 to 4280 Å. In addition to confirming previous line and band identifications, Wallace provided the first identification of the CN violet band–system in the lightning spectrum. Further, Wallace estimated the temperature necessary to provide the relative intensities observed for two regions of the CN band spectrum. Wallace suggested a "rotational temperature" in the neighborhood of 6000 to 30,000°K.

Hu (1960) reported on spectra obtained with a slit spectrometer covering the wavelength range 2750 to 7000 Å. Hu presents reproductions of his lightning spectra as well as a spectrum of a high-frequency steady-state air discharge generated by a Tesla coil.

5.5 LIGHTNING SPECTROSCOPY AFTER 1960

5.5.1 General

Prior to 1960, all recorded lightning spectra were integrated over the time of at least one flash. In 1961, Salanave (1961) reported obtaining the first stroke-integrated spectra. Salanave made use of a revolving film drum (see Fig. 5.3) to separate the spectra of the strokes constituting a flash. The time resolution was of the order of milliseconds. During the summers of 1960 and 1961, approximately 200 slitless spectra of lightning flashes were obtained in the wavelength region 3850 to 6900 Å by Salanave and co-workers. Typical spectra and wavelength identifications from these spectra are given by Salanave et al. (1962). They reported that no outstanding differences exist between the spectra of the strokes making up a flash, with the exception that some strokes were followed by a continuing luminosity. Spectral lines due to ionized argon were identified. No lines due to doubly ionized atoms could be definitely identified. Molecular bands attributed to CN and N_2^+ in the wavelength region 3800 to 4000 Å were reported to appear infrequently and with highly variable intensity. In approximately 95 percent of the spectra there were no clearly recognizable molecular bands between 3850 and 5700 Å. The $O_2(1,0)$ absorption band at 6883 Å was recorded in a few cases, as was the water vapor absorption in the region 5900 to 6000 Å. Salanave et al. (1962) observed, as had others previously, that the degree of excitation in the discharge channel decreases from the ground up.

Wallace (1964a) published slit spectra of lightning in the wavelength range 3100 to 9600 Å. Wallace's data represent the best slit spectra presently available. Some of these data are shown in Fig. 5.2. The principal emission features are listed in Table 5.1. Wallace (1964a) also gives the wavelengths of 35 absorption features in the spectra, most, but not all, of which are probably due to water vapor. Band spectra due to N_2, N_2^+, CN, and NH are identified. It was noted that some of the lines due to neutral oxygen and nitrogen are Stark broadened. Wallace stated that the majority of previous line and band identification, made from lower resolution spectra than his, are correct, and that although new multiplets of NI and OI have been identified, these were expected on the basis of previously identified lines. Unidentified features in the lightning spectrum are listed by Wallace (1960, 1964a).

The ultraviolet spectrum of lightning has received relatively little attention. Slit spectra in the ultraviolet have been obtained by Dufay (1949), Hu (1960), and Wallace (1964a). The first slitless ultraviolet spectra of lightning were reported by Salanave (1962). Among the ultraviolet spectra obtained by Salanave there is one which extends to the ozone absorption cutoff which appears at about 2850 Å. Part of this spectrum has been reproduced by Meinel and Salanave (1964). The entire spectrum from the ozone cutoff to 4000 Å has been reproduced by Orville (1967). Dufay (1949) reported the appearance of the Huggins ozone absorption band in the 3200 to 3250 Å region of his spectrum and noted a sharp cutoff in the 3020 to 3060 Å region. This cutoff, according to Orville (1967), is due to absorption by the Hartley band of ozone. In Salanave's spectrum the Huggins band observed by Dufay does not appear. Further, the wavelength cutoff due to the Hartley absorption band is at 2850 Å, rather than in the 3020 to 3060 Å region. Dufay (1949) measured the rate of decrease of the transmitted intensity through the 40-Å cutoff region and, assuming the absorption was due to ozone, he calculated a reduced ozone thickness of 0.3 cm along the path between spectrograph and the several flashes that produced the slit spectrum. (The amount of ozone is expressed as the thickness of an equivalent layer of pure gas at standard temperature and pressure.) Orville (1967) has used Salanave's spectrum showing a Hartley band absorption but no Huggins band absorption to determine the amount of ozone occurring between the slitless spectrograph and a single flash at a distance of 0.85 km. A value of 0.01 to 0.05 cm of ozone was found, this value being one order of magnitude greater than the clear air value. Orville (1967) suggests that Dufay's ozone values are larger than his because of the greater distance from Dufay's spectrometer to the flashes observed. This argument assumes a relatively homogeneous distribution of ozone along the line of observation to the discharge. This ozone is presumably created by corona processes at the ground and in the cloud. It may be, however, that the lightning flash itself creates ozone and that the measured ozone exists as a mantle around the lightning channel. Orville (1967) has outlined further experiments to determine which, if either, of the two proposed processes of ozone generation is dominant in producing the observed spectra absorption.

The infrared spectrum of lightning has been observed with slit spectrometers by Jose (1950), Petrie and Small (1951), Knuckles and Swensson (1952), and Wallace (1964). The first slitless infrared spectra of lightning were reported by Salanave (1966). A reproduction of one of Salanave's infrared spectra is given in Fig. 5.4.

Lightning spectra obtained by various observers often show conspicuous differences. Probably the most apparent difference is in the strength of the molecular bands. The spectra obtained with slit spectrometers by Israel and Wurm (1947), Dufay (1949), and Wallace (1960, 1964a) show strong molecular features whereas spectra obtained with a slitless spectrometer by Salanave (1961, 1962) and Salanave et al. (1962) show only occasional traces of molecular bands. Salanave (1964) and Meinel and Salanave (1964) have published papers in which slit and slitless lightning spectra are compared. There are two explanations generally advanced for the observation that lightning spectra obtained with slit spectrometers would appear to be associated with discharges of lower energy than the discharges observed with slitless spectrometers: (1) The observations are fortuitous and are merely due to the effects of geographical location and variation in the individual storms; (2) the observations are influenced by the method of observation in that the slit spectrometer usually records the time integral of luminosity from ground discharges, cloud discharges, and interstroke processes in the cloud and between cloud and ground whereas the slitless spectrometer usually records the spectrum of only the cloud-to-ground discharge channel. The latter point of view is adopted by Meinel and Salanave (1964) who have examined the N_2^+ bands in slit and in slitless spectra. They suggest that the N_2^+ emission recorded on slit spectra occurs at some distance from the visible lightning channel, probably in discharges within the cloud, and that the occasional occurrence of N_2^+ in slitless spectra probably arises from channel luminosity due to continuing current. Meinel and Salanave (1964) list seven possible sources of the N_2^+ band emission other than the return-stroke channel.

The desirability of time resolving the spectral features of a lightning stroke is evident. The first attempt at time resolving the stroke spectrum was made by Israel and Fries (1956), who constructed a scanning spectrometer with a time resolution of 20 μsec. No data obtained with this spectrometer have been published although a spectrum was recorded. Krider (1965a,b) reported the first time resolution of spectral emissions from individual return strokes. Krider used narrow-passband interference filters and photoelectric detectors to measure the time history of several spectral lines and of the continuum in the 3700 to 3900 Å region. Unfortunately, Krider's photoelectric system failed to isolate a narrow section of the lightning channel and therefore the true time history of light emission was obscured by the variation in luminosity due to the propagating return stroke (Krider, 1966). Krider (1965a,b) did show, however, that the NII lines reached maximum intensity first, followed by the continuum maximum, which in turn was followed by the peaking of the neutral hydrogen lines. Lines of lower excitation potential therefore reach a maximum later in time, consistent with a channel temperature that decreases with time. The indication is that the effective excitation potential of the continuum lay between that of the ions and the neutrals, consistent with a continuum due to radiative recombination. Orville and Uman (1965) have shown that the time-integrated continuum on Salanave's spectra is not the result of blackbody radiation or electron-ion bremsstrahlung emitted at constant temperature.

Using a calibrated photodiode, Krider (1966) and Krider et al. (1968) have made absolute estimates of the spectral power radiated by lightning in the wavelength region from about 4000 to about 11,000 Å. In addition, by comparing the broadband optical radiation from lightning with similar radiation from a long laboratory spark of known input power and input energy, Krider et al. (1968) have deduced values of input power and input energy for lightning. For a single-stroke lightning flash the peak radiated power (4000 to 11,000 Å) was found to be 6.2 × 10⁶ watts/m, the radiated energy 870 joules/m. The computed value of peak input power was 7.8 × 10⁸ watts/m, of input energy 2.3 × 10⁵ joules/m. The value of input energy is in good agreement with values of that quantity deduced from thunder measurements (Sec. 6.3.1) and from electrical measurements (Sec. 7.3.3). According to Krider et al. (1968) the values for peak radiated and input power should be considered lower limits to the actual values characteristic of the channel. Additional comments on the input power to lightning are given in Sec. 7.3.3.

The first time-resolved spectral studies of lightning strokes with both good spatial resolution and good temporal resolution were reported by Orville (1966a,b). Further results of time-resolved spectroscopic studies have been published by Orville (1968). A schematic diagram of Orville's spectrometer is shown in Fig. 5.5. Typical time-resolved return-stroke spectra are shown in Fig. 5.6a and b. Orville's spectra represent the radiation from about a 10-m length of channel. The time resolution is 2 to 5 μsec and the wavelength resolution about 10 Å. Orville (1968) found that in general the time for both the NII line intensities and the continuum intensity to rise to peak was less than about 10 or 15 μsec, the line intensities reaching peak first, and that the singly ionized species emitted radiation before the continuum spectrum is evident. The NII lines were found to remain intense for 30 to 60 μsec. The NII lines with the lowest excitation potential appeared first followed by NII lines with higher excitation potential. The NII lines with lower excitation potentials also persisted longer than did the NII lines with higher excitation potentials. Thus, NII 5680 Å (upper energy level, 20.6 ev) appears before NII 5942 Å (23.2 ev) and persists longer (see Fig. 5.6a). The hydrogen line H_α (Balmer series) was found to be absent or very faint in the initial microseconds of the discharge, but later became intense (see Fig. 5.6b). The H_α line reached maximum intensity in 20 to 40 μsec and often could be detected for more than 150 μsec. Information about the return-stroke channel derived from these measurements is given in Sec. 5.5.2.

Orville (1968) reported spectra from one flash composed of at least five strokes. The five recorded spectra could be grouped into two general classes. In one class, the NII emission was short lived and H_α was long lasting. The continuum emission was relatively weak. In the other class, the NII emission persisted for a relatively long time and the continuum emission was relatively intense. The H_α emission was very intense but short lived.

In the Section to follow, Sec. 5.5.2, we shall examine how the theory presented in Sec. 5.3 can be applied to the spectral data of Salanave, Orville, and others to determine the temperature, particle densities, and pressure in the return stroke.

5.5.2 Return-stroke properties

Return-stroke properties have been determined from the flash-integrated spectra of Wallace (1960) and of Zhivlyuk and Mandel'shtam

(1961), from the stroke-integrated spectra of Salanave and co-workers, and from the stroke-resolved spectra of Orville. We consider first the analyses of the data of Salanave.

The first step in analyzing a spectrum is the opacity determination. In Fig. 5.8 are shown measured line profiles for an NII multiplet whose component members originate from approximately the same upper energy level and have approximately the same wavelength. It follows from Eq. (5.6) and Eq. (5.10) that for an isothermal channel which is optically thin the intensity ratios among the various lines composing the multiplet should depend only on the ratios of the *gf* products. Uman and Orville (1965) have shown theoretically that even on time-integrated spectra the intensity ratios should depend only on the *gf* ratios if the channels are optically thin. In Table 5.2 a comparison is given between measured intensity ratios within two NII lightning multiplets and the results of theory and laboratory experiment. From this data Uman and Orville (1965) conclude that the lightning channel is optically thin to the NII

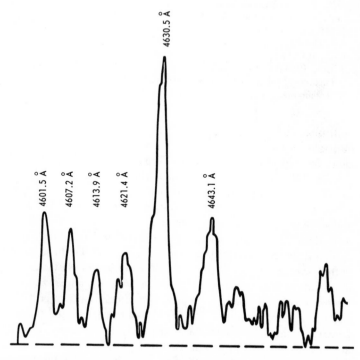

Fig. 5.8 *Relative intensity vs. wavelength within the* NII(5) *multiplet for lightning stroke A3, no. 2. Dashed line indicates the line base chosen.* (*Adapted from Uman and Orville (1965).*)

TABLE 5.2

Intensity ratios within NII(3) multiplet and NII(5) multiplet for five lightning strokes. Adapted from Uman and Orville (1965).

Stroke destination	Average excitation temperature,* °K	NII(3)		NII(5)				
		A	B	C	D	E	F	G
a†		2.6	8.0	3.0	3.8	5.0	3.8	3.0
b‡		2.8	9.9	3.2	4.1	7.2	4.8	2.9
A1	24,400 ± 800	2.8		2.6	2.7	4.1	3.1	2.3
A2	27,800 ± 400	2.4		2.1	3.1	4.4	3.8	1.9
A3, no. 1	24,900 ± 400	3.0		2.9	3.1	3.5	3.1	2.3
A3, no. 2	26,600 ± 600	3.1		2.7	3.6	4.6	3.1	2.1
B1, no. 1	26,900 ± 600	2.4	6.8	2.9	2.5	4.2	3.4	2.0
B1, no. 2	28,400 ± 1,000	2.7	7.6	2.4	2.8	4.3	3.5	2.2
B1, no. 3	27,500 ± 900	2.8	6.5	3.0	2.8	4.0	3.5	2.2
C1, no. 1	24,200 ± 900	2.5	6.7	2.8	3.8	4.6	3.5	2.4
C1, no. 2	24,700 ± 500	2.8	7.2	3.1	3.2	6.1	3.3	2.3

Letters designate the following measured line-intensity ratios: $A = (5,679.6 + 5,686.2 + 5,676.0)/5,666.6$, $B = (5,679.6 + 5,686.2 + 5,676.0)/5,710.8$, $C = 4,630.5/4,601.5$, $D = 4,630.5/4,607.2$, $E = 4,630.5/4,613.9$, $F = 4,630.5/4,621.4$, $G = 4,630.5/4,643.1$

* From Prueitt (1963).

† Theoretical calculation of gf ratio by Griem (1964).

‡ Measurement of gf ratio by Mastrup and Wiese (1958).

multiplets of interest, at least over the greater part of the time that those multiplets are integrated on film. From the calculated absorption coefficient for NII 4630.5 Å and the measured optical thinness, Uman and Orville (1965) have determined that the effective diameter within which the radiating NII atoms are contained must be less than a few millimeters.

Since the lightning channel is optically thin to the NII lines of interest, and since the energy levels which yield those lines are populated according to Boltzmann statistics on a 0.1-μsec time scale (see Sec. 5.3), a lightning temperature can be calculated from Eq. (5.7). To do so from time-integrated spectra involves the implicit assumption that the return-stroke temperature is constant in time and is constant across the channel cross section. Since this is not the case, the calculated temperature will represent some average value. Prueitt (1963) has calculated this average temperature, which he terms the excitation temperature, for nine different strokes using a variation of Eq. (5.7) applied to five NII multiplets. These data are included in Table 5.2. The average temperatures range from 24,200 to 28,400°K.

Although it is not possible to determine the error incurred by assuming the channel properties to be uniform as a function of channel radius, it is possible to make an estimate of the error incurred by assuming the return-stroke temperature to be constant as a function of time. Uman (1964a) integrated Eq. (5.6) by computer for two NII multiplets (4041 and 3995 Å) for various assumed temperature vs. time characteristics. The ratio of integrated multiplet intensities was matched with the measured ratio by adjusting the peak temperature for a given variational form of temperature vs. time. Using this technique it was found that if the average temperature was near 24,000°K, the peak temperature was within 10 percent of the average. This result is reasonable since in the temperature range of interest the NII line intensities increase rapidly with temperature, thus strongly weighting the average temperature toward the peak temperature. The method of analysis is only valid for peak temperatures below about 30,000°K. Additional comments regarding the peak temperature calculation are presented later in this section. It should be noted that the lightning temperature may rise to a very high value for a very short time without leaving a measurable line intensity on the time-integrated spectrum; thus, peak temperature really means the highest temperature that the return-stroke channel attains for a few microseconds or longer.

Once the return-stroke temperature has been determined, the techniques involving use of the Saha equation can be applied to determine electron density, channel pressure, percentage ionization in the channel, and other channel properties. Three return strokes, found by Prueitt to have average temperatures near 24,000°K, have been analyzed by Uman et al. (1964a,b). It was assumed that the spectral lines of OI, NI, and NII considered in the analysis were primarily emitted at temperatures near 24,000°K. The existence of LTE, optical thinness, and uniformity of cross-sectional properties was also assumed. The electron density calculated was of the order of 3×10^{24} electrons per cubic meter. A fundamental uncertainty in the analysis is due to the assumption that the spectral lines of interest are emitted primarily at temperatures near 24,000°K. Although it is not unreasonable to expect the OI, NI, and NII line intensities to be near maximum at 24,000°K, the OI and NI lines will contribute more to the time-integrated spectrum at temperatures below 24,000°K than will the NII lines. Since we wish to know the return-stroke properties near 24,000°K, which is the temperature determined from an analysis of NII lines, the measured intensities radiated by the OI and NI lines at temperatures below 24,000°K should be subtracted from the total OI and NI radiation. The result of overestimating the OI and NI contribution at 24,000°K is an overestimation of the electron density in the channel at that temperature. Thus, $3 \times$

10^{24} m^{-3} can be considered an upper limit to the electron density. A further error may occur if the NII lines are emitted primarily from a hot central portion of the channel while the NI and OI lines originate from a cooler part of the channel.

If the temperature and particle densities in a lightning channel are known, the electric conductivity of the channel can be calculated. Uman (1964b) has found a channel conductivity of 1.8×10^4 mhos/m for the lightning stroke analyzed by Uman et al. (1964a). For temperatures above about 13,000°K, the channel conductivity varies directly as $T^{3/2}$ and inversely as $\ln n_e$. Thus, in the high-temperature range the channel conductivity is relatively insensitive to temperature and electron density. It follows that the lightning conductivity should be between about 10^4 and 2×10^4 mhos/m during the high-temperature phase of the lightning discharge.

Electron densities have been determined from time-integrated spectra by a comparison of the Stark profiles of H$_\alpha$ (Balmer series) with theory (Uman and Orville, 1964). Figure 5.9 shows an example of that comparison. Electron densities of between 1×10^{23} and 5×10^{23} m^{-3} have been

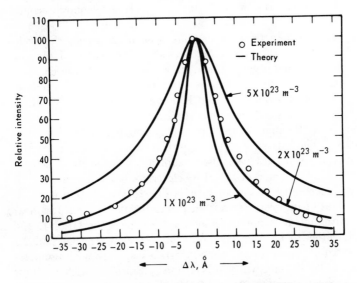

Fig. 5.9 *Comparison of the measured and calculated line profiles for* H$_\alpha$ *in lightning stroke A1. Theoretical profiles are given for three electron densities at 20,000°K. The excellent agreement between theory and experiment must be considered fortuitous in light of the various sources of experimental error. (Adapted from Uman and Orville (1964).)*

found for three return strokes. According to Drellishak (1964), by calculation and assuming LTE, the electron density in a nitrogen plasma at atmospheric pressure is between 1×10^{23} m^{-3} and 2×10^{23} m^{-3} for temperatures between 13,000 and 35,000°K. Since the Stark profiles are strong functions of electron density and only weak functions of temperature, the H_α profiles emitted from the lightning channel would not be expected to change much from the time when the channel has attained a near-atmospheric pressure until the time at which the temperature falls below about 13,000°K. Thus the value of electron density determined from Stark broadening of H_α would appear to be characteristic of a return-stroke channel at atmospheric pressure with temperature above 13,000°K. The electron density determined from time-integrated spectra using Stark broadening of H_α is considerably more reliable than the electron density determined using the Saha equation. Some data regarding an electron density determination from a time-resolved spectrum are given later in this section.

In addition to return-stroke properties determined from the stroke-integrated slitless spectra of Salanave and co-workers, measurements on

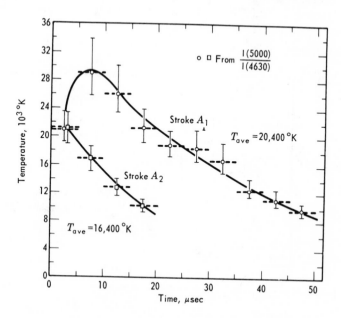

Fig. 5.10a *Stroke temperature as a function of time for flash A. Horizontal dashed lines indicate the time resolution, vertical bars the error limits. (Adapted from Orville (1966a, 1968).)*

time-integrated slit spectra have been made by Wallace (1960) and by Zhivlyuk and Mandel'shtam (1961). Wallace, as mentioned in Sec. 5.4, has determined a lightning temperature of 6000 to 30,000°K from a study of the optically thin intensities within the N$_2^+$ bands near 3900 Å. Zhivlyuk and Mandel'shtam measured the relative intensities at the centers of several spectral lines and, assuming the channel to be *optically thick* to those line centers, used Eq. (5.8) to calculate an average blackbody temperature of 21,000°K. The return-stroke spectra obtained by Zhivlyuk and Mandel'shtam would appear to be significantly different from those obtained by Salanave and co-workers.

We consider now the results of the analysis of the time-resolved spectra of Orville. Orville (1966a, 1968) has determined the temperature vs. time characteristics of 10-m vertical sections of 10 lightning channels. Two of the stroke temperatures were measured with 2-μsec time resolution, the remaining eight with 5-μsec resolution. Several lightning temperature vs. time curves are shown in Fig. 5.10. In Fig. 5.10, the horizontal lines about the data points indicate the time resolution and the vertical bars the error limits. Typical peak temperatures determined by Orville were of the order of 28,000 to 31,000°K. No temperatures exceeded 36,000°K. Two of the stroke temperatures measured appear

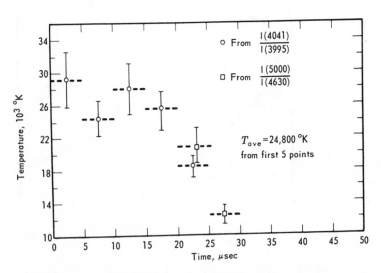

Fig. 5.10*b* *Stroke temperature as a function of time for flash B. Horizontal dashed lines indicate the time resolution, vertical bars the error limits. (Adapted from Orville (1966a, 1968).)*

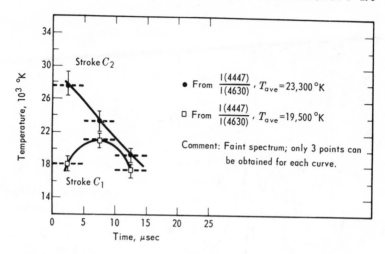

Fig. 5.10c *Stroke temperature as a function of time for flash C. Horizontal dashed lines indicate the time resolution, vertical bars the error limits. (Adapted from Orville (1966a, 1968).)*

to rise to peak during the first 10 μsec; the remaining eight strokes (including the two measured with 2-μsec time resolution) have temperatures which are maximum during the first temperature-determination period and monotonically decrease during their total time history. In one case Orville (1968) measured the temperature of a branch of the main channel. For the initial 5-μsec period, the branch temperature was about 22,000°K while the main-channel temperature was 31,000°K. Orville (1968) has compared his data with the average (time-integrated) temperatures of Prueitt (1963) (Table 5.2) by integrating, with pencil and paper, the time-resolved time intensities, and from these integrated values calculating an average temperature. Orville's average temperatures so computed are shown in Fig. 5.10. They are in good agreement with Prueitt's temperatures. Orville (1968) has verified Uman's (1964a) calculation that the peak lightning temperature is within 10 percent of the average temperature if both temperatures are determined using the (NII 4041 Å)/(NII 3995 Å) intensity ratio. Orville has shown, however, that the average temperature is more than 10 percent below the peak temperature if NII lines with low excitation potentials are used in the average-temperature determination. The reason for this effect is that NII lines with low excitation potentials radiate appreciably more at lower temperatures than do NII lines with high excitation potential. The measured radiation from the time-integrated low-excitation-potential lines thus appears

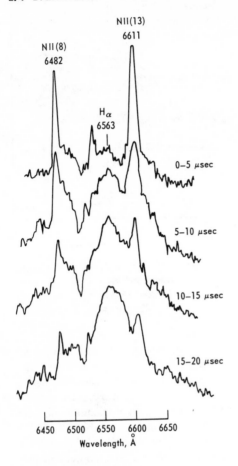

NII(8)
6482

NII(13)
6611

H$_\alpha$
6563

0–5 μsec

5–10 μsec

10–15 μsec

15–20 μsec

6450 6500 6550 6600 6650
Wavelength, Å

Fig. 5.11*a   Microdensitometer traces in the* H$_\alpha$ *region for times between 0 and 20 μsec. The unidentified line to the left of 6563 in the 0 to 5 μsec period is the NII 6611 emission from a branch of the main channel. The* H$_\alpha$ *emission from the branch is similarly displaced to the left and produces an asymmetry in the NII 6482 line profile. Traces are uncorrected for film nonlinearity. (Adapted from Orville (1968).)*

to be from a cooler channel than the time-integrated radiation from high-excitation-potential lines

Some microdensitometer traces of the photographic spectrum of lightning as a function of time in the H$_\alpha$ region are shown in Fig. 5.11 (see also Fig. 5.6b). An electron density vs. time curve obtained by Orville (1968) is presented in Fig. 5.12. The electron density vs. time data could be obtained for only one stroke. As can be seen, an electron density of the order of 10^{24} m^{-3} existed during the first 5 μsec of the discharge. The electron density decreased to a value between 1×10^{23} and 1.5×10^{23} m^{-3} in about 25 μsec. The electron density was then reasonably constant to 50 μsec and beyond 50 μsec was less than 1.5×10^{23} m^{-3}. Errors in electron density measurement are of the order of 50 percent. Since the effective instrument width of the spectrometer was about 10 Å, electron

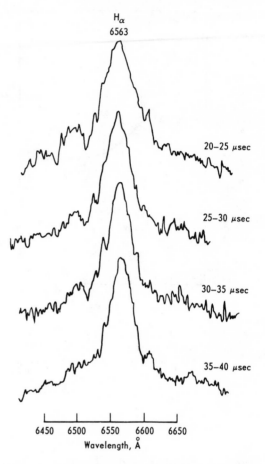

Fig. 5.11b *Microdensitometer traces in the H_α region for times between 20 and 40 μsec. Continuation of Fig. 5.11a. Traces are uncorrected for film nonlinearity. (Adapted from Orville (1968).)*

densities below about 1.0×10^{23} m^{-3} could not be determined. Initially the H_α intensity appears to be very low (see Fig. 5.6b). This is apparently due to the high temperature and high electron density present in the early stages of the discharge. The high temperature allows few neutral hydrogen atoms to be present in the discharge. The high electron density, via the Stark effect, causes the H_α line profile to be broadened to a width of perhaps 100 Å. The broadened profile blends into the back-

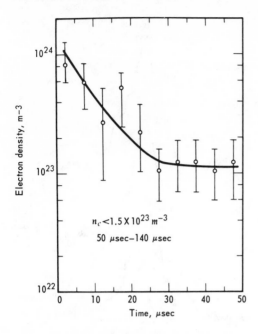

$n_e < 1.5 \times 10^{23} \, m^{-3}$

50 μsec–140 μsec

Fig. 5.12 *Electron density as a function of time as calculated from the measured Stark broadening of* H_α*. Vertical bars indicate error limits. (Adapted from Orville (1968).)*

ground continuum and is obscured by nearby lines. Orville's electron density determination is consistent with previous determinations made from time-integrated spectra. From a Saha-equation analysis, Uman et al. (1964a,b) found an upper limit to the electron density of $3 \times 10^{24} \, m^{-3}$ at a peak temperature near 24,000°K. Uman and Orville (1964) found electron densities between 1×10^{23} and $5 \times 10^{23} \, m^{-3}$ from an analysis of time-integrated profiles of H_α (Fig. 5.9). Since the peak intensity of H_α occurs when the profile corresponds to an electron density of the order of $1 \times 10^{23} \, m^{-3}$ and since the profile remains relatively constant at high intensity for tens of microseconds, it is not too surprising that the time-integrated profiles yielded an electron density of the order of $1 \times 10^{23} \, m^{-3}$.

It is of some interest to note that the temperature and electron density variation measured for lightning are similar to the temperature and electron density variation occurring in the 5-m air spark studied by Orville et al. (1967). The electron density in the initial 2-μsec period of the spark was about 10^{24} m^{-3} and the temperature was about 34,000°K. The electron density fell monotonically to about 2×10^{23} m^{-3} at 5 μsec and remained constant at that value for an additional 10 μsec (which was as long as an electron-density measurement could be made). At 15 μsec from spark initiation, the channel temperature had fallen to 19,000°K. Thus the spark would appear similar to lightning but on a time-scale shorter by a factor of 3 to 5.

If the lightning channel temperature and electron density are known as a function of time and if LTE exists in the channel as a function of time, then many channel properties (e.g., pressure, percent ionization, various particle densities) can be calculated as a function of time. These calculations can be made directly for a known temperature and electron density by solving simultaneously a series of Saha equations, an equation of charge conservation, and an equation of percentage composition. Of particular importance is the channel pressure and percent ionization as a function of time. For a temperature of about 30,000°K and an electron density of about 10^{24} m^{-3}, the channel pressure is of the order of 10 atm and the channel contains between one and two electrons per air atom (i.e., some atoms are twice ionized). Since the channel pressure exceeds the pressure of the surrounding air, the channel will expand until its pressure is equal to the ambient pressure (see Sec. 7.6). That this is the case is indicated by the spectral data. An electron density of 1 to 2×10^{23} m^{-3} is probably indicative of atmospheric pressure if the channel temperature is above about 13,000°K (Drellishak, 1964). Thus, for the data given in Fig. 5.12, the channel apparently reaches atmospheric pressure in 10 to 20 μsec. At 20,000°K and an electron density of about 2×10^{23} m^{-3}, there is about one electron per air atom. Information regarding the variation of channel pressure, percent ionization, and particle constituents as a function of time is given by Orville (1968).

References

1. Cannon, J. B.: A Note on the Spectrum of Lightning, *J. Roy. Astron. Soc. Can.*, **12**:95–97 (1918).

2. Drellishak, K. S.: Partition Functions and Thermodynamic Properties of High Temperature Gases, *AEDC-TDR-64-22*, January 1964. (Defense Documentation Center, AD-428210, unclassified.)

3. Dufay, M.: Spectres des éclairs, *Compt. Rend.*, **182**:1331–1333 (1926).

4. Dufay, M.: Sur le spectre des éclairs dans les régions violette et ultraviolette *Compt. Rend.*, **225**:1079–1080 (1947).

5. Dufay, M.: Recherches sur les spectres des éclairs, deuxième partie: Étude du spectre dans les régions violette et ultraviolette, *Ann. Geophys.*, **5**:255–263 (1949).

6. Dufay, M., and J. Dufay: Spectres des éclairs photographies au prisme objectif, *Compt. Rend.*, **229**:838–841 (1949).

7. Dufay, J., and M. Tcheng: Spectres des éclairs, de 3830 à 6570 Å, *Compt. Rend.*, **228**:330–332 (1949a).

8. Dufay, J., and M. Tcheng: Recherches sur les spectres des éclairs, première partie: Étude des spectres, de 3830 à 6570 Å, au moyen de spectrographes à fente, *Ann. Geophys.*, **5**:137–149 (1949b).

9. Eriksson, K. B. S.: The Spectrum of the Singly-ionized Nitrogen Atom, *Arkiv Fysik*, **13**:303–329 (1958).

10. Fox, P.: The Spectrum of Lightning, *Astrophys. J.*, **14**:294–296 (1903).

11. Gilmore, F. R.: Equilibrium Composition and Thermodynamic Properties of Air to 24,000°K, *RAND Corp., Res. Memo RM*-1543, August, 1955. (Defense Documentation Center, AD-84502, unclassified.) Thermal Radiation Phenomena. Volume 1, The Equilibrium Thermodynamic Properties of High Temperature Air, Lockheed Missiles and Space Company, Palo Alto, California, May, 1967. (Defense Documentation Center, AD-654054, unclassified.)

12. Griem, H. R.: "Plasma Spectroscopy," McGraw-Hill Book Company, New York, 1964.

13. Herschel, J.: On the Lightning Spectrum, *Proc. Roy. Soc. (London)*, **15**:61–62 (1868).

14. Herzberg, G.: "Molecular Spectra and Molecular Structure. 1. Spectra of Diatomic Molecules," 2d ed., D. Van Nostrand Company, Inc., Princeton, N.J., 1950.

15. Hilsenrath, J., and M. Klein: Tables of Thermodynamic Properties of Air in Chemical Equilibrium Including Second Virial Corrections from 1500°K to 15,000°K, *AEDC-TR-65-58*, March, 1965. (Defense Documentation Center, AD-612301, unclassified.)

16. Hu, R.: The Lightning Spectra in the Visible and Ultra-violet Regions with Grating Spectrograph, *Sci. Rec. (Peking)*, **4**:380–383 (1960).

17. Israel, H., and G. Fries: Ein Gerät zur spektroskopischen Analyse verschiedener Blitzphasen, *Optik*, **13**:365–368 (1956).

18. Israel, H., and K. Wurm: Das Blitzspektrum, *Naturwissenshaften*, **52**:778–779 (1941).

19. Israel, H., and K. Wurm: Das Spektrum der Blitz, *Wiss. Arb. DMD-ZFO*, **1**:48–57 (1947).

20. Jose, P. D.: The Infrared Spectrum of Lightning, *J. Geophys. Res.*, **55**:39–41 (1950).

21. Kayser, H., "Tabelle der Hauptlinien der Linienspektra," Springer-Verlag OHG, Berlin, 1939.

22. Knuckles, C. F., and J. W. Swensson: The Spectrum of Lightning in the Region λ 6159–λ7157, *Ann. Geophys.*, **8**:333–334 (1952).

23. Krider, E. P.: "The Design and Testing of a Photoelectric Photometer for Selected Lines in the Spectrum of Lightning," M.S. thesis, Department of Physics, University of Arizona, Tucson, Ariz., 1965a.

24. Krider, E. P.: Time-resolved Spectral Emissions from Individual Return Strokes in Lightning Discharges, *J. Geophys. Res.*, **70**:2459–2460 (1965b).

25. Krider, E. P.: Some Photoelectric Observations of Lightning, *J. Geophys. Res.*, **71**:3095–3098 (1966).

26. Krider, E. P., G. A. Dawson, and M. A. Uman: Peak Power and Energy Dissipation in a Single-stroke Lightning Flash, *J. Geophys. Res.*, **73**:3335–3339 (1968).

27. Larsen, A.: Photographing Lightning with a Moving Camera, *Smithsonian Inst. Rept.*, **60** (pt. 1):119–127 (1905).

28. Mastrup, F., and W. Wiese: Experimentelle Bestimmung des Oszillatorenstarken einiger NII und OII Linien, *Z. Astrophys.*, **44**:259–279 (1958).

29. Meinel, A. B., and L. E. Salanave: N_2^+ Emission in Lightning, *J. Atmospheric Sci.*, **21**:157–160 (1964).

30. Meyer, G.: Ein Versuch, das Spectrum des Blitzes zu photographieren, *Ann. Physik Chem.*, **51**:415–416 (1894).

31. Moore, C. E.: A Multiplet Table of Astrophysical Interest, rev. ed., *Contrib. Princeton Univ. Obs.*, no. 20, 1945.

32. Nicolet, M.: Le spectre des éclairs, *Ciel Terre*, **59**:91–98 (1943).

33. Orville, R. E.: "A Spectral Study of Lightning Strokes," Ph.D. thesis, Department of Meteorology, University of Arizona, Tucson, Ariz., 1966a.

34. Orville, R. E.: High-speed, Time-resolved Spectrum of a Lightning Stroke, *Science*, **151**:451–452 (1966b).

35. Orville, R. E.: Ozone Production during Thunderstorms, Measured by the Absorption of Ultraviolet Radiation from Lightning, *J. Geophys. Res.*, **72**:3557–3562 (1967).

36. Orville, R. E.: A High-speed Time-resolved Spectroscopic Study of the Lightning Return Stroke, pts. 1, 2, 3, *J. Atmospheric Sci.*, **25**:827–856 (1968).

37. Orville, R. E., and M. A. Uman: The Optical Continuum of Lightning, *J. Geophys. Res.*, **70**:279–282 (1965).

38. Orville, R. E., M. A. Uman, and A. M. Sletten: Temperature and Electron Density in Long Air Sparks, *J. Appl. Phys.*, **38**:895–896 (1967).

39. Petrie, W., and R. Small: The Near Infrared Spectrum of Lightning, *Phys. Rev.*, **84**:1263–1264 (1951).

40. Pickering, E. C.: Spectrum of Lightning, *Astrophys. J.*, **14**:367–369 (1901).

41. Prueitt, M. L.: The Excitation Temperature of Lightning, *J. Geophys. Res.*, **68**:803–811 (1963).

42. Salanave, L. E.: The Optical Spectrum of Lightning, *Science*, **134**:1395–1399 (1961).

43. Salanave, L. E.: The Ultraviolet Spectrum of Lightning: First Slitless Spectra down to 3000 Angstroms, *Trans. Am. Geophys. Union*, **43**:431–432 (1962).

44. Salanave, L. E.: The Optical Spectrum of Lightning, *Advan. Geophys.*, **10**:83–98 (1964).

45. Salanave, L. E.: The Infrared Spectrum of Lightning, *1966 IEEE Reg. Six Conf. Record*, 1966.

46. Salanave, L. E., R. E. Orville, and C. N. Richards: Slitless Spectra of Lightning in the Region from 3850 to 6900 Angstroms, *J. Geophys. Res.*, **67**:1877–1884 (1962).

47. Sampson, D. H.: Approximate Method for Determining When any Gas is Collision-dominated, *Astrophys. J.*, **144**:96–102 (1966).

48. Schuster, A.: On Spectra of Lightning, *Proc. Phys. Soc. (London)*, **3**:46–52 (1880).

49. Slipher, V. M.: The Spectrum of Lightning, *Lowell Obs. Bull. 79*, Flagstaff, Ariz., 55–58 (1917).

50. Steadworthy, A.: Spectrum of Lightning, *J. Roy. Astron. Soc. Can.*, **8**:345–348 (1914).

51. Uman, M. A.: The Peak Temperature of Lightning, *J. Atmospheric Terrest. Phys.*, **26**:123–128 (1964a).

52. Uman, M. A.: The Conductivity of Lightning, *J. Atmospheric Terrest. Phys.*, **26**:1215–1219 (1964b).

53. Uman, M. A., and R. E. Orville: Electron Density Measurement of Lightning from Stark-broadening of H_α, *J. Geophys. Res.*, **69**:5151–5154 (1964).

54. Uman, M. A., and R. E. Orville: The Opacity of Lightning, *J. Geophys. Res.*, **70**:5491–5497 (1965).

55. Uman, M. A., R. E. Orville, and L. E. Salanave: The Density, Pressure, and Particle Distribution in a Lightning Stroke near Peak Temperature, *J. Atmospheric Sci.*, **21**:306–310 (1964a).

56. Uman, M. A., R. E. Orville, and L. E. Salanave: The Mass Density, Pressure, and Electron Density in Three Lightning Strokes near Peak Temperature, *J. Geophys. Res.*, **69**:5423–5424 (1964b).

57. Vassy, A.: Comparaison des spectres d'étincelles de grande longueur dans l'air et du spectre de l'éclair, *Compt. Rend.*, **238**:1831–1833 (1954).

58. Wallace, L.: Note on the Spectrum of Lightning in the Region 3670 to 4280 Å, *J. Geophys. Res.*, **65**:1211–1214 (1960).

59. Wallace, L.: The Spectrum of Lightning, *Astrophys. J.*, **139**:944–998 (1964a).

60. Wallace, L.: private communication (1964b).

61. Wiese, W. L.: Line Broadening, in R. H. Huddlestone and S. L. Leonard (eds.), "Plasma Diagnostic Techniques," pp. 265–317, Academic Press Inc., New York, 1965.

62. Zhivlyuk, Yu. N., and S. L. Mandel'shtam: On the Temperature of Lightning and the Force of Thunder, *Soviet Phys. JETP (English Trans.)*, **13**:338–340 (1961).

6

Thunder

6.1 EARLY HISTORY

As we have said, there are two kinds of exhalation, moist and dry; and their combination (air) contains both potentially. It condenses into clouds, as we have explained before, and the condensation of clouds is thicker toward their farther limit. . . . But any of the dry exhalation that gets trapped when the air is in the process of cooling is forcibly ejected as the clouds condense and in its course strikes the surrounding clouds, and the noise caused by the impact is what we call thunder.

Aristotle (384–322 B.C.): "Meteorologica" (translated by H. P. D. Lee), pp. 223–225, Loeb Classical Library, Harvard University Press, Cambridge, Mass., 1951.

In the first place the blue of heaven is shaken with thunder, because the ethereal clouds clash together as they fly aloft when the winds combat from opposite quarters.

T. Lucretius (98–55 B.C.): "On the Nature of Things," book VI, H. A. J. Munro, trans., Great Books of the Western World, p. 81, William Benton, Chicago, 1952.

. . . the sound which is known as thunder is due simply to the fact that the air traversed by an electric spark, that is, a flash of lightning, is suddenly raised to a very high temperature, and has its volume, moreover, considerably increased. The column of gas thus suddenly heated and expanded is sometimes several miles long, and as the duration of the flash is not even a millionth of a second, it follows that the noise bursts forth at once from the whole column,

though for an observer in any one place it commences where the lightning is at the least distance.

. . . the beginning of the thunder clap gives us the minimum distance of the lightning, and the length of the thunder clap gives us the length of the column.

M. Hirn: The Sound of Thunder, *Sci. Am.*, **59**:201 (1888).   Copyright 1888 by Scientific American, Inc..   All rights reserved.

Aristotle was of the opinion that thunder preceded lightning, the lightning being a burning wind produced after the impact of the dry exhalation on a cloud.   Lucretius thought that lightning and thunder were created simultaneously, both being due to a collision between clouds.   Both Aristotle and Lucretius were aware that an observer sees lightning before he hears thunder.   This was the case, according to Lucretius, "because things always travel more slowly to the ears than those which excite vision travel to the eyes."   A similar view was held by Aristotle who noted that "the oars are already drawing back again when the sound of the stroke which they have made first reaches us."

In the year 1637, Descartes (Haldane, 1905) suggested that thunder was due to higher clouds descending on lower clouds, the sound coming from the resonance of the air between the clouds.

According to Remillard (1960), it was the English physicist Robert Hooke (discoverer of Hooke's law which relates the deformation of an elastic body directly to the applied force) who, in the middle of the seventeenth century, first suggested that the duration of the thunder was due to the difference in path lengths between the observer and various portions of the lightning stroke.

In 1738, De L'Isle (1738) published the results of his measurements of the time interval between the lightning flash and the arrival of the first sound of thunder and the duration of thunder.   These data have been reproduced and analyzed by Remillard (1960).   Of particular importance was the observation by De L'Isle that thunder is seldom heard from lightning strokes that are more than 25 km away.   An explanation for this observation has been given by Fleagle (1949) and is discussed in Sec. 6.3.2.

A review of the information regarding thunder available prior to the mid-nineteenth century is given by Arago (1854).   Apparently, the generally accepted theory of thunder in the mid-nineteenth century was the so-called "vacuum theory"; the lightning discharge was supposed to produce a vacuum along its path, the thunder being due to the subsequent motion of air into the vacuum.   Arago did not support this view.   He held that the explanation of thunder was yet to be found.

Various theories of thunder were published during the latter half of the nineteenth century and the early twentieth century.   Mershon (1870) suggested that

The electricity in passing from one cloud to another, or to the earth, decomposes the water in the cloud into its component gases; and the great heat of the electricity ignites and explodes these gases, and reforms them into water.

The sound of the explosion was supposedly the thunder. A similar theory was proposed by Bates (1903). Reynolds (1903) suggested that thunder was due to "steam explosions" created when the water along the lightning path was heated by the lightning discharge. Lyon (1903) criticized both the chemical explosion and the steam explosion theories and pointed out that laboratory sparks in atmospheres devoid of water or explosive gases produce sound. Lyon's theory of thunder is similar to that given by Hirn (1888), quoted at the beginning of this section. Hirn's explanation of the mechanism of thunder, with the exception of an error in the stated time duration of a flash, remains today the consensus view.

The early history of thunder research can be considered to end with the work of Veenema (1917, 1918, 1920) who was principally interested in determining how far thunder could be heard. Veenema made observations on nearly every thunderstorm which occurred in his area during the years 1895 to 1916. He confirmed the conclusions of De L'Isle (1738) that generally thunder could not be heard more than about 25 km from the lightning discharge. He recorded, however, isolated instances when thunder was heard from distances up to and over 100 km.

6.2 THE EXPERIMENTAL DATA

The first recordings of the pressure variations due to thunder were published by Schmidt (1914). Recent recordings of the pressure variations due to thunder are shown in Fig. 6.1a and b. In order to discuss properly the pressure measurements, it is necessary to consider first the terms *peal*, *clap*, *roll*, and *rumble* commonly used to describe the audible

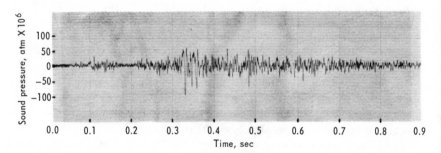

Fig. 6.1a *A portion of the pressure variation due to thunder. (Adapted from Few et al. (1967).)*

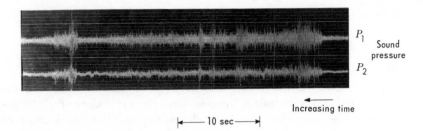

P_1 Sound pressure

P_2

Increasing time

|←——— 10 sec ———→|

Fig. 6.1b   *A complete thunder record thought to be of a cloud flash. The time delay from the flash to the arrival of the first sound was about 14 sec. The total thunder duration was about 36 sec. P_1 is from a microphone whose frequency response is essentially flat between 60 and 1,250 Hz; P_2 from a microphone flat between 0.08 and 450 Hz. Maximum thunder pressure is about 10^{-5} atm. (Courtesy of A. A. Few.)*

components of thunder. Unfortunately, these terms are not used consistently in the literature nor in everyday speech. Often peal and clap are used synonymously, as are roll and rumble. Peals or claps are the sudden loud sounds which occur against a background of prolonged roll or rumble. The peals or claps are generally separated by time intervals of 1 sec or more. A clap is sometimes considered to be of less intensity and shorter duration than a peal. The term roll is sometimes used to describe irregular sound variations whereas rumble is used to describe relatively weak sound of long duration. In this chapter, unless stated otherwise, we shall arbitrarily use the term clap to describe all sudden loud sounds and the term rumble to describe all other sounds of thunder.

Schmidt (1914) used two different instruments to measure the pressure variations due to thunder. The first instrument was essentially a box with a single hole in one side in which was suspended a thin metal plate. The plate was free to move in or out in response to the sound pressure impinging on it. The motion of the plate was recorded via levers attached to the plate. The instrument was capable of recording pressure variation between about 0.2 and 3 Hz and was calibrated in this range for absolute pressure-change measurements. The system exhibited a resonance near 0.4 Hz, plate oscillations of decreasing amplitude being recorded for pressure oscillations of fixed amplitude as the frequency difference from the resonance point increased. The second instrument consisted of a megaphone at whose throat was a smoky turpentine flame which moved in response to the impinging pressure variations. A motor-driven strip of paper was moved near the flame. The pressure record was impressed on the paper as darker or lighter soot. Using this instrument, Schmidt was able to measure pressure-variation frequencies from

25 to over 100 Hz. The instrument was not absolutely calibrated. Hence the data from the "high-frequency" instrument could not be compared with the data from the "low-frequency" instrument.

Schmidt recorded low-frequency pressure variations with periods between 0.2 and 0.54 sec and amplitudes up to 10^{-5} atm. The strongest recorded pressure variations were rarefactions. Preceding each strong rarefaction a short compression was recorded. Schmidt stated that it was possible, owing to the poor high-frequency response of his instrument, that the compressions were as strong as the rarefactions. The infrasonic pressure vibrations measured by Schmidt corresponded in time with the audible thunder claps. Schmidt stated that the infrasonic pressure variations were considerably stronger than those corresponding to the loudest audible sounds (presumably of cannons, etc., since the audible pressure of thunder was not known and was not measured by Schmidt). Using his crude high-frequency instrument Schmidt found that frequencies between 25 and 40 Hz and between 75 and 120 Hz were most common.

It is possible that Schmidt's interpretation of his low-frequency data was in error. In particular, his instrument may have recorded the modulation envelope of an audio-frequency "carrier" rather than a true low-frequency component,* or it may have been set in resonant oscillation by the successive peaks (claps) of the modulation envelope. Similar comments are applicable to the work of Arabadzhi (1952), considered next.

Almost 40 years elapsed after Schmidt's work until the next measurement of the thunder pressure was reported. Arabadzhi (1952) constructed a device similar to Schmidt's low-frequency instrument and obtained results which essentially confirm those of Schmidt. Arabadzhi reported that the maximum energy in thunder corresponded to a frequency between 0.25 and 2 Hz, the most frequently observed value being 0.5 Hz. (No audio-frequency pressure measurements were made.) Arabadzhi reported low-frequency pressure variations of about 10^{-3} atm for discharges about 1 km distant. This value is two orders of magnitude greater than the value reported by Schmidt. Some of the discrepancy may be due to the fact that the discharges studied by Arabadzhi were at relatively close range. Schmidt stated that the distances to the strokes studied by him were probably less than 5 km. Arabadzhi found that the strongest (and apparently the initial) low-frequency pressure variations were compressions.

In a later review paper, Arabadzhi (1957) reported that absolute-pressure measurements were made with a microphone for frequencies up to

* A sound wave consisting of a 100-Hz carrier wave modulated at 1 Hz has a frequency spectrum composed of three discrete frequencies, 99, 100, and 101 Hz, that is, the carrier frequency and two sidebands displaced from the carrier by the modulation frequency.

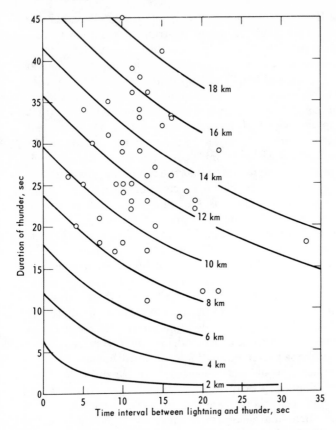

Fig. 6.2 *Data points of De L'Isle (1738) showing observed relation between duration of thunder and the time interval between lightning and thunder. Solid curves are due to Remillard (1960) and represent expected values if thunder is due to the difference in path lengths from the observer to the two extremes of a vertical lightning channel to ground of length 2 to 18 km. (Adapted from Remillard (1960).)*

about 10 kHz, that thunder contains frequencies from 0.25 to 500 Hz, and that the fundamental thunder energy is in the very low frequencies. An audio-frequency thunder spectrum with amplitudes expressed in arbitrary units has been given by Arabadzhi (1965). Data are given between 64 and 2,000 Hz. The peak amplitude occurs at 200 Hz.

A thesis presenting a theory of thunder was completed in 1960 by

Remillard (1960) and two theses containing measurements of thunder were completed in 1964 (Latham, 1964; Bhartendu, 1964).

Remillard's thesis contains a great deal of background data regarding thunder and many valuable references. Remillard's mathematical theory of thunder would appear to be of little value since he assumes as a starting point for his calculations that the lightning channel emits a small-amplitude sound wave. It is now the consensus view, based on theory and experiment, that the channel emits a relatively strong shock wave. Remillard has analyzed the data of De L'Isle (1738) concerning the time interval between the lightning flash and the arrival of thunder and the duration of thunder. Some of these data are shown in Fig. 6.2. Remillard concluded that the measured thunder duration was twice as long as it should be for channels to ground of the usual height. That is, the average thunder records analyzed by Remillard appeared to be due to channels to ground of 12-km height, the sound duration being determined by the difference between the times taken for sound to reach the observer from the bottom of the channel and from the top of the channel. Remillard, therefore, postulated that there exists a graupel (soft ice) layer in

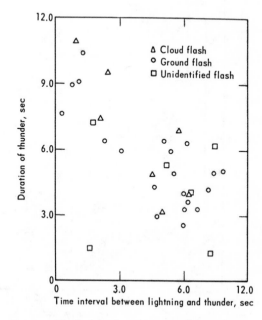

Fig. 6.3 *Total time duration of thunder vs. time interval between lightning and thunder. (Adapted from Latham (1964).)*

the cloud at about 6-km height which reflects the thunder and prolongs its duration. Latham (1964), on the other hand, has presented data to show that the duration of thunder corresponds very well to the usual length of the channel and that it is therefore not necessary to postulate a reflecting surface in the cloud. Latham's data are shown in Fig. 6.3. The discrepancy between Latham's and Remillard's conclusions can be resolved in favor of Latham if it is assumed that De L'Isle's thunder records are not of ground discharges but are primarily of vertical cloud discharges in clouds that were nearly overhead. For example, the thunder record shown in Fig. 6.1*b* would appear similar to De L'Isle's typical data.

Latham (1964), using a condenser microphone, made measurements of the pressure variations due to thunder. He found that a low-intensity sound (thunder leader) with duration between 0.1 and 2.2 sec existed in almost every case prior to the main thunder, and that thunder in general consists of three or four discrete pulses or claps, the pressure oscillations within these claps being about 100 Hz. The time duration of the claps

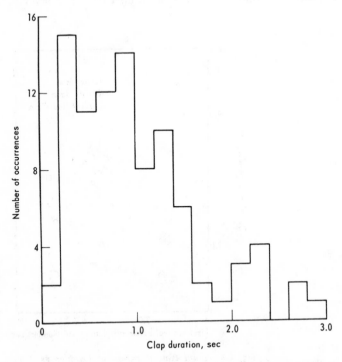

Fig. 6.4 *Histogram of clap duration.* (*Adapted from Latham (1964).*)

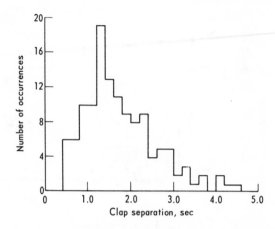

Fig. 6.5 *Histogram of time interval between the beginning of successive claps. (Adapted from Latham (1964).)*

was 0.2 to 1 sec (Fig. 6.4), and the time separation between claps (the time from the beginning of one clap until the beginning of the next clap) generally was 1.2 to 1.4 sec (Fig. 6.5). As shown in Fig. 6.6, the amplitudes of the claps did not vary much with clap order. Latham found that the initial portion of the thunder leader as well as the initial portions of the claps was a compression. Thunder from cloud strokes was found to have the same general characteristics as from ground strokes although thunder from the latter was usually stronger.

Bhartendu (1964) (see also Bhartendu and Currie, 1963) used a hot-wire microphone (infrasonic frequencies) and a crystal microphone and a woofer (audible frequencies) to record thunder pressure. In the hot-wire microphone, the resistance of an electrically heated wire which is subjected to air-pressure variations is recorded. The change in resistance of the hot wire gives a measure of the pressure oscillations in the air. The response of a hot-wire microphone to infrasonic-pressure variations (compressions or rarefactions) is unidirectional. Thus, the device acts, in some sense, like a full-wave rectifier in an electric circuit. Bhartendu states that the maximum thunder energy is infrasonic but does not give absolute pressure data for the low-frequency variations. The average audible pressure as measured with the crystal microphone was about 10^{-5} atm, the maximum 10^{-4} atm. The hot-wire microphone data were analyzed for relative-frequency content for frequencies between about 1 and 20 Hz. A typical spectrum is shown in Fig. 6.7. In the 1 to 20 Hz range the power content was found generally to increase with decreasing

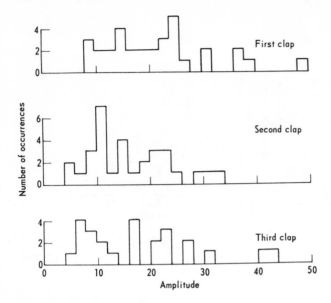

Fig. 6.6 *Histograms of relative amplitudes of thunder claps vs. clap order. (Adapted from Latham (1964).)*

frequency. Primary maxima were in the frequency range 0.75 to 6 Hz. It is not clear how the distortions in the recorded data due to the non-linear properties of the hot-wire microphone were separated out from the recorded data in the frequency analysis. The frequency spectrum of the audible frequencies recorded by a crystal microphone were found to vary irregularly within a factor of 2 or 3 between 20 and 100 Hz.

Bhartendu defines peals to be of longer duration and higher intensity than claps. He states that claps often contribute to rumbling in the last phases of thunder. Using a three-microphone system Bhartendu was able to determine the direction and height of various components of the thunder. He found that the first sound was generally due to the base of the channel, that peals arrive directly from the discharge channel, and that successive peals arrive from different (often higher) parts of the channel. Bhartendu states that claps in general arrive from different directions than peals, and thus could be due to reflections or due to higher and more distant parts of the flash that those parts which produce peals. Bhartendu found that the initial impulses from most of the large pressure variations were compressions, that normally thunder could be heard no more than 25 km, and that most thunder durations were in the range 5 to 20 sec.

Recently, Few et al. (1967) have reported on measurement of the thun-

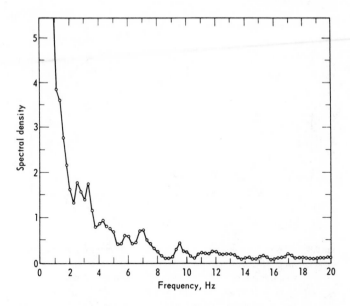

Fig. 6.7   *The infrasonic frequency content of thunder as determined by Bhartendu (1964).   (Adapted from Bhartendu (1964).)*

der spectrum made in two independent studies, one in Texas and one in New Mexico.   The microphone system used in Texas had a flat frequency response from about 0.1 to over 400 Hz; the system used in New Mexico was flat from about 0.2 Hz to 20 kHz.   Typical maximum overpressures of 10^{-4} atm were measured.   No absolute spectral power measurements were reported.   The Texas data were analyzed by counting zero crossings on the pressure vs. time record.   The dominant frequencies so obtained were in the range 180 to 260 Hz.   The New Mexico data were analyzed by playing magnetic tape records of the pressure wave into an acoustic-spectrum analyzer.   An average acoustic power spectrum obtained in this way is shown in Fig. 6.8.   The acoustic spectra obtained all showed broad dominant peaks in the 200-Hz region.   There were no infrasonic frequency peaks.   In neither study was any evidence found that there was appreciable energy carried by the infrasonic frequencies.   These findings are contradictory to those of Schmidt (1914), Arabadzhi (1952, 1957), and Bhartendu (1964).   Few et al. (1967) present a theory to explain the observed broad frequency maximum around 200 Hz.   This theory will be presented in Sec. 6.3.1.

Newman et al. (1967a) have made pressure measurements 0.3 m from several lightning discharges.   The results of these measurements are given at the end of Sec. 6.3.1.

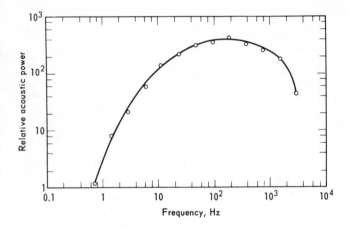

Fig. 6.8 *The acoustic spectrum of thunder as determined by playing magnetic tape records of thunder from 12 cloud-to-ground and 11 intracloud flashes into an acoustic-spectrum analyzer. Data shown represent the average of the 23 individual spectra. (Adapted from Few et al. (1967).)*

6.3 IDEAS AND THEORIES

6.3.1 The shock wave

The energy per unit length delivered to a *stroke* is probably of the order of 10^5 joules/m (Sec. 7.3.3). This energy appears as dissociation, ionization, excitation, and kinetic energy of the channel particles, as radiation, and as the energy of expansion of the channel. The fraction of the input energy appearing as dissociation, ionization, excitation, and kinetic energy of the channel particles and as radiation is apparently small (Sec. 7.3.3). It is probable, therefore, that an appreciable fraction of the input energy appears as the energy of the channel expansion. That is, the rapidly heated channel of initially very high pressure creates a roughly cylindrical shock wave which propagates radially outward. Several investigators (Lin, 1954; Sakurai, 1953, 1954, 1955a,b, 1959; Swigart, 1960; Rouse, 1959) have calculated the properties of strong cylindrical shock waves. All of these calculations give similar results. According to Lin (1954) the radial position of a strong cylindrical shock wave is given by

$$r = S \left(\frac{W}{\rho_0} \right)^{1/4} t^{1/2} \tag{6.1}$$

where S is a factor depending on the properties of the gas and which is of order of magnitude unity for air, W the energy release (assumed instantaneous) per unit length, t the time, and ρ_0 the atmospheric density. The shock-wave pressure (pressure immediately behind the shock front) is given by

$$P = 0.2 \frac{W}{r^2} \quad \text{newtons/m}^2 \tag{6.2}$$

Equations (6.1) and (6.2) are only valid if the shock-wave pressure significantly exceeds the ambient pressure. References to papers containing experimental verification of Eqs. (6.1) and (6.2) and of similar equations for the case of spherical geometry are given in the review by Pain and Rogers (1962).

The theory of Lin (1954) assumes that the energy input to the channel is instantaneous, that is, fast compared with the time in which the shock wave can move appreciably. Drabkina (1951) and Braginskii (1958) have considered the case of a non-instantaneous energy input to the discharge channel. The results of a calculation of lightning-channel radius vs. time using the Braginskii theory are given in Sec. 7.3.3. It is probable that the energy input to the lightning channel is sufficiently fast that Eqs. (6.1) and (6.2) are reasonable approximations. For this to be the case, the bulk of the energy input to a short length of channel must take place before the shock wave created by expansion of the high pressure channel can break away from the channel.

Since the shock-front pressure in the strong-shock regime decreases with the square of the shock-front radius, the strong shock will at some radius transform into a weak shock which in turn will eventually become a sound wave. There is no theory available for the weak cylindrical-shock regime, the regime in which the shock overpressure is between about 0.1 and 10 atm. Weak-shock calculations have been performed for the spherical case by Brode (1956). Brode's calculations extend well into the sound wave region.

According to Few et al. (1967) the dominant frequencies present in thunder are determined by the value of the energy input per unit length to the lightning channel.* The energy input produces a shock wave which relaxes into a sound wave whose length is found by Few et al. to be directly proportional to the square root of the energy input per unit length. The frequency characterizing the sound wave is given, to order of magnitude, by dividing the length of the sound wave into the speed of sound. Few et al. have used the results of cylindrical shock wave theory to describe the strong cylindrical shock and a scaled version of the spher-

---

* The form of the measured frequency spectrum of thunder at a given distance from the channel is also determined in part by the preferential atmospheric absorption of the higher sound frequencies, as discussed in Sec. 6.3.2.

ical shock wave theory (Brode, 1956) to describe the weak cylindrical
shock and sound waves. The length of the sound wave resulting from
relaxation of the cylindrical shock wave was found to be

$$l = 1.47 \left(\frac{W}{p} \right)^{\frac{1}{2}} \tag{6.3}$$

where p is the ambient pressure. The radius at which the weak shock
wave becomes a sound wave is apparently of the same order of magnitude
as l (see also Kitagawa, 1965). Few et al. (1967) assumed a maximum
energy input to the channel of 1.7×10^6 joules/m and found a maximum
value for l of 6 m and a minimum dominant frequency of about 57 Hz at
ground level. The minimum frequency at the top of a 6-km channel was
found to be 39 Hz due to the decreased pressure at that altitude. An
energy input of about 10^5 joules/m yields a dominant frequency of about
200 Hz, in good agreement with the experimental data presented by Few
et al. Dawson et al. (1968) have shown that use of the empirical
expression

$$f = c \left(\frac{p}{W} \right)^{\frac{1}{2}} \tag{6.4}$$

where f is the dominant frequency and c the speed of sound, gives good
results for both the lightning discharge ($W \sim 10^5$ joules/m) and for a
4-m laboratory spark (for which $W = 5 \times 10^3$ joules/m and the observed
sound pressure variations were between 1,350 and 1,650 Hz). The fact
that the thunder theory works well for energy inputs differing by a factor
of twenty can be considered reasonable confirmation of the validity of
the theory.

We consider now the pressure due to the strong shock wave. From
Eq. (6.2) the pressure immediately behind the shock front is for
$W = 10^5$ joules/m

$$P = \frac{0.2 \times 10^5}{r^2} \quad \text{newtons/m}^2$$

$$= \frac{0.2}{r^2} \quad \text{atm} \tag{6.5}$$

Thus the predicted overpressure at a radius of 0.02 m is about 500 atm,
at $r = 0.05$ m about 80 atm, and at $r = 0.14$ m about 10 atm. Zhivlyuk
and Mandel'shtam (1961) have used the theory of Braginskii (1958) to
calculate the shock wave pressure as a function of the rate of rise of stroke
current. They find that for a typical first-stroke current growth rate of
3 ka/μsec, the shock wave overpressure at $r = 0.05$ m is about 91 atm,
at $r = 0.5$ m about 8.9 atm, and at $r = 5.0$ m about 0.87 atm. These
overpressure values are in rough agreement with those derived from Eq.

(6.5) for small radii, but the Braginskii theory predicts a slower shock-front pressure decrease with increasing radius than does the strong shock theory (for overpressures above 10 atm) or the theory of Few et al.

As noted in Section 5.5.2, values of channel pressure can be derived from spectroscopic measurements. It is the high initial channel pressure which launches the shock wave. The initial channel pressure determined spectroscopically is *averaged* over about 5 μsec. The value thus determined is of the order of 10 atm. The averaging process, however, tends to weight preferentially pressures near and below 10 atm. This is the case because (1) the H_α line profile is very broad and the line is difficult to detect for electron densities above 10^{24} m^{-3} so that such densities, if they exist, are not recorded, and (2) the NII lines used to determine temperature are brightest in the 30,000°K range, and thus if the channel temperature were, say, 50,000°K for 0.1 μsec and 30,000°K for 4.9 μsec, analysis of the integrated radiation emitted in the 5-μsec interval would yield a temperature value near 30,000°K. It is probable, in light of the available shock wave data, that the initial pressure in the lightning channel is considerably in excess of 10 atm.

The only experimental data regarding the shock wave pressure within a few meters of the lightning channel would appear to be that reported by Newman et al. (1967a). Newman and co-workers artificially initiated lightning discharges by firing thin wires into the air under thunderclouds. Most of the discharges so initiated had relatively slow current wavefronts (millisecond rise times) and thus were apparently initiated by upward-moving leaders (see Sec. 2.4.2, 2.5.4, and 4.4). For four lightning discharges, peak overpressures at 0.3 m from a spark gap through which the lightning current was forced to flow were measured to be 1.2 atm, 1.2 atm, 2.0 atm, and 0.3 atm. Since the current rise times were relatively slow, it might be expected that the measured sound overpressures would be lower than the overpressures emitted by the usual cloud-to-ground discharge. Newman et al. (1967b) state that the accuracy of the blast gauges used in obtaining the overpressure data is still to be evaluated. According to the theory of Few et al. (1967), the overpressure 0.3 m from a discharge with an energy input of 10^5 joules/m is about 4 atm.

6.3.2 The sound wave

As we have seen in the previous section, the cylindrical shock wave emitted by the return-stroke channel relaxes to a sound wave at a distance of a few meters from the channel. It is apparently this sound wave originating from all along the channel which, after modification by the medium through which it travels, is the thunder we hear. When heard at a distance of about 100 m or less from the discharge channel, thunder consists of one loud bang, although low-intensity hissing and clicking sounds are often reported to occur just prior to the bang. When heard at a distance of 1 km from the discharge channel, thunder generally consists of a rumbling noise punctuated by several loud claps. In this section we first consider the theory advanced to explain why thunder can be heard only for relatively short distances. We then consider the available explanations for the observed thunder pressure variations, in particular the origin of the dominant thunder frequency and of the thunder claps.

Generally, thunder cannot be heard at distances from the channel more than about 25 km. Isolated reports of thunder heard at greater distances have been given by Veenema (1917, 1918, 1920), Brooks (1920), Cave (1920), Page (1944), and Taljaard (1952). On the other hand, Ault (1916), captain of the research ship *Carnegie*, reported that a thunderstorm at sea became inaudible beyond a distance of about 8 km.

Fleagle (1949) has suggested that the inaudibility of thunder beyond 25 km is due to the upward curving of sound rays resulting from temperature gradients and wind shear. Since the velocity of sound is proportional to the square root of the temperature and since the temperature in general decreases as a function of height, Snell's law indicates that sound waves will be refracted upward. Fleagle has shown that in the presence of a linear lapse rate (temperature decreasing linearly with height) those sound rays which leave the channel and at some point become tangent to the ground plane exhibit a trajectory which is very nearly parabolic. For a typical lapse rate of $7.5°K/km$, sound which originates at a height of 4 km has a maximum range of audibility of 25 km if wind shear is ignored. That is, the sound rays from a 4-km height are tangent to the ground 25 km distant from the discharge channel. All sounds originating from heights below 4 km will not be heard at 25 km; sounds originating from above 4 km will be heard. Only the very close observer can hear the sound from the base of the discharge channel. Fleagle has shown that wind shear, a change in wind velocity with height, can provide a refraction of the thunder which is of the same order of magnitude as that due to temperature gradients. Sound rays may be refracted upward or downward depending on the relation between the wind direction and the sound ray direction. A wind shear of 4 m/sec per km can yield a sound ray trajectory almost equivalent to a lapse rate of $7.5°K/km$.

Fleagle cautions that factors other than the lapse rate and wind shear may affect the audibility of thunder. For example, a region of temperature inversion will tend to increase the range of audibility; and features of the terrain which hinder the essentially horizontal propagation of the critical sound ray in its final several kilometers will decrease the range of audibility.

It is to be expected that the base of the lightning channel is a powerful source of sound, since it is the base which is the brightest on channel photographs. Perhaps the loud bang of thunder heard at a distance of 100 m or less from a lightning discharge is due to the strong sound wave from the channel base; when the observer is at a distance of 1 km from the channel the initial loud bang is refracted over his head and the discharge begins with a rumble.

We consider now the frequency content of the thunder. We have seen in Sec. 6.2 that there is some discrepancy between the results of different investigators regarding the dominant frequency or frequencies present in thunder. Schmidt (1914), Arabadzhi (1952, 1957), and Bhartendu (1964) contend that most of the thunder energy is in the infrasonic frequencies around 1 Hz, while Latham (1964) and Few et al. (1967) have obtained data which show the dominant frequency to be in the 100-Hz range. Schmidt did not measure the absolute thunder pressure in the audible range, but rather arrived at his conclusion from a measurement of the infrasonic pressure variations and a comparison of these with the known pressure variations from common loud sources. Since Few et al. (1967) and Bhartendu (1964) have measured stronger thunder pressure variations in the sonic regions than Schmidt did in the infrasonic, Schmidt's conclusions would appear not to be valid. Further, there is, as discussed in Sec. 6.2, some question regarding the validity of Schmidt's interpretation of his data. Arabadzhi (1957) reports that he has made absolute pressure measurements over both the infrasonic and the sonic ranges and that the dominant frequencies are infrasonic. He presents no detailed data to support this claim. Bhartendu (1964) has given data regarding absolute thunder pressure for the sonic range but not for the infrasonic range. The justification for his statement that the dominant thunder energy is in the infrasonic is not apparent. Evidence that the pressure variations due to thunder are mainly near 100 Hz is derived from the data of Latham (1964), from the two independent studies reported by Few et al. (1967), and from the theory presented in Sec. 6.3.1. In view of all the evidence, it would appear likely that the dominant thunder frequency is in the 100-Hz range. As pointed out by Few et al., in order for the shock wave emitted by the channel to produce a 1-Hz dominant frequency, the radius at which the shock wave must relax to a sound wave is about 130 m; the input energy needed to produce this dimension is about 5.5×10^8 joules/m, three orders of magnitude greater than other estimates of the input energy (Sec. 7.3.3).

In the acoustic spectrum shown in Fig. 6.8, there is a sharp decrease in the acoustic power with increasing frequency for frequencies above several hundred hertz. Some, if not all, of this decrease is due to the fact that the attenuation of sound waves in air increases with increasing frequency. For example, a 2,000-Hz sound wave is attenuated to about 0.3 of its

original amplitude in traveling through 400 m of air at standard temperature and pressure and 50 to 100 percent relative humidity (Harris, 1967).
A 300-Hz signal is similarly attenuated in traveling through about 4 km
of air (Harris, 1967). Thus one would expect the measured acoustic
spectrum of thunder to have a greater high-frequency content than that
shown in Fig. 6.8 if measurements were made close to the discharge
channel.

We consider now the waveshape of thunder. According to Latham
(1964) the thunder we hear at 1 km or more from the discharge channel
begins with a low-intensity sound (thunder leader) whose duration is
between 0.1 and 2.2 sec. The initial portion of the thunder leader is
compressional. The main thunder consists of three or four discrete claps
superimposed on a rumbling noise. The origin of the thunder leader is
unknown. A reasonable guess can be made, however, about the origin
of the claps: The claps are probably to be associated with sound emitted
by sections of the main channel and channel branches which are approximately perpendicular to the line of sight from the observer to those sections. As shown independently by Brook and McCrory (1968) and by
Uman et al. (1968), the sound received from a section of the channel is
strongly dependent on the orientation of that section with respect to the
observer. When a channel or branch section is perpendicular to the
observer's line of sight, all points on the section produce sound that arrives
at the observer in a minimum time period, resulting in a clap. It is
reasonable to expect branches of first strokes to be powerful sources of
sound since, according to the data of Malan and Collens (1937), branches
may be instantaneously brighter than the channels above those branches.
Further, it is interesting to note that there are roughly the same number
of claps per thunder (Latham, 1964) as there are branches in a first stroke
(Schonland et al., 1935), so that the branches may well account for a
significant fraction of the claps.

Brook and McCrory (1968) and Uman et al. (1968) have shown that a
long, perfectly straight section of channel which generates sound uniformly from all portions of the section will yield a much lower sound level
far from the channel than will an approximately straight section of channel of similar length exhibiting small-scale tortuosity. The relatively low
levels of sound characteristic of the perfectly straight section are due to
interference effects; except for phenomena associated with the ends of the
channel, for every air compression that arrives at the observer, an air
rarefaction will arrive simultaneously from a slightly nearer section of the
channel, and the compression and rarefaction will add to yield a small
signal. If the channel is tortuous, these pronounced interference effects
do not occur.

When thunder is heard at very close range, low-intensity hissing and clicking sounds are often reported to occur just prior to the main thunder. The origin of these sounds has not been determined. It is possible that the hissing sound is due to corona discharge at the ground in the presence of strong electric fields induced by the leader, and that the clicking sound is to be associated with an upward-moving discharge (Sec. 7.6).

References

1. Arabadzhi, V.: Certain Characteristics of Thunder, *Dokl. Akad. Nauk SSSR*, **82**:377–378 (1952). (Translation available as RJ-1058 from Associated Technical Services, Inc., Glen Ridge, New Jersey.)
2. Arabadzhi, V.: Some Characteristics of the Electrical State of Thunderclouds and Thunderstorm Activity, *Uch. Zap. Minsk. Gos. Ped. Inst., im A. M. Gorkogo Yubil. Vypusk, Ser. Fiz.-Mat.*, No. 7, 1957. (Translation available as RJ-1315 from Associated Technical Services, Inc., Glen Ridge, New Jersey.)
3. Arabadzhi, V.: The Spectrum of Thunder, *Priroda (Moscow)*, **54**(7):74–75 (1965). In Russian.
4. Arago, F.: "O'Euvres complètes, notices scientifiques," tome 1, Legrand, Pomey et Crouzet Libraries, Paris, 1854.
5. Aristotle, (384–322 B.C.): "Meteorologica," H. P. D. Lee, trans., pp. 223–225, Loeb Classical Library, Harvard University Press, Cambridge, Mass., 1951.
6. Ault: Thunder at Sea, *Sci. Am.*, **218**:525 (1916).
7. Bates, E. L.: The Cause of Thunder Again, *Sci. Am.*, **88**:115 (1903).
8. Bhartendu: "Acoustics of Thunder," Ph.D. thesis, Physics Department, University of Saskatchewan, Saskatoon, Saskatchewan, 1964.
9. Bhartendu and B. W. Currie: Atmospheric Pressure Variations from Lightning Discharges, *Can. J. Phys.*, **41**:1929–1933 (1963).
10. Braginskii, S. I.: Theory of the Development of a Spark Channel, *Sov. Phys. JETP (English Transl.)*, **34**:1068–1074 (1958).
11. Brode, H. L.: The Blast Wave in Air Resulting from a High Temperature, High Pressure Sphere of Air, *Rand. Corp. Rept., RM-1825-AEC*, 1956.
12. Brook, M., and R. McCrory: private communication, 1968.
13. Brooks, C. F.: Another Case, *Monthly Weather Rev.*, **48**:162 (1920).
14. Cave, C. J. P.: The Audibility of Thunder, *Nature*, **104**:132 (1919).
15. Dawson, G. A., C. N. Richards, E. P. Krider, and M. A. Uman: The Acoustic Output of a Long Spark, *J. Geophys. Res.*, **73**:815–816 (1968).
16. De L'Isle, J. N.: "Memoires pour servir à l'histoire et au progrès de l'astronomie de la géographie et de la physique," L'Imprimerie de l'Académie des Sciences, St. Petersbourg, 1738.
17. Drabkina, D. I.: The Theory of the Development of the Spark Channel, *J. Exptl. Theoret. Phys. (USSR)*, **21**:473–483 (1951). (English translation, AERE Lib/Trans. 621, Harwell, Berkshire, England.)
18. Few, A. A., A. J. Dessler, D. J. Latham, and M. Brook: A Dominant 200 Hz Peak in the Acoustic Spectrum of Thunder, *J. Geophys. Res.*, **72**:6149–6154 (1967).
19. Fleagle, R. G.: The Audibility of Thunder, *J. Acoust. Soc. Am.*, **21**:411–412 (1949).

20. Harris, C. M., Absorption of Sound in Air versus Humidity and Temperature, *NASA-CR-647*, Columbia University, New York, 1967.

21. Haldane, E. S.: "Life of René Descartes," p. 181, John Murray (Publishers), Ltd., London, 1905.

22. Hirn, M.: The Sound of Thunder, *Sci. Am.*, **59**:201 (1888).

23. Kitagawa, N.: Discussion, in S. C. Coroniti (ed.), "Problems of Atmospheric and Space Electricity," pp. 350–351, American Elsevier Publishing Company, New York, 1965.

24. Latham, D. J.: A Study of Thunder from Close Lightning Discharges, M.S. thesis, Physics Department, New Mexico Institute of Mines and Technology, Socorro, N.M., 1964.

25. Lin, Shao-Chi: Cylindrical Shock Waves Produced by Instantaneous Energy Release, *J. Appl. Phys.*, **25**:54–57 (1954).

26. Lucretius, T. (98–55 B.C.): "On the Nature of Things," book VI, H. A. J. Munro, trans., Great Books of the Western World, p. 81, William Benton, Chicago, 1952.

27. Lyon, J. A.: The Cause of Thunder Again, *Sci. Am.*, **88**:191 (1903).

28. Malan, D. J., and H. Collens: Progressive Lightning, pt.3, The Fine Structure of Return Lightning Strokes, *Proc. Roy. Soc. (London)*, **A162**:175–203 (1937).

29. Mershon, R. S.: A Theory of Thunder, *Sci. Am.*, **23**:68 (1870).

30. Newman, M. M., J. R. Stahmann, and J. D. Robb: Experimental Study of Triggered Natural Lightning Discharges, *Rept. DS-67-3, Project 520-002-03X*, Federal Aviation Agency, Washington, D.C., March, 1967a.

31. Newman, M. M., J. R. Stahmann, J. D. Robb, E. A. Lewis, S. G. Martin, and S. V. Zinn: Triggered Lightning Strokes at Very Close Range, *J. Geophys. Res.*, **72**:4761–4764 (1967b).

32. Page, D. E.: Distance to Which Thunder Can Be Heard, *Bull. Am. Meteorol. Soc.*, **25**:366 (1944).

33. Pain, H. J., and E. W. E. Rogers: Shock Waves in Gases, *Rept. Progr. Phys.*, **25**:287–336 (1962).

34. Remillard, W. J.: The Acoustics of Thunder, *Tech. Mem.* 44, Acoustics Research Laboratory, Division of Engineering and Applied Physics, Harvard University, Cambridge, Mass., September, 1960.

35. Reynolds, R. V.: The Cause of Thunder, *Sci. Am.*, **88**:41 (1903).

36. Rouse, C. A.: Theoretical Analysis of the Hydrodynamic Flow in Exploding Wire Phenomena, in W. G. Chace and H. K. Moore, (eds.), "Exploding Wires," Plenum Press, Inc., New York, 1959.

37. Sakurai, A.: On the Propagation and Structure of the Blast Wave (1), *J. Phys. Soc. Japan*, **8**:662–669 (1953).

38. Sakurai, A.: On the Propagation and Structure of a Blast Wave (2), *J. Phys. Soc. Japan*, **9**:256–266 (1954).

39. Sakurai, A.: On Exact Solution of the Blast Wave Problem, *J. Phys. Soc. Japan*, **10**:827–828 (1955a).

40. Sakurai, A.: Decrement of Blast Wave, *J. Phys. Soc. Japan*, **10**:1018 (1955b).

41. Sakurai, A.: On the Propagation of Cylindrical Shock Waves, in W. G. Chace and H. K. Moore, (eds.), "Exploding Wires," Plenum Press, Inc., New York, 1959.

42. Schmidt, W., Über den Donner, *Meteorol. Z.*, 487–498, (1914).

43. Schonland, B. F. J., D. J. Malan, and H. Collens: Progressive Lightning, pt. 2, *Proc. Roy. Soc. (London)*, **A152**:595–625 (1935).

44. Swigart, R. J.: Third Order Blast Wave Theory and Its Application to Hypersonic Flow Past Blunt-nosed Cylinders, *J. Fluid Mech.*, **9**:613–620 (1960).

45. Taljaard, J. J.: How Far Can Thunder Be Heard? *Weather*, **7**:245–246 (1952).

46. Uman, M. A., D. K. McLain, and F. Myers: Sound from Line Sources with Application to Thunder, *Westinghouse Res. Lab. Rept. 68-9E4-HIVOL-R1*, 1968. Portions of this report to be published.

47. Veenema, L. C.: Die Hörweite des Gewitterdonner, *Z. Angew. Meteorol., Wetter*, 127–130, 187–192, 258–262 (1917).

48. Veenema, L. C.: Die Hörweite des Gewitterdonner, *Z. Angew. Meteorol., Wetter*, 56–68 (1918).

49. Veenema, L. C.: The Audibility of Thunder, *Monthly Weather Rev.*, **48**:162 (1920).

50. Zhivlyuk, Yu., and S. L. Mandel'shtam: On the Temperature of Lightning and Force of Thunder, *Soviet Phys. JETP (English Transl.)*, **13**:338–340 (1961).

7

Theory: The Discharge Processes

7.1 INTRODUCTION

Each of these successive strokes or partial discharges which make up a complete lightning discharge takes place in two stages, the downward-moving leader stage being followed by an upward-moving return stage. These processes will be described as the leader and the return streamer, since they have the same properties as electrical streamers produced in the laboratory. Such a streamer is a conducting filament of ionized gas which extends its length by virtue of ionizing processes occurring in the strong field in front of its tip. It is electrically charged throughout its length, but is not at the same potential as the electrode from which it started, for there is a drop of potential along it. This drop of potential provides a field which drives a current through the stem of the streamer and this current serves to charge up newly formed sections of the stem to the potential necessary for further progress.

B. F. J. Schonland: Progressive Lightning, pt. 4, The Discharge Mechanism, *Proc. Roy. Soc. (London)*, **A164**:132–150 (1938).

It already appears that the condition for the occurrence of a flash is not that the breakdown potential be obtained over the whole intervening space, but that a concentration of field should occur of sufficient intensity to initiate what may be called a "self-propagating" discharge in a field which elsewhere may only reach a few hundred volts per cm. . . .

Such an arc channel can be regarded simply as an extension of the initiating conductor, and the voltage gradient along it is only of the order of 10 volts per cm. . . .

The condition for the occurrence of a spark in a non-homogeneous field would thus appear to be that the current in the brush discharge from an electrode should be sufficient to cause glow-arc transition to occur, after which the leader progresses by the above mechanism.

C. E. R. Bruce: The Lightning and Spark Discharges, *Nature,* **147**:805–806 (1941).

To reconcile a natural phenomenon with an experiment in the laboratory is a difficult task, as is shown by the history of lightning research. Evaluations and knowledge obtained by model tests on a small scale are often incorrect, but survive for a remarkably long time.

D. Müller-Hillebrand: The Protection of Houses by Lightning Conductors— An Historical Review, *J. Franklin Inst.,* **274**:34–54 (1962).

Considerable experimental data have been accumulated regarding the leader and return-stroke processes. From these data information has been derived regarding the relative intensities and the propagation velocities of various luminous phenomena and the charges and currents associated with these phenomena. Unfortunately, the physical models derived from the experimental data or from the information determined directly from experimental data have often been obtained more on the basis of intuition than on the basis of detailed quantitative analysis. Lightning research has, in fact, been characterized by a marked absence of quantitative theoretical work. To some extent, this lack of quantitative theory is excusable. There is, for example, no quantitative theory for laboratory electrical breakdown due to *nonuniform* electric fields, although considerable experimental data for this type of breakdown have been collected. To muddle further the literature on lightning "theory," the laboratory data, much of which is conflicting, have frequently been extrapolated in an effort to "explain" lightning phenomena. The whole lamentable situation is well characterized by the various theories of the stepped leader, some of which we shall discuss in Sec. 7.4. In much of the lightning literature the words *pilot leader* and *streamer* have attained the status of explanations or theories. To name is not to explain. Further, at a given time different researchers have used these words to mean different things, and the consensus view of the meaning of these words has changed with time. In the previous chapters we have not used the words pilot leader and have made only limited use of the word streamer. We shall try in this chapter to make clear the difficulties involved in using these words.

7.2 DISCHARGE MECHANISMS

7.2.1 Introduction

When a potential difference is applied between two metal electrodes, an electric field is created in the space between the electrodes. In general, the gas in which the electrodes are immersed can be considered an insulator unless the applied voltage exceeds a well-defined value known as the breakdown voltage. For parallel-plane electrodes (uniform electric field) in air at atmospheric pressure and room temperature the electric field intensity necessary for breakdown is about 3×10^6 volts/m. The breakdown voltage is found by multiplying the breakdown field by the gap spacing. The breakdown voltage of a nonuniform gap is always less than the breakdown voltage of a uniform gap with the same gap spacing. In order for breakdown to occur, a few initiating electrons must be present in the gap between the electrodes. In practice, these electrons are always present due to cosmic-ray bombardment or natural radioactivity. The initiating electrons are accelerated in the breakdown field to a sufficient energy that they can produce ionization in colliding with gas atoms. The electrons produced by ionization in turn produce new electrons by the same process and a growth of electron and ion densities ensues. The discharge process may also be critically dependent on so-called secondary processes, the primary process being direct electron-impact ionization. Possible secondary processes include: (1) Secondary electron emission at the cathode due to the impact of positive ions, (2) photoelectric emission at the cathode due to photons emitted by the gas, (3) electron emission from the cathode due to the incidence of metastable atoms or ions, (4) ionization of the gas by positive ions, and (5) photoionization of the gas due to photon emission by the gas. Electrons may also be lost from the gas by attachment to a neutral atom or molecule, by recombination to an ion, and by diffusion. For a discussion of the various particle interactions occurring in gases the reader is referred to Massey and Burhop (1952), Loeb (1961), and McDaniel (1964). The fundamental physical problem is to describe quantitatively the development of electron and ion densities in the breakdown gap as a function of position and time in terms of the interaction processes taking place in the gap and at the electrodes.

7.2.2 Discharge phenomena in the absence of appreciable space charge

We consider now the formation of ionization in uniform field geometry prior to the onset of space-charge distortion of the applied field. We consider parallel-plane electrodes and allow n_0 electrons to be emitted per unit time from the cathode at $x = 0$. The emitted electrons drift in the uniform applied electric field E and produce additional electrons by ionizing collisions. Let α be the number of new electrons created per centimeter by a drifting electron. The coefficient α is called the first Townsend ionization coefficient or the coefficient of primary ionization by electrons. For simplicity, we neglect electron loss processes. If n_x electrons are moving through a plane at x, in going a distance dx they generate by ionization $\alpha n_x \, dx$ new electrons, so that

$$dn_x = \alpha n_x \, dx \tag{7.1}$$

The solution to Eq. (7.1) is

$$n_x = n_0 e^{\alpha x} \tag{7.2}$$

That is, there is an exponential growth of electrons. The group of electrons resulting from cumulative ionizations started by a few initial electrons at the same position in space is known as an *avalanche*. It can be shown (Townsend, 1947) that

$$\frac{\alpha}{p} = f\left(\frac{E}{p}\right) \tag{7.3}$$

where p is the gas pressure at constant temperature. A simple expression, derived by Townsend, for the function α/p at large values of E/p is

$$\frac{\alpha}{p} = A \exp\left(-\frac{B}{E/p}\right) \tag{7.4}$$

where A and B are constants for the gas under consideration. An analytical approximation to α/p for air at room temperature for E/p between 10^2 and 10^5 volts/m-torr is given by Kontaratos (1965). For air at room temperature α/p has a value of about 6 (m-torr)$^{-1}$ for E/p about 5×10^3 volts/m-torr, about 50 for E/p about 10^4, and about 5×10^2 for E/p about 3.5×10^4 (Frommhold, 1964; Loeb, 1961). For E/p between 3×10^3 and 10^4, the electron drift velocity in air at room temperature is between 1 and 3×10^5 m/sec (Frommhold, 1965; Raether, 1964).

We have, thus far, allowed only n_0 electrons per unit time to leave the cathode. These may be considered to be due to photoelectric emission caused by an external light source. Additional electrons may be emitted from the cathode due to various secondary processes, as mentioned in Sec. 7.2.1. (We neglect secondary processes which may occur within the gas although they may be easily included in the analysis.) Let

$n_0' = $ total electron emission from the cathode per unit time

$n_0'' = $ electron emission from the cathode per unit time due to secondary processes

then

$$n_0' = n_0 + n_0'' \tag{7.5}$$

If n_x electrons in traversing dx at x have the secondary effect of causing the emission of $\omega n_x \, dx$ electrons from the cathode, then *in the steady state*

$$n_0'' = \int_0^d \omega n_0' \exp \alpha x \, dx \tag{7.6}$$

$$= \frac{\omega}{\alpha} n_0'[(\exp \alpha d) - 1]$$

where d is the x coordinate of the anode. Combining Eqs. (7.5) and (7.6), we obtain

$$n_0' = n_0 + \frac{\omega}{\alpha} n_0'[(\exp \alpha d) - 1] \tag{7.7}$$

or

$$n_0 = n_0' \left\{ 1 - \frac{\omega}{\alpha} [(\exp \alpha d) - 1] \right\} \tag{7.8}$$

The total number of electrons at the anode is $n_d = n_0' \exp \alpha d$ so that

$$n_d = \frac{n_0 \exp \alpha d}{1 - \frac{\omega}{\alpha} [(\exp \alpha d) - 1]} \tag{7.9}$$

Equation (7.9) is a steady-state expression. Since all electric current at the anode is carried by electrons and since in the steady state current must be continuous across the gap, Eq. (7.9) is proportional to the discharge current. It is possible to verify the validity of the theory given above by experiment. Such experiments usually involve the steady-state measurement of gap current vs. gap distance at constant E/p and a comparison of the data thus obtained with a theoretical expression similar to Eq. (7.9), or the measurement of the temporal growth of gap current [the theory of which we have not considered; see, for example, Phelps (1960)]. To some extent, it is possible by experiment to distinguish between the various secondary processes.

Even though Eq. (7.9) is only valid in the absence of space-charge distortion of the applied field and even though complete electrical breakdown (collapse of the applied gap voltage) does not occur without space-charge distortion, for many experimental conditions an adequate breakdown criterion can be obtained from Eq. (7.9). The Townsend criterion for breakdown can be expressed as follows: Breakdown occurs if the denominator of Eq. (7.9) goes to zero, that is, if

$$1 - \frac{\omega}{\alpha} [(\exp \alpha d) - 1] = 0 \tag{7.10}$$

For a given gap spacing and gas pressure there will be an electric field intensity and hence a static breakdown potential for which Eq. (7.10) is satisfied. Equation (7.10) is equivalent to the condition that for every electron leaving the cathode and arriving at the anode a secondary process will regenerate a replacement electron from the cathode. From a control-engineering point of view, it is the positive feedback mechanism (regenerative replacement of electrons from the cathode) which leads to the unstable condition which eventually results in breakdown. Only a very small increase in voltage above the sustaining voltage (static breakdown potential) is needed to yield complete breakdown and hence the criterion for a self-sustaining discharge is essentially a breakdown criterion.

For a detailed discussion of the generation of ionization in uniform gaps without space-charge distortion of the applied field including a consideration of the various secondary processes active both at the cathode and in the gas, the reader is referred to Little (1956), Loeb (1961), Raether (1964), and Llewellyn-Jones (1957, 1967). It is shown in these references that for all secondary processes which have been considered, the form of the breakdown criterion given in Eq. (7.10) remains essentially unchanged.

7.2.3 Discharge phenomena in the presence of space charge

The theory for the generation of ionization in a uniform electric field as given in Sec. 7.2.2 is valid as long as the electric field due to the electrons and ions of the discharge is small compared to the applied field. For complete breakdown to occur, however, appreciable space-charge fields are needed. These fields play an important part in the creation of the high-conductivity region between cathode and anode that is characteristic of complete breakdown.

We consider now space-charge-influenced phenomena in plane-parallel gaps occurring when step-function voltages are applied to those gaps. Raether (1964) has stated that there are two mechanisms by which complete breakdown can occur: (1) A relatively slow mechanism, the theory of which is a time-dependent version of the theory of Sec. 7.2.2 including space-charge distortion. In the slow mechanism (which Raether calls the generation or Townsend mechanism) the time to breakdown is such that many successive electron avalanches cross the gap before breakdown occurs. (2) A rapid mechanism in which the first avalanche develops directly into a thin channel of high conductivity via streamer formation. The idea of the streamer mechanism was developed independently in the 1930s by Loeb and by Raether (Loeb and Meek, 1941). A streamer is defined by Loeb and Meek (1941) and by Raether (1964) as a filamentary discharge with strong space-charge fields at and ahead of the filament tip. These fields are presumed to lead to enhanced collisional ionization and to appreciable photoionization at the tip of the filament with the result that the filament propagates. Photoionization is an important part of the streamer mechanism and is thought to be essential for the propagation of cathode-directed positive streamers.

There is some controversy as to whether two distinct breakdown mechanisms actually do exist in uniform gap geometry. It is the view of some researchers (e.g., Ward, 1965a,b) that the streamer mechanism is merely a variation of the generation or Townsend mechanism; that the observed "streamer" characteristics are due primarily to collisional ionization in the presence of space-charge distortion of the applied field and that photo-ionization in the gas is not important.

Calculations of ionization growth in parallel-plane gaps with space-charge distortion of the applied field have been published, for example, by Miyoshi (1956, 1960), Davies et al. (1964), Köhrmann (1964a,b) and Ward (1965a,b). Recent experimental studies of streamer breakdown in parallel-plane gaps have been published by Wagner (1966, 1967).

Thus far, we have not considered breakdown in a nonuniform applied electric field. The electric fields which produce the lightning discharge are strongly nonuniform: They are strong at the cloud base and weak at the ground. Since the breakdown must initially occur in the region of strong field, the lightning discharge must propagate from the high-field region into the low-field region in order to reach the ground.

Breakdown in nonuniform gaps is generally described in terms of a streamer model. Details of non-uniform gap breakdown are given by Loeb (1965a,b). Typically the discharge begins with the propagation of luminous pulses of ionization from the high-field electrode into the low-field region. These initial pulses have been termed by Loeb (1965a) *primary streamers*. As a result of the action of numerous primary streamers, a comparatively narrow luminous channel is propagated from the high-field electrode into the gap. This luminous channel is termed by Loeb (1965a) a *secondary streamer*. Immediately after the secondary streamer bridges the gap between the electrodes, a return stroke propagates from the low-field electrode to the high-field electrode rendering the discharge channel highly conducting. In most of the older literature on nonuniform-gap breakdown, only the secondary streamer and return stroke were observed.

If one considers the secondary streamer to be a good conductor, then it can be self-propagating since it carries the potential of one electrode and hence a high electric field at its tip. Thus the streamer mechanism can account qualitatively for the propagation of the discharge into the low field region. The detailed physics of the situation, however, remains unknown.

A streamer-like mechanism can be invoked to explain the propagation of first leaders, dart leaders, and return strokes of lightning, as shown diagrammatically in Fig. 7.1a, b, and c. The propagation velocities of lightning leaders and return strokes are apparently greater than the drift velocity of the electrons at the tips of the propagating discharges. Cravath and Loeb (1935) first pointed out that this was the case for leaders and proposed the following physical explanation: In the intense field in front of the leader tip, the initial free electrons rapidly multiply by collisional ionization. If the intense field extends a distance d in front of the leader tip, the discharge will have advanced a distance d when the

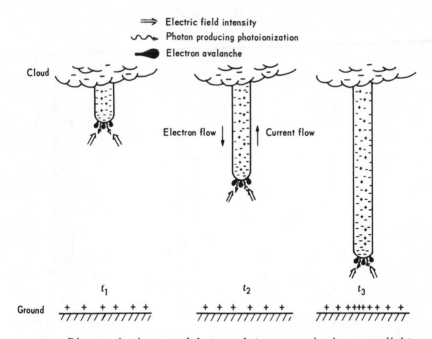

⇒ Electric field intensity

〜〜 Photon producing photoionization

◖ Electron avalanche

Cloud

Electron flow ↓ ↑ Current flow

t_1 t_2 t_3

Ground + + + + + + +   + + + + + + +   + + +++++ + + +

Fig. 7.1a   *Diagram showing general features of streamer mechanism as applied to lightning leader into virgin air in absence of step mechanism. The times $t_3 > t_2 > t_1$.*

electron density in front of the tip is high enough to allow sufficient current to flow ahead of the tip so that the necessary charging of the leader tip can occur. Cravath and Loeb (1935) presented a calculation to show that the theory yielded reasonable results. Their technique of calculation is reasonable but the values chosen for some of the physical parameters are probably in error. On the basis of the Cravath-Loeb theory, Schonland (1938) proposed an analytical expression for the propagating wavefront velocity v_w. The derivation of Schonland's expression is as follows: Assume that the electron density in front of the wave is n_i before cumulative ionization is produced and that the wavefront of strong electric field extends over a distance d. Schonland (1938) assumed, "as the simplest possible hypothesis," that the time t for the wavefront to advance d was the time necessary for each initial electron to cover the mean distance from one electron to the next. This time is $1/n_i^{1/3}v_e$, where v_e is the electron drift velocity and thus the wave velocity is

$$v_w = \frac{d}{t} = n_i^{1/3}v_e\, d \tag{7.11}$$

The propagation velocity v_w thus depends on the preexisting electron density and on the electric field intensity associated with the wavefront, the

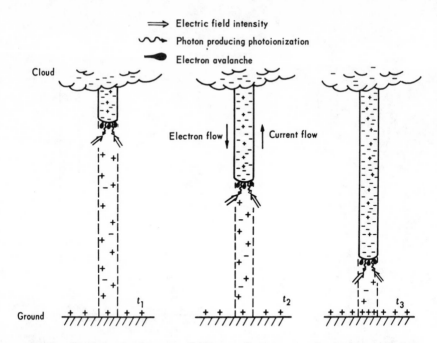

$\Longrightarrow$ Electric field intensity

$\rightsquigarrow$ Photon producing photoionization

Electron avalanche

Fig. 7.1b *Diagram showing general features of streamer mechanism as applied to dart leader. The times $t_3 > t_2 > t_1$.*

wavefront field determining both v_e and d. Schonland's expression for the propagation velocity is erroneously conceived as we shall see in Sec. 7.5.

The wave velocity for the Cravath-Loeb theory has been derived by Loeb (1965a) as follows: If n_f is the final electron density after the wave has traversed a distance d equal to the wavefront thickness, then the final electron density due to many avalanches within d is

$$n_f = n_i \exp \alpha v_e t \tag{7.12}$$

where the length of each avalanche, $v_e t$, must be less than d. Since the time t for the wavefront to advance d is $t = d/v_w$, the wave velocity is

$$v_w = \frac{\alpha v_e d}{\ln n_f/n_i} \tag{7.13}$$

According to Loeb (1966a), if the streamer advance is into un-ionized air, n_i is given by the density of photoelectrons created within d by photons from the wavefront. The wavefront velocity is only weakly dependent on the ratio of the initial to final electron density. We will consider an application of Eq. (7.13) in Sec. 7.5.

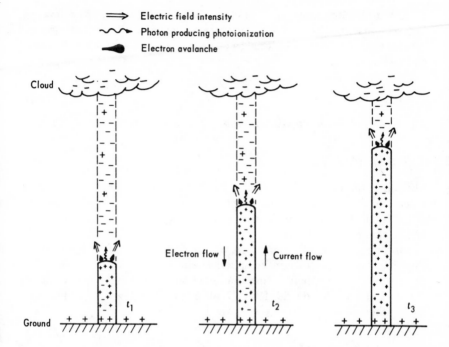

Fig. 7.1c *Diagram showing general features of streamer mechanism as applied to return stroke. Times $t_3 > t_2 > t_1$. Note that all current is carried by electrons since the mobility of positive ions is low.*

Winn (1965) has made the following valid comment regarding the two expressions for wave velocity, Eq. (7.11) and Eq. (7.13):

Both formulas require knowledge of the extent of the wavefront d, a quantity which is not usually known and which may not need to be known, because it probably is not independent of the more fundamental parameters—voltage, channel diameter, initial electron density, pressure, and properties of the gas. The same can be said of n_f. The formula of Cravath and Loeb has the attractive feature that it depends upon Townsend's coefficient α, a quantity which should make some difference. Unfortunately, neither formula constitutes a complete or workable theory.

7.3 SOME ROUGH CALCULATIONS

7.3.1 Some electrostatic relations

In this section we present analytical expressions from which rough calculations for the cloud-charge dimensions, the leader diameter, and the cloud-to-ground potential difference can be made. In the following two sections we shall derive numerical data from these expressions. A general reference for the material to be presented in this section is Plonsey and Collin (1961).

The integral form of Maxwell's equation

$$\nabla \cdot \mathbf{D} = \rho_v \tag{7.14}$$

where $\mathbf{D} = \epsilon_0 \mathbf{E}$ is the electric flux density and ρ_v the electric charge density is

$$\oiint \mathbf{D} \cdot \mathbf{n} \, dA = Q_{\text{encl}} \tag{7.15}$$

where $\mathbf{n}$ is the normal to the closed surface A over which the integral is taken and Q_{encl} is the net charge enclosed by the surface A. Equation (7.15) is easily applied in the case of a spherically or cylindrically symmetric charge distribution.

Consider a spherically symmetric charge distribution of total charge Q contained within an imaginary spherical surface of radius a. Since for the assumed charge distribution the direction of the vector $\mathbf{D}$ is radially outward, the left-hand side of Eq. (7.15) can be integrated over an imaginary spherical surface at $r \geq a$ to yield

$$D4\pi r^2 = Q \tag{7.16}$$

or

$$E = \frac{Q}{4\pi\epsilon_0 r^2} \qquad r \geq a \tag{7.17}$$

a result we have stated in Eq. (3.1). If the charge density inside the sphere is uniform at $Q/\frac{4}{3}\pi a^3$ coul/m³, then for $r \leq a$ Eq. (7.15) yields

$$D4\pi r^2 = \frac{Q}{\frac{4}{3}\pi a^3} \frac{4}{3}\pi r^3 \tag{7.18}$$

or

$$E = \frac{Qr}{4\pi a^3} \qquad r \leq a \tag{7.19}$$

Consider now a cylindrically symmetric charge distribution of total charge per unit length ρ contained within an imaginary cylindrical surface of radius a. The electric field intensity outside the radius a is, from Eq. (7.15),

$$E = \frac{\rho}{2\pi\epsilon_0 r} \qquad r \geq a \tag{7.20}$$

The potential difference V between two points b and c in an electrostatic field is

$$V = -\int_b^c \mathbf{E} \cdot d\mathbf{l} \tag{7.21}$$

where l is the path along which the integral is taken. The potential difference between two points r_b and r_c in the field of a spherically symmetric charge distribution is, for $r \geq a$, found by combining Eq. (7.17) with Eq. (7.21)

$$V = \frac{Q}{4\pi\epsilon_0}\left(\frac{1}{r_c} - \frac{1}{r_b}\right) \tag{7.22}$$

The potential difference between the surface, $r = a$, of a uniform spherical charge distribution and the center, $r = 0$, of the charge distribution is

$$V = -\int_a^0 \frac{Qr}{4\pi\epsilon_0 a^3}\, dr = \frac{Q}{8\pi\epsilon_0 a} \tag{7.23}$$

7.3.2 Cloud-charge dimensions and leader radius

The N region of the thundercloud provides the negative charge which flows to ground in the usual cloud-to-ground lightning flash. Let us assume for the sake of rough calculations that the N charge is isolated from all other charges, is spherically symmetric, and is contained within a radius a. We can estimate the radius a as follows: The strongest electric field produced by the spherical charge is found at $r = a$, as can be seen by comparing Eq. (7.17) with Eq. (7.19). At $r = a$, the electric field should not exceed the breakdown field, which for dry air at atmospheric pressure is about 3×10^6 volts/m. If we assume that the N charge is 40 coul and that the field at $r = a$ is less than 3×10^6 volts/m, then from Eq. (7.17) we find that $a \geq 350$ m. Due to decreased pressure at cloud height the breakdown field in the cloud will be about a factor of 2 less than at atmospheric pressure. In addition, the breakdown field may be lowered by a factor of about 3 by the presence of water drops in the cloud (Macky, 1931; English, 1948). Thus we might expect the effective radius for the N charge to be of the order of 2 km. This result is in reasonable agreement with the cloud-charge model shown in Fig. 1.2 and in Fig. 3.3. It is worth noting that the lightning flash can drain an appreciable fraction of the N charge in less than 1 sec even though that charge occupies a relatively large volume.

A similar analysis to that given in the preceding paragraph can be applied to determine the minimum leader radius. A reasonable value for the charge per unit length along the stepped-leader path is 10^{-3} coul/m (Sec. 3.7.2). Assuming that the electric field at the leader surface is less than a breakdown field of 3×10^6 volts/m, we find from Eq. (7.20) that the minimum stepped-leader radius is 6 m. The effective radius within which the charge of the stepped leader is contained is not necessarily the photographed radius since the presence of charge does not insure the presence of luminosity. Schonland (1953) reports luminous stepped leader radii of between 0.5 and 5.0 m (Sec. 2.3.2). Further, Brook (1967) reports that the luminous steps have no apparent structure as a function of radius, that photographs of steps appear to be taken with the camera lens out of focus although this is not the case; the images are fuzzy. The charge lowered by dart leaders is perhaps a factor of about 5 less than that lowered by stepped leaders (Sec. 3.7.3). Thus on the basis of the isolated charge model a minimum effective dart-leader radius is about 1 m. A photographic determination of the dart-leader radius has apparently not been published.

7.3.3 The cloud-to-ground potential difference and energy available for lightning

We estimate now the potential difference between the bottom of the N-charge region and ground. We will ignore the possible presence of the p charge and consider the bottom of the N charge to be at 3 km. A *maximum* value for the voltage difference can be obtained by assuming that the dry-air atmospheric-pressure breakdown field of 3×10^6 volts/m is necessary everywhere between cloud and ground to effect the breakdown, although this is certainly not the case for the highly nonuniform gap. Use of Eq. (7.21) yields for the maximum potential difference between cloud and ground a value of about 10^{10} volts. A *minimum* value for the voltage difference can be obtained by assuming that the field magnitudes observed at the ground in the vicinity of thunderclouds extend up into the cloud (where, in fact, the fields must become much greater). Using the usual field observed at ground, about 10^4 volts/m, we obtain a minimum cloud-to-ground potential difference of about 3×10^7 volts. The actual potential difference must fall between these calculated maximum and minimum values.

We can calculate the actual cloud-to-ground potential difference by assuming a model distribution of charge in the thundercloud and by using the expressions for electric field and potential difference presented in the previous section. In order to get a feeling for the order of magnitudes involved, we first determine the potential for an isolated spherical charge of radius 1 km and charge 40 coul. From Eq. (7.22) we find that the potential difference from the surface of the sphere to infinity is about 4×10^8 volts. The potential difference between the surface and a point 3 km away from the surface is 3×10^8 volts. The potential difference between the surface of the sphere of charge and the center of the sphere is about 2×10^8 volts. Thus we find that potential differences of the order of 10^8 volts are created over distances of the order of kilometers by charge distributions which approximate the N or P regions of the thundercloud. If we use the cloud-charge model shown in Figs. 1.2 and 3.3 (excluding the p charge), we can determine the potential difference between ground and any point above ground exactly. For a vertical path directly beneath the charges, the potential difference between the ground and 3 km above the ground is 1.1×10^8 volts. The value is not much different from that of the isolated spherical charge because one charged region, the N region, provides the dominant electric field beneath the cloud. We conclude therefore that the potential difference between the cloud base and ground is probably of the order of 10^8 volts.

The energy stored in the thundercloud and available for dissipation in the lightning channel is of the order of QV, or about 4×10^9 joules. The energy dissipated in a single lightning stroke in which 5 coul are brought to ground is of the order of 5×10^8 joules. If the lightning channel is 5 km long, the energy dissipated per meter is 10^5 joules/m. This value is in good agreement with values of input energy deduced from acoustic measurements (Sec. 6.3.1) and from spectroscopic measurements (Sec. 5.5.1). If the bulk of the input energy of 10^5 joules/m is dissipated in, say, 10 μsec, the effective power input for this time is about 10^{10} watts/m. Measurements of the light output of the channel indicate that the lower limit on the peak input power is 7.8×10^8 watts/m (Sec. 5.5.1). For a power input of 10^{10} watts/m and current of 10^4 amp, an electric field in the channel of 10^6 volts/m would be required. A high electric field along the channel (probably of the order of 10^7 volts/m; see calculation in Section 7.5) is initially provided by the return-stroke wavefront which traverses several meters of the channel in 0.1 μsec and which is probably about 1 m in extent. On the basis of laboratory experiments (Somerville, 1959), it is thought that relatively high electric fields (probably between 10^4 and 10^5 volts/m) are characteristic of the discharge channel during the initial stages of the transition from spark to arc (during the shock-wave phase). In time, the lightning channel becomes similar to a steady-state arc in air. An arc channel in air with a current of 10^4 amp has an electric field intensity of about 10^3 volts/m (King, 1962), and thus has a power dissipation of 10^7 watts/m. This level of dissipation could be maintained for many milliseconds with the available input energy. In reality the lightning current is usually less than 10^4 amp after 100 μsec and has a total duration of a few milliseconds. It appears then that the input energy calculated to be available for a lightning stroke, 10^5 joules/m, is a reasonable value and is capable of accounting for the high electric fields of the return-stroke wavefront and the channel-expansion stage as well as for the power dissipation during the arc phase of the discharge. It is to be noted that some of the input energy available must be dissipated in propagating the leader to ground. Even if half the available energy goes to the leader, the rough calculations given above are not strongly affected.

The energy input to the return-stroke channel is converted to dissociation, ionization, excitation, and kinetic energy of the channel particles, to the energy of expansion of the channel, and to radiation. We can make a rough estimate of the energy delivered to the channel particles: Assume that the leader channel has initially 10^{24} particles per cubic meter (corresponding to a temperature of about 3000°K at atmospheric pressure) and that soon after the return-stroke wavefront passes a given section of the channel all the particles within a radius of 2 cm are once ionized and at a temperature of about 30,000°K. Thus about 10^{21} particles per meter will be involved. If the energy delivered to each particle is about 15 ev $(2.4 \times 10^{-18}$ joule), the total energy stored in the particles is of the order of 10^3 joules/m, 1 percent of the available energy. Krider et al. (1968) have shown that the radiated energy in the 4000 to 11,000 Å spectral range in a single-stroke lightning flash amounted to about 870 joules/m, or about 0.4 percent of the input energy. The stroke energy radiated in the frequencies between 1 kHz and 100 MHz can be estimated from the data of Horner (1964) to be less than 10 joules/m. Unless large amounts of energy are radiated in the wavelengths below 4000 Å and in the infrared, it would appear that a sizable portion of the input energy must go into the energy of channel expansion. That this is probably the case is evidenced by the good agreement between measured and calculated thunder frequencies (Sec. 6.3.1 and 6.3.2). Note that the channel expansion must initially be driven by a high channel pressure and hence that the initial channel temperature is probably much higher than 30,000°K (see Sec. 6.3.1).

7.4 STEPPED-LEADER THEORIES

7.4.1 The experimental data

Any theory devised to describe the stepped leader must necessarily take account of (and predict) the observed stepped-leader characteristics. Some of the salient stepped-leader characteristics are the following: (1) The minimum average velocity for negatively charged, downward-moving stepped leaders is about 1×10^5 m/sec (Table 2.1). (2) The pause times between steps are relatively constant, and, in general, the longer pause times are followed by longer step lengths (Sec. 2.3.2). (3) Step lengths for leaders far above ground are of the order of 50 m; pause times of the order of 50 μsec (Sec. 2.3.2). (4) The luminous step process takes place in 1 μsec or less (Sec. 2.3.2). (5) The electrostatic field change due to the stepped leader is relatively smooth and the implication of this observation is that the step process itself does not lower appreciable charge (Sec. 3.7.2). (6) A reasonable order-of-magnitude estimate for the charge lowered by the stepped leader is 5 coul. For a charge of this magnitude to be lowered in tens of milliseconds requires an average current flow of the order-of-magnitude of 100 amp (Sec. 3.7.2). (7) Current in upward-moving stepped leaders is generally between twenty and several hundred amperes and is relatively smooth (Sec. 4.4). (8) Both upward- and downward-moving negatively charged leaders exhibit stepping. Positively charged leaders are of a more continuous nature (Sec. 2.5.4). (9) Photographs have been obtained of upward-going, negatively charged stepped leaders which show a faint corona discharge extending for about one step length in front of the bright leader step. The luminosity of the advance corona does not appear to develop between steps, but rather occurs simultaneously with the creation of the bright step behind it. The corona luminosity is impressed on film in a time less than 5 μsec, the time resolution of the camera used (Sec. 2.5.4).

7.4.2 The two original theories of the stepped leader

The two original, and probably still most quoted, theories of the stepped leader are due to Schonland (1938, 1953, 1956) and to Bruce (1941, 1944). Briefly, Schonland's model of the leader consists of a channel of uniform cross-sectional properties whose radius is of the order of a meter. On the other hand, Bruce pictures the leader channel as composed of a central arc core of small radius surrounded by a large corona envelope. We consider now the details of these two theories.

Schonland (1938) considered the observed minimum average velocity of the stepped leader and the relatively constant ratio of the step length to pause time sufficient grounds for postulating the existence of a *pilot leader* or *pilot streamer*. The pilot leader is defined by Schonland (1938) as "a true negative virgin air streamer (that) travels continuously downward in front of the step streamer processes with a velocity equal to the effective velocity of these." The observed minimum effective velocity of the stepped leader process, 1×10^5 m/sec, was identified with the minimum velocity of the advance of the pilot. Since the pilot velocity could be no slower than the electron drift velocity at the pilot tip and since the electric field at the tip must be of the order of 10^6 volts/m or greater, a minimum pilot velocity of the order of 1×10^5 m/sec is quite reasonable. [For E/p between 3×10^3 and 10^4 volts/m-torr, the electron drift velocity in air is between 1 and 3×10^5 m/sec according to Frommhold (1964).]

Further support for the idea of the pilot leader was provided (Schonland, 1953) by the observation that the stepping process did not yield appreciable electrostatic field changes and hence that it was apparently not the agent for the lowering of negative charge (Sec. 3.7.2). The measurement of smooth currents in upward-moving, negatively charged stepped leaders adds weight to this conclusion (Sec. 4.4).

According to Schonland (1953) the average radius of both the pilot leader and the step process is about 2.5 m; and the average stepped-leader current, arrived at by considerations similar to those given in Sec. 3.7.2, is most frequently 320 amp. Schonland (1953) presents several arguments for the channel charge being contained within a radius of 2.5 m, all of which are essentially equivalent to the calculation of leader radius given in Sec. 7.3.2. Using the values of radius and current given above, Schonland (1953) calculates the current density in the pilot leader to be 16 amp/m^2, the current density being uniformly distributed across the channel cross section. For this current density and an assumed electric field of 6×10^5 volts/m, the electron density is 1.8×10^{15} m^{-3}.

Two qualitative explanations of the mechanism of stepping are proposed by Schonland (1938) and a third by Schonland (1953, 1956). In all of the explanations the pilot leader is thought to proceed toward ground with the visible step intermittently catching up. In other words, the pilot prepares the way for the step process. The various theories of stepping invoke space charge, recombination, electron capture, and ionization processes in order to explain the stepping phenomenon. Little is gained by reiterating the various intuitive arguments since the physics of each is muddled and inexact. Nevertheless, what is clear, according to Schonland (1962), is that the pilot cannot move forward unless the channel behind it is capable of carrying the current to sustain it. Thus the function of the step process is to render the channel suitably conducting.

As is apparent, relatively little is known about the details of the stepping process. Various estimates, including those of Schonland, of the current present in a step yield values of the order of 1 ka. The details of these estimates have been considered in Sec. 3.7.2.

In 1962, Schonland (1962) stated that it was the general consensus that the stepped-leader channel must carry at its center a thin conducting core similar to an electric arc and that the leader current flows in the core. The core is surrounded by a corona sheath several meters in radius which carries the leader charge. It is not clear whether in 1962 this view was shared by Schonland, but the implication is that it was.

The word *pilot leader* was originally coined by Schonland (1938) to mean a downward-moving cylinder of ionization with a radius of the order of a meter which propagates itself into virgin air. Schonland (1938) supposed the pilot motion to be continuous; Schonland (1953) suggested that the pilot stopped during the step process. Schonland (1962) proposed that there may be several variants of the pilot process in the laboratory spark. Apparently, the pilot which Schonland (1962) refers to as occurring in the laboratory spark is Loeb's (1965a,b) primary streamers, whereas the laboratory leader referred to by Schonland is the secondary streamer. In view of the various different existing concepts of the pilot and in view of the great hazard of extrapolating laboratory data to the lightning leader, a reasonable definition for the pilot leader in lightning would be "the sum of all processes which precede the bright steps." In some of the lightning literature the term pilot is used in this loose and general sense. In other of the literature the original definition of the term pilot is accepted as dogma. The literature would be somewhat more coherent if the word pilot were used with more discretion.

We consider now the stepped-leader theory due to Bruce (1941, 1944). Bruce's view of the leader channel, first proposed in 1941, is similar to that recently stated by Schonland (1962), and presented in the paragraph before last, except that Bruce attributes considerable significance to radial-flowing corona currents. Bruce considers the arc channel at the center of the stepped leader to be of sufficiently high conductivity that the center of the leader is at a potential of about 10^8 volts with respect to ground. Radial corona currents are thought to flow because of the large potential difference between the core and the surrounding air, and these currents are thought to play a significant role in determining the properties of the leader as well as in controlling the stepping process. According to Bruce, the region ahead of the leader channel, because of the high voltage at the leader tip, is in the state of a brush or glow discharge. [Bruce (1944) identified this glow discharge with Schonland's pilot leader.] The current required to maintain the corona currents in the glow increases as the glow discharge lengthens. According to Bruce, when the current to the glow discharge exceeds a value of the order of 1 amp, a sudden transition from glow to arc discharge occurs. The resulting elongation of the arc channel appears as a luminous step. Once arc conditions are attained, the channel will not revert back to glow conditions even if the current is lowered considerably below 1 amp due to the characteristics of the arc-glow transition.

In Bruce's view the current flowing in the arc core of the stepped leader channel decreases with distance from the source of the channel since the current flowing through a given channel cross section must supply both radial corona currents associated with the channel ahead of the cross section and the current necessary to extend the channel. Using a value of the corona constant extrapolated from laboratory experiment, Bruce (1944) calculated the source current and step length to be expected for upward-going leaders. His values for step length agree with those found in the Empire State Building study, but the calculated source current, of the order of 5,000 amp, would appear to be too large.

It is appropriate now to make a few comments regarding the Schonland and Bruce theories. There would appear to be general agreement at present that in order for the leader channel to conduct currents of the order of a hundred amperes or more, that channel must contain at its center a highly conducting core similar to a laboratory arc. No such arc core is apparent on stepped-leader photographs. An envelope of negative charge of meter radius must surround the arc core in order to account for the charge lowered by the leader. Whether radial corona currents are significant or not will depend on the extent and behavior of the negative space charge which must carry the corona currents (Pierce, 1955). This question remains unresolved. Both the Schonland and Bruce theories assume that a continuously developing, low-luminosity corona is present ahead of the last leader step. The continuous development is necessary in order to account for the continuous nature of the leader electric field change as well as for the continuous nature of the measured leader currents. The observation of a pulsed corona occurring simultaneously with and ahead of the last leader step (Sec. 2.5.4) would appear to be difficult to explain on the basis of either theory.

7.4.3 Other stepped-leader theories

In addition to the work of Schonland and of Bruce, stepped leader theories have been proposed by Komelkov (1947, 1950), by Wagner and Hileman (1958, 1961), by Griscom (1958), and by Loeb (1966). All of these authors except the latter consider the leader channel to be composed of a small-diameter conducting core surrounded by a corona sheath. Loeb (1966) proposes that the stepped leader is composed of an invisible (low-luminosity) pilot-leader channel which is similar in characteristics to Schonland's pilot and which is formed by and on which are superimposed various streamer and traveling-wave phenomena.

Komelkov (1947, 1950) proposes that the leader tip is fed by current from a thermally ionized channel core with an internal electric field intensity of about 6×10^3 volts/m and a current density of 10^7 amp/m^2, and that a corona discharge consisting of weakly luminous streamers is produced around and ahead of the tip in a field of 6×10^5 to 10^6 volts/m. Halting of the corona and the advance of the arc are attributed to physical processes supposedly initiated by a fall of potential at the arc tip. As in the case of Schonland's stepping theories, the physics is sufficiently murky so that no detailed presentation of the theory is justified.

Wagner and Hileman (1958, 1961) propose a leader channel structure very similar to the model of Komelkov.   They state that the leader steps "are formed by the alternate advance of the central channel and its surrounding corona sheath," and that "the actual transition from the glow to the arc is of the nature proposed by Bruce but on a filamentary basis." That is, the corona region in front of the arc-core tip is presumed to be fed by a multitude of filamentary channels.   The current at the base of each channel increases as the channel length increases.   When one filament develops sufficient current, it transforms into an arc, short-circuits the other filaments, and becomes the leader step.   Wagner and Hileman (1958, 1961) present a number of valuable numerical calculations regarding values of charge density, potential, and electric field intensity in and around their model leader.

Griscom (1958) has presented a stepped leader theory which he terms the *prestrike* theory.   According to this theory, the head of the leader is a giant bulbous volume of charge many times greater in diameter than the cylindrical envelope of charge around the arc core of the leader channel.   The leader head contains a disproportionately large amount of charge compared to a similar length of the leader channel proper.   Griscom predicts that the transfer of the leader-head charge to a grounded object will result in a high-amplitude short-duration prestrike current preceding the actual return-stroke current.   According to the prestrike theory, a leader step originates at the surface of the bulbous leader head and develops back to the leader channel.   The point of origin on the leader-head surface is determined by chance.   As the new step is formed, the bulbous corona head collapses, its charge draining into the step.   As the old head is collapsing, a new leader head begins to grow from the tip of the new leader step.   Criticism of the prestrike theory is given by Wagner (1967).

Loeb (1966) has proposed a stepped-leader theory based on the laboratory studies of breakdown in air by Wagner (1966).   In Loeb's view, a nonluminous step is formed by positive and negative streamers being launched earthward and cloudward, respectively, from a Schonland-like pilot, the pilot being created out of previous streamers.   The two streamers each travel about half a step length when the upward-going streamer initiates an ionizing wave which sweeps down the new step.   It is this latter wave which illuminates the new step.   The channel thus created has, according to Loeb, an electron density of 10^{16} m^{-3} or more.   The excited molecules in the channel radiate their light in 10^{-7} sec or so and the channel is dark but still full of electrons.   The conductivity of the channel above the step is maintained by various traveling-wave phenomena associated with the step formation.   The leader channel is originally of the order of 1 m in radius but it slowly shrinks in size due to the action of successive passing waves in rendering the core conducting.   Thus, when the leader reaches the ground, the return stroke has a relatively small, highly conducting channel up which to propagate, despite the initial large radii of the steps.

7.4.4 Stepped-leader initiation

The cloud-to-ground lightning discharge is thought by many investigators (Malan and Schonland, 1951; Loeb, 1953, 1966; Clarence and Malan, 1957) to begin with a local discharge between the p and the N regions of the cloud. This p-N discharge would serve to make the N charge mobile as well as to lower the N charge closer to the ground (as a byproduct of the neutralization of the lower and smaller p charge). Experimental data relating to the p-N discharge are given by Clarence and Malan (1957) and are considered in Sec. 3.7.2. Clarence and Malan suggest, on the basis of their experiments, that the p-N discharge is just as often started from the p charge as from the N charge, and that, following the initial breakdown stage, a second stage occurs during which the breakdown channel is being negatively charged. The most detailed physical (theoretical) explanation of the stepped-leader initiation process is given by Loeb (1966). Loeb considers that a small body of positively charged raindrops of some hundred meters or so diameter (the p charge) is suddenly carried upward toward the N charge whose cross-sectional area is some square kilometers. An electric field of 7 to 10×10^5 volts/m is produced at the top of the positive p charge. The raindrops of the p-charge body elongate and yield corona streamers directed upward toward the N charge. Loeb calculates the shape and size of the cloud volume which the streamers bridge and the time required for the p-N discharge, once initiated, to launch a stepped leader downward from the cloud base.

7.5 DART-LEADER THEORIES

The dart leader traverses a pre-ionized trail, the remains of the previous-stroke channel, from cloud to ground in about 1 msec and in that time lowers about 1 coul of charge. It follows that the current which must flow in the dart leader channel is of the order of 1 ka. The length of the luminous dart which travels to ground is about 50 m. The radius within which the dart-leader charge is contained is, according to the calculation given in Sec. 7.3.2, of the order of meters. Dart leaders traversing previous-stroke channels after long interstroke periods have slower propagation velocities than dart leaders which are initiated following short interstroke periods (see Table 2.5 and Fig. 2.10). Dart-leader characteristics have been considered in Secs. 2.3.3 and 3.7.3.

The general features of dart-leader propagation can be explained from a streamer-theory point of view. A schematic diagram of the dart-leader propagation according to the streamer theory is given in Fig. 7.1b. The value of preexisting electron density in front of the dart-leader tip is not known. Evidence will be presented in Sec. 7.7 to show that the electron density is between 10^{13} and 10^{19} m^{-3}. The dart-leader velocity might be expected to decrease with decreasing initial electron density, so that one might expect slower dart leaders after longer interstroke intervals (Schonland et al., 1935). Additional factors which may influence the dart-leader velocity are the heavy-particle density in the channel and the channel radius. These factors will be considered in Sec. 7.7.

It is instructive to consider the application of Eqs. (7.11) and (7.13) to the case of the dart leaders. According to Schonland (1956), a dart leader with tip radius of 0.1 m and potential of 10^7 volts will be able to produce significant ionization in front of the tip for a distance of about 0.3 m. (Schonland's stated conditions can be produced by a charge of about 10^{-4} coul contained within a roughly spherical tip of radius 0.1 m.) At 0.15 m the electric field is, according to Schonland, about 7.2×10^6 volts/m and the electron-drift velocity about 1.6×10^5 m/sec. The use of this value of drift velocity assumes that the channel pressure is atmospheric and the channel temperature is about 300°K. If the channel temperature is higher than 300°K, which it probably is, and the pressure is atmospheric, the drift velocity will be higher than stated. A reasonable value for α is about 4×10^4 m^{-1} for the electric field given above. From Schonland's formula, Eq. (7.11), we can calculate n_i for a typical dart leader since the dart velocity is known and other needed quantities are given above. For the conditions given above, Schonland (1956) finds an electron density of 10^9 m^{-3} for a dart velocity of 4.8×10^7 m/sec. Schonland's formula is open to criticism from two points of view. First, there are no physical grounds for the assumption used in the derivation that the time for the wavefront to go a distance d is the time for each electron to travel the average distance between electrons; second, for initial electron densities greater than about 10^{12} m^{-3}, the formula predicts dart velocities greater than the speed of light.

Loeb's formula, Eq. (7.13), can be used to determine the dart velocity if the ratio n_f/n_i is chosen. Let us assume that ln (n_f/n_i) is about 10, corresponding to n_f/n_i of about 10^5. Then for an electric field of 7.2×10^6 volts/m, $d = 0.3$ m, $\alpha = 4 \times 10^4$ m^{-1}, and $v_e = 3 \times 10^5$ m/sec, we find a wave velocity of about 3×10^8 m/sec. The total potential difference across the wavefront is about 2×10^6 volts. For the parameters given above the dart-leader velocity is the speed of light. A value of electric field smaller than that chosen will yield a reduced α, d, and v_e and hence yield a reasonable wave velocity for dart leaders. It would appear, then, that according to the Loeb formula, Eq. (7.13), the dart leader wave-front can propagate with quite moderate electric field intensities. A calculation similar to that given above has been presented by Loeb (1958) for the case of the return stroke.

It follows from the foregoing that the extent of the dart-leader wave-front electric field is probably less than a meter. It is this strong field which is responsible for the dart-leader luminosity. According to Schonland (1938), the wavefront field produces a channel luminosity which is visible for a time of the order of the lifetime τ of the longest significant atomic or molecular excited states involved in the luminosity. For a dart leader of velocity 10^7 m/sec, the luminous dart length is observed to be about 50 m. Thus, the effective lifetime τ of the excited states involved would be 5×10^{-6} sec. Most of the spectral lines observed in the arc and spark spectra of air have lifetimes between 10^{-6} and 10^{-8} sec. For example, OI 4368 Å has $\tau = 1.36 \times 10^{-6}$ sec, NI 4935 Å has $\tau = 4.46 \times 10^{-7}$ sec, and NII 3995 Å has $\tau = 9.5 \times 10^{-9}$ sec (Uman et al., 1964). A measurement of the spectral features constituting the dart luminosity would provide a check on Schonland's suggestion that dart length is determined by excited-state lifetime. No dart leader spectra have thus far been obtained. In the event that the significant spectral features in the dart-leader spectrum have lifetimes much less than 10^{-6} sec, a circumstance which would appear not unlikely, the long-lasting luminosity can be explained on the basis of a heat-transfer model (see Sec. 7.7). That is, we can consider the dart-leader channel to be heated by the wavefront field so that the channel will emit light characteristic of some elevated temperature until that temperature has been lowered by thermal-conductive, thermal-convective, and radiative energy loss.

Some of the properties of dart leaders have been simulated in the laboratory by Winn (1965). He applied pulsed voltages to defunct 0.1-m spark channels in air and thereby produced luminous waves which traversed the defunct channels. The waves traveled more slowly along spark channels that had been allowed to decay for longer times.

7.6 RETURN-STROKE THEORIES

The lightning return stroke and its properties have been considered in numerous sections throughout this book. Of particular importance are the discussions of return-stroke photography in Secs. 2.3.4 and 2.3.5, of return-stroke electric field changes in Sec. 3.7.4, of return-stroke currents in Sec. 4.3, and of return-stroke spectroscopy in Sec. 5.5.2. In the present section we consider in a sequential fashion the physical events occurring in the return stroke.

When the negatively charged lightning leader approaches ground, a high electric field is produced between the leader head and ground. It is possible, therefore, for an upward-moving discharge to be initiated from the ground or from a sharp object. Such an upward-moving discharge would meet the negative leader at some point above ground and at this point the return stroke would be initiated. The existance of the positively charged, upward-moving discharge appears probable in view of the fact that positive "streamer" discharges can be initiated in the laboratory at lower applied electric fields that can negative discharges. Further, it is observed in the laboratory that the electrical breakdown between a negative rod and a positive plane takes place via a filamentary discharge which proceeds from the rod toward the plane and which is met near the plane by another filamentary discharge initiated from the plane. Direct observational evidence of the upward-moving discharge in lightning is relatively scanty. Golde (1967) presents two still photographs of close lightning discharges on which the probable meeting point of the upward-moving discharge and the downward-moving leader can be discerned. The two upward-moving discharges appear to have been about 15 m and about 30 m in length. Golde (1947) has reproduced and discussed a sketch of a Boys camera photograph due to Malan which shows an upward-moving discharge from the ground meeting the downward-moving stepped leader about 50 m above the flat earth. Golde also references two papers containing still-camera photographs which show evidence of upward-moving discharges of a few meters length above the flat earth. Observational evidence for the existence of upward-moving discharges of about 20 to 70 m in length in lightning strokes to towers has been presented in Sec. 2.5.4. It is probably reasonable to assume that for an electric field intensity at the ground of between 10^6 and 3×10^6 volts/m an upward-moving discharge will be initiated. The electric field intensity at the ground as a function of the height of the downward-moving leader above the ground can be calculated if a charge and charge distribution for the leader are assumed. Golde (1945, 1947) has used a charge-distribution model which assumes the charge on the leader channel decreases exponentially upward (decay distance 1 km) from a maximum at the leader tip. Golde considers a field of 10^6 volts/m sufficient to launch an upward-moving discharge from the flat earth. For 1 coul on the channel, Golde finds that the field at the ground is 10^6 volts/m when the leader tip is 17 m above the ground. For 4 coul on the channel, a field of 10^6 volts/m is produced when the leader is 60 m above the ground. A smaller field is probably necessary to launch an upward-moving discharge from a nonflat conductor such as a lightning rod, a TV antenna, or a transmission-line tower than from a flat conductor.

Whether the lightning leader strikes ground (as the dart leader may do) or whether it is met by a discharge carrying ground potential upward, the negatively charged leader whose tip potential may be of the order of 10^7 to 10^8 volts with respect to ground is effectively short-circuited to ground. Ground potential is then propagated up the channel. The propagating electric field present between the ground potential level and the channel potential above that level is the return-stroke wavefront. The return stroke probably propagates in the manner shown in Fig. 7.1c. Its propagation characteristics are similar to those of the dart leader in that it propagates into a pre-ionized region. Its propagation velocity has been calculated by Loeb (1958) using a theory identical to that presented in Sec. 7.5 for the dart leader. On the other hand, the return-stroke is a positively charged discharge while the dart leader is negatively charged. It is to be emphasized that although the return stroke is positively charged, there is effectively no motion of the positive ions within the channel during the return-stroke propagation phase due to their low mobility. For practical purposes, all current is carried by electrons. In order for the return-stroke wavefront to become charged positively, electrons must be removed from the regions where ionization processes have created equal numbers of negative electrons and positive ions. These electrons flow to ground, as shown in Fig. 7.1c, and constitute the "positive" stroke current measured at the ground.

As noted in Secs. 2.3.4 and 2.3.5, the velocity of the first return stroke decreases as it moves upward whereas the velocity of subsequent return strokes is relatively constant. Further, return strokes travel more rapidly and are brighter along branches than along portions of the main channel which are being traversed by it at the same time as the branches. Schonland (1938) has suggested that the observations stated above can be explained if the return-stroke wavefront propagation is slower in older sections of the channel. Schonland suggests that it is the electron density which controls the return-stroke velocity and that the electron density decreases with the age of the leader. The top of a stepped-leader channel is of the order of 10 msec old when the return stroke reaches it, whereas the top of a dart-leader channel is of the order of 1 msec old. Further, as the return stroke travels along a branch, it encounters a leader channel of increasingly recent age. Schonland (1938) nevertheless cautions that the high branch velocities and luminosities may also be due to the effects of a greater negative charge being present in the branches than in the main channel. It should also be noted that the heavy-particle density in the leader and the leader radius change with time, and that these changes may influence the return-stroke velocity. A discussion of these effects for the case of dart-leader propagation is given in Sec. 7.7

If the leader channel is composed of a conducting arc core surrounded by a corona envelope, the return stroke can be expected to propagate up the conducting arc core. The question then arises as to how the charge in the corona is transported into the stroke channel which will probably have a radius of the order of a few centimeters. Two points of view have been expressed regarding this matter. Wagner and Hileman (1958) suggest that filamentary discharges might be projected from the core into the corona and that the effective charge-transfer time due to these discharges would be microseconds. Thus, Wagner and Hileman envision a rapidly collapsing corona charge. An objection to this view is to be found in the fact that the charge transferred to ground during the return-stroke propagation is only a fraction of the charge on the leader. On the other hand, Pierce and Wormell (1953) suggest that the collapse of the corona envelope takes a time of the order of milliseconds (Sec. 3.7.4). In their view, the core is tied to ground potential while the corona envelope is at essentially cloud potential causing a radial corona current with a peak value of about 10^3 amp and a duration of milliseconds to flow. Calculations of the corona current to be expected according to this model have been made by Pierce (1958) and by Rao and Bhattacharya (1966). A slow corona-current flow can be used to explain the R_c field change (Section 3.7.4).

It is observed that the luminous stepped-leader radius is of the order of meters or more and that the return-stroke radius is at least an order of magnitude less and is probably only a few centimeters. Many authors (e.g., Schonland, 1956; Malan, 1963) attribute the rapid decrease in luminous radius to the magnetic pinch effect. Apparently, the return-stroke current is thought to flow initially within a meter radius and then to constrict itself. That this is very unlikely is apparent if a few numbers are inserted in the simple formulas for the quasi-equilibrium pinch effect (Uman, 1964). For the magnetic pressure at a radius of 1 m to exceed, say, 1 atm (to balance or overcome the kinetic overpressure induced by the return-stroke ionization and temperature), the current flowing within that radius must exceed 10^6 amp, an unreasonably high current. The magnetic pressure at the surface of the channel varies directly as the square of the current and inversely as the square of the channel radius. A more likely explanation for the apparent collapse of the leader channel is simply that the corona brightness decreases when the core potential is discharged. From this point of view there is only the appearance of a collapse as determined by the change in corona luminosity. The leader current and the return-stroke current both flow in a comparatively narrow channel. An alternative view of the leader collapse has been expressed by Loeb (1966) and is presented in Sec. 7.4.3.

The return-stroke wavefront traverses several meters of the leader channel in 0.1 μsec. We will consider now the physical events taking place along this short length of lightning channel. We assume that the leader channel has a conducting core and that the return stroke follows this core. The leader core is probably in equilibrium with the air around it and thus is at about atmospheric pressure. In a time short compared to the time in which the heavy gas atoms and ions can move appreciably, the return stroke supplies considerable energy to the leader core (or to the remains of the leader core). The core temperature, therefore, rises. The core (or stroke) pressure must rise also since the mass density in the core cannot change appreciably. The particle density in the core increases due to dissociation and ionization. If the core temperature increases an order of magnitude, say from 3000 to 30,000°K, the pressure will increase to above 10 atm. It is possible that at this time the magnetic pinch effect may become important. For a channel radius of 1 cm, a current of 8×10^4 amp must flow before the magnetic pressure at 1 cm exceeds a kinetic pressure of 10 atm; for a channel radius of 0.1 cm, the current must exceed 8×10^3 amp. We shall assume that magnetic effects are not important. Since the channel kinetic pressure exceeds the pressure of the surrounding air, the channel will expand. This expansion apparently takes place with supersonic speed and hence produces a roughly cylindrical shock wave which eventually becomes the thunder we hear (Secs. 6.3.1 and 6.3.2). The shock wave phase of the channel expansion lasts about 5 to 10 μsec. As the shock wave expands, the gas density in the current-carrying channel behind it decreases, and late in the shock wave phase, as we have seen in Sec. 5.5.2, the channel temperature is near 30,000°K. Some speculation regarding the electric field and power dissipation in the lightning channel during the shock wave phase is given in Sec. 7.3.3. Evidence for the channel model described above is derived from spectroscopic measurement of pressure in lightning and in long air sparks (Sec. 5.5.2; Orville et al., 1967), from thunder measurements (Sec. 6.3.1), from numerous laboratory measurements of spark-channel expansion (e.g., Flowers, 1943; Higham and Meek, 1950a,b; Norinder and Karsten, 1948), and from theory (Drabkina, 1951; Braginskii, 1958).

After the shock wave phase of the channel expansion is completed, the high-temperature low-density channel approaches, in microseconds or a few tens of microseconds, a state of approximate pressure equilibrium with the surrounding air (Somerville, 1959). Laboratory experiments by Flowers (1943) suggest that at this time the current density in the channel stabilizes at 10^7 amp/m², which is about two orders of magnitude greater than the current density in a free-burning arc in air. The channel now slowly expands and the current density slowly decreases to the value characteristic of a stable arc (Somerville, 1959). The electric field and power dissipation during the arc phase are considered in Sec. 7.3.3.

We can obtain a rough idea of the variation of the channel radius with time from the experiments of Norinder and Karsten (1948) and of Flowers (1943) and from the theory of Braginskii (1958). A typical lightning current measured at the ground rises to a peak of 10,000 to 20,000 amp in about 1 μsec or so and falls to half-value in about 40 μsec. It is not clear that the return-stroke current waveform above the ground is similar to that measured at the ground, but we shall assume that it is. Norinder and Karsten have measured the radius of a long oscillatory discharge in air whose current rose to a peak of 10,000 amp in 4 μsec (one-fourth period). Initially the luminous radius was very small. At current maximum it had expanded to about 0.5 cm, at the first current zero (8 μsec) to about 0.75 cm. Flowers found that a nonoscillatory spark, which reached a peak current of 20,000 amp in 3.3 μsec, exhibited a radius of 0.65 cm at that time. The theory of Braginskii, which relates radius to current, follows the experimental data quite closely during the shock wave phase if it is assumed that the conductivity of the channel is constant at 1.5×10^4 mhos/m (see Sec. 5.5.2) and that Braginskii's ξ factor is 4.5 (Hanlon, 1964). According to the theory of Braginskii, the typical lightning current just described would be characterized by a radius of about 1 cm at 10 μsec. The channel should reach a luminous radius of about 2 cm when the current is half peak value. (The 2-cm value comes from extrapolation of data of Norinder and Karsten and of Flowers.) It is of some interest to compare the lightning radii postulated with the radii determined from direct measurements made on lightning (Sec. 2.5.2). Some comments on the measurements are given by Loeb (1964). Additional evidence that the channel radius is of the order of centimeters is given in the following section and in Appendix D.

7.7 THE LIGHTNING CHANNEL BETWEEN STROKES

In Sec. 3.7.5 (see also Sec. 3.8) we considered the available information regarding the J and K processes occurring in the cloud between strokes. In the present section we will focus our attention on the channel between cloud base and ground during the interstroke period. In particular, we wish to know why the strokes in a flash generally follow the same channel even though that channel may be photographically dark for periods up to 100 msec between strokes.

The fact that most records of flashes to the Empire State Building indicate continuous current flow between current peaks led to the belief among some investigators (e.g., McEachron, 1940) that this was a necessary condition for the multiple-stroke mechanism. It is certainly reasonable from a physical point of view to expect that the aged stroke channel must maintain some degree of ionization if subsequent strokes are to follow the same channel. The most straightforward mechanism for the maintenance of this ionization would appear to be a low-level current flow between strokes. Brook et al. (1962) have discussed this possibility and have concluded that a current of the order of 10 amp might flow between strokes and not be detected by photographic or electric field measuring techniques. On the other hand, McCann (1944) on the basis of direct measurements of lightning currents states that in 21 out of 24 multiple-stroke flashes recorded the current fell below 0.1 amp between strokes. Loeb (1966) has presented an alternate mechanism for keeping the channel conducting between strokes. He suggests that nonluminous ionizing waves associated with K field changes traverse the channel between strokes and thereby keep it conducting.

Uman and Voshall (1968) have presented calculations to show that no special mechanism is needed to keep the lightning channel conducting between strokes. They have shown that in the absence of input energy to the channel the channel temperature will decay sufficiently slowly so that conditions conducive to the initiation and propagation of a dart leader will exist in the channel after a typical interstroke period. The

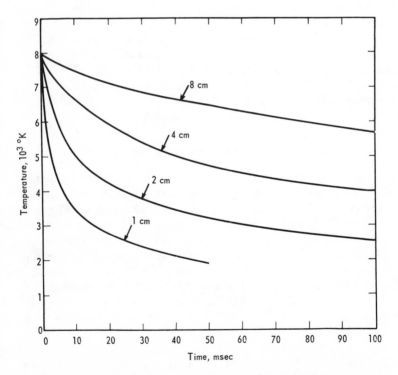

Fig. 7.2 *Decay of the central temperature for channel radii of 1, 2, 4, and 8 cm. The initial central temperature is 8,000°K. (From Uman and Voshall (1968).)*

channel temperature decay is viewed as a heat transfer problem. Energy must leave the channel before the channel temperature can decay. The channel temperature determines the degree of ionization and hence the conductivity. The cooling rate is found to be appropriately slow for channels of centimeter radius. In Fig. 7.2 the central temperature vs. time is shown for channels of several initial radii and an assumed temperature of 8000°K at current cessation. It was found that a given initial radius was maintained approximately constant during a typical interstroke time period. Larger channels take longer to decay because of their greater volume-to-surface ratio. The volume is a measure of the stored energy, the surface area a measure of the rate at which heat can be conducted away. At 4000°K, dry air is a reasonable conductor with a conductivity of about 1 mho/m and an electron density of about 10^{19} m^{-3}; at 2000°K, dry air is essentially an insulator with a conductivity of about 10^{-6} mho/m and an electron density of about 10^{13} m^{-3} (Yos, 1963). It would thus appear that between 2000 and 4000°K the channel is ripe for a dart leader since this is the conductor-insulator transition range and since channels with radii between 1 and 2 cm decay into this temperature range during a typical interstroke period. The centimeter channel radius so deduced is in good agreement with other estimates of the stroke radius (Secs. 7.6 and 2.5.2).

The behavior of the dart leader is influenced by the properties of the defunct stroke channel. The salient properties of the aged channel are its conductivity, its radius, and its heavy particle density. The channel conductivity and heavy particle density are probably the controlling factors in dart leader behavior since the channel radius changes little with time. Estimates of how the channel conductivity (electron density) could determine the dart leader velocity have been given in Sec. 7.5. The channel heavy-particle density near 3000°K at atmospheric pressure is about an order of magnitude less than that existing outside the channel. In the presence of a strong electric field, ionization rates and charged-particle velocities are determined by the heavy-particle density. For a given electric field provided by the dart leader, the low value of heavy-particle density in the channel relative to regions outside the channel may serve to make the channel a preferred path for the dart leader.

References

1. Braginskii, S. I.: Theory of the Development of a Spark Channel, *Sov. Phys. JETP* (*English Transl.*), **34**:1068–1074 (1958).

2. Brook, M.: private communication (1967).

3. Brook, M., N. Kitagawa, and E. J. Workman: Quantitative Study of Strokes

and Continuing Currents in Lightning Discharges to Ground, *J. Geophys. Res.*, **67**:649–659 (1962).

4. Bruce, C. E. R.: The Lightning and Spark Discharges, *Nature*, **147**:805–806 (1941).

5. Bruce, C. E. R.: The Initiation of Long Electrical Discharges, *Proc. Roy. Soc.* (*London*), **A183**:228–242 (1944).

6. Clarence, N. D., and D. J. Malan: Preliminary Discharge Processes in Lightning Flashes to Ground, *Quart. J. Roy. Meteorol. Soc.*, **83**:161–172 (1957).

7. Cravath, A. M., and L. B. Loeb: The Mechanism of the High Velocity of Propagation of Lightning Discharges, *Physics* (now *J. Appl. Phys.*), **6**:125–127 (1935).

8. Davies, A. J., C. J. Evans, and F. Llewellyn-Jones: Electrical Breakdown of Gases: The Spatio-temporal Growth of Ionization in Fields Distorted by Space Charge, *Proc. Roy. Soc.* (*London*), **A281**:164–183 (1964).

9. Drabkina, D. I.: The Theory of the Development of the Spark Channel, *J. Exptl. Theoret. Phys.* (*USSR*), **21**:473–483 (1951). (English translation, AERE Lib/Trans. 621, Harwell, Berkshire, England.)

10. Dutton, J., S. C. Haydon, and F. Llewellyn-Jones: Photo-ionization and the Electrical Breakdown of Gases, *Proc. Roy. Soc.* (*London*), **A218**:206–223 (1953).

11. English, W. M.: Corona from Water Drops, *Phys. Rev.*, **74**:179–189 (1948).

12. Flowers, J. W.: The Channel of the Spark Discharge, *Phys. Rev.*, **64**:225–235 (1943).

13. Frommhold, L.: Über verzögerte Elektronen in Elektronenlawinen, insbesondere in Sauerstoff und Luft, durch Bildung und Zerfall negativer Ionen (O^-), *Fortschr. Physik*, **12**:597–643 (1964).

14. Golde, R. H.: The Frequency of Occurrence and the Distribution of Lightning Flashes to Transmission Lines, *Trans. AIEE*, **64**:902–910 (1945).

15. Golde, R. H.: Occurrence of Upward Streamers in Lightning Discharges, *Nature* **160**:395–396 (1947).

16. Golde, R. H., "The Lightning Conductor," *J. Franklin Inst.*, **283**:451–477 (1967).

17. Griscom, S. B.: The Prestrike Theory and Other Effects in the Lightning Stroke, *Trans. AIEE*, **77**(pt. 3):919–933 (1958).

18. Hanlon, J.: private communication (1964).

19. Higham, J. B., and J. M. Meek: The Expansion of Gaseous Spark Channels, *Proc. Phys. Soc.* (*London*), **B63**:649–661 (1950a).

20. Higham, J. B., and J. M. Meek: Voltage Gradients in Long Gaseous Spark Channels, *Proc. Phys. Soc.* (*London*), **B63**:633–648 (1950b).

21. Horner, F.: Radio Noise from Thunderstorms, in J. A. Saxton (ed.), "Advances in Radio Research," vol. 2, pp. 121–204, Academic Press Inc., New York, 1964.

22. King, L. A.: The Voltage Gradient of the Free Burning Arc in Air or Nitrogen, *Proc. Intern. Conf. Ionization Phenomena Gases, 5th, Munich, 1961*, pp. 871–877, North Holland Publishing Company, Amsterdam, 1962.

23. Köhrmann, W.: Der Stromanstieg einer Townsend-entladung unter dem Einfluss Raumladung, *Z. Naturforsch.*, **19a**:245–253 (1964a).

24. Köhrmann, W.: Die zeitliche Entwicklung der Townsend-entladung bis zum Durchslag, *Z. Naturforsch.*, **19a**:926–933 (1964b).

25. Komelkov, V. S.: Structure and Parameters of the Leader Discharge, *Bull. Acad. Sci. USSR, Tech. Sci. Sect.*, no. 8, pp. 955–966, 1947. In Russian.

26. Komelkov, V. S.: The Development of Electric Discharges in Long Gaps, *Bull. Acad. Sci. USSR, Tech. Sci. Sect.*, no. 6, pp. 851–865, 1950. In Russian.

27. Kontaratos, A. N.: On the Functional Dependence of Townsend's First Ionization Coefficient, *Appl. Sci. Res., Sect. B*, **12**:27–32 (1965).

28. Krider, E. P., G. A. Dawson, and M. A. Uman: Peak Power and Energy Dissipation in a Single-stroke Lightning Flash, *J. Geophys. Res.*, **73**:3335–3339 (1968).

29. Little, P. F.: Secondary Effects, "Handbuch der Physik," vol. 21, pp. 574–663, Springer-Verlag OHG, Berlin, 1956.

30. Llewellyn-Jones, F.: "Ionization and Breakdown in Gases," John Wiley & Sons, Inc., New York, 1957.

31. Llewellyn-Jones, F.: "Ionization Avalanches and Breakdown," Methuen & Co., Ltd., London, 1967.

32. Loeb, L. B.: Experimental Contributions to the Knowledge of Charge Generation, in H. R. Byers (ed.), "Thunderstorm Electricity," pp. 150–192, University of Chicago Press, Chicago, 1953.

33. Loeb, L. B.: The Positive Streamer Spark in Air in Relation to the Lightning Stroke, in H. G. Houghton (ed.), "Atmospheric Explorations," pp. 46–75, John Wiley & Sons, Inc., New York, 1958.

34. Loeb, L. B.: "Basic Processes of Gaseous Electronics," University of California Press, Berkeley, Calif., 1961.

35. Loeb, L. B.: Discussion of Paper by M. A. Uman, 'The Diameter of Lightning,' *J. Geophys. Res.*, **69**:587–589 (1964).

36. Loeb, L. B.: Ionizing Waves of Potential Gradient, *Science*, **148**:1417–1426 (1965a).

37. Loeb, L. B.: "Electrical Coronas, Their Basic Physical Mechanisms," University of California Press, Berkeley, Calif., 1965b.

38. Loeb, L. B.: The Mechanisms of Stepped and Dart Leaders in Cloud-to-ground Lightning Strokes, *J. Geophys. Res.*, **71**:4711–4721 (1966).

39. Loeb, L. B., and J. M. Meek: "The Mechanism of the Electric Spark," Stanford University Press, Stanford, Calif., 1941.

40. Macky, W. A.: Some Investigations on the Deformation and Breaking of Water Drops in Strong Electric Fields, *Proc. Roy. Soc. (London)*, **A133**:565–587 (1931).

41. Malan, D. J.: "Physics of Lightning," The English Universities Press Ltd., London, 1963.

42. Malan, D. J., and B. F. J. Schonland: The Distribution of Electricity in Thunderclouds, *Proc. Roy. Soc. (London)*, **A209**:158–177 (1951).

43. Massey, H. S. W., and E. H. S. Burhop: "Electronic and Ionic Impact Phenomena," Oxford University Press, London, 1952.

44. McCann, G. D.: The Measurement of Lightning Currents in Direct Strokes, *Trans. AIEE*, **63**:1157–1164 (1944).

45. McDaniel, E. W.: "Collision Phenomena in Ionized Gases," John Wiley & Sons, Inc., New York, 1964.

46. McEachron, K. B.: Wave Shapes of Successive Lightning Current Peaks, *Elec. World*, 56–59, (January 10, 1940).

47. Miyoshi, Y.: Theoretical Analysis of Buildup of Current in Transient Townsend Discharge, *Phys. Rev.*, **103**:1609–1618 (1956).

48. Miyoshi, Y.: Development of Space Charge and Growth of Ionization in the Transient Townsend Discharge, *Phys. Rev.*, **117**:355–365 (1960).

49. Müller-Hillebrand, D.: The Protection of Houses by Lightning Conductors— A Historical Review, *J. Franklin Inst.*, **274**:34–54 (1962).

50. Norinder, H., and O. Karsten: Experimental Investigations of Resistance and Power within Artificial Lightning Current Paths, *Arkiv Mat.*, **36**:1–48 (1948).

51. Orville, R. E., M. A. Uman, and A. M. Sletten: Temperature and Electron Density in Long Air Sparks, *J. Appl. Phys.*, **38**:895–896 (1967).

52. Phelps, A. V.: Role of Molecular Ions, Metastable Molecules, and Resonance Radiation in the Breakdown of Rare Gases, *Phys. Rev.*, **117**:619–632 (1960).

53. Pierce, E. T.: The Development of Lightning Discharges, *Quart. J. Roy. Meteorol. Soc.*, **81**:229–240 (1955).

54. Pierce, E. T.: Some Topics in Atmospheric Electricity, in L. G. Smith (ed.), "Recent Advances in Atmospheric Electricity," pp. 5–16, Pergamon Press, New York, 1958.

55. Pierce, E. T., and T. W. Wormell: Field Changes due to Lightning Discharges, in H. R. Byers (ed.), "Thunderstorm Electricity," pp. 251–266, University of Chicago Press, Chicago, 1953.

56. Plonsey, R., and R. E. Collin: "Principles and Applications of Electromagnetic Fields," McGraw-Hill Book Company, New York, 1961.

57. Raether, H.: "Electron Avalanches and Breakdown in Gases," Butterworths, Washington, D.C., 1964.

58. Rao, M., and H. Bhattacharya: Lateral Corona Currents from the Return Stroke Channel and the Slow Field Change after the Return Stroke in a Lightning Discharge, *J. Geophys. Res.*, **71**:2811–2814 (1966).

59. Schonland, B. F. J.: Progressive Lightning, pt. 4, The Discharge Mechanism, *Proc. Roy. Soc. (London)*, **A164**:132–150 (1938).

60. Schonland, B. F. J.: The Pilot Streamer in Lightning and the Long Spark, *Proc. Roy. Soc. (London)*, **A220**:25–38 (1953).

61. Schonland, B. F. J.: The Lightning Discharge, "Handbuch der Physik," vol. 22, pp. 576–628, Springer-Verlag OHG, Berlin, 1956.

62. Schonland, B. F. J.: Lightning and the Long Electric Spark, *Advan. Sci.*, November, 1962, pp. 306–313.

63. Schonland, B. F. J., D. J. Malan, and H. Collens: Progressive Lightning, pt. 2, *Proc. Roy. Soc. (London)*, **A152**:595–625 (1935).

64. Somerville, J. M.: "The Electric Arc," pp. 101–119, John Wiley & Sons, Inc., New York, 1959.

65. Townsend, J. S.: "Electrons in Gases," Hutchinson & Co. (Publishers), Ltd., London, 1947.

66. Uman, M. A.: "Introduction to Plasma Physics," pp. 184–193, McGraw-Hill Book Company, New York, 1964.

67. Uman, M. A., R. E. Orville, and L. E. Salanave: The Density, Pressure and Particle Distribution in a Lightning Stroke near Peak Temperature, *J. Atmospheric Sci.*, **21**:306–310 (1964).

68. Uman, M. A., and R. E. Voshall: Time-interval between Lightning Strokes and the Initiation of Dart Leaders, *J. Geophys. Res.*, **73**:497–506 (1968).

69. Wagner, C. F.: Lightning and Transmission Lines, *J. Franklin Inst.*, **283**:558–594 (1967).

70. Wagner, C. F., and A. R. Hileman: The Lightning Stroke, *Trans. AIEE*, **77** (pt. 3):229–242 (1958).

71. Wagner, C. F. and A. R. Hileman: The Lightning Stroke (2), *Trans. AIEE*, **80**(pt. 3):622–642 (1961).

72. Wagner, K. H.: Die Entwicklung der Electronenlawine in den Plasmakanal, untersucht mit Bildverstärker und Wischverschluss, *Z. Physik*, **189**:465–515 (1966).

73. Wagner, K. H.: Vorstadium des Funkens, untersucht mit dem Bildverstärker, *Z. Physik*, **204**:177–197 (1967).

74. Ward, A. L.: Calculations of Electrical Breakdown in Air at Near-atmospheric Pressure, *Phys. Rev.*, **138**:A1357–A1362 (1965a).

75. Ward, A. L.: The Purported Transition from Avalanche to Streamer Breakdown, *Physics*, **1**:215–217 (1965b).

76. Winn, W. P.: A Laboratory Analog to the Dart Leader and Return Stroke of Lightning, *J. Geophys. Res.*, **70**:3265–3270 (1965).

77. Yos, J. M.: Transport Properties of Nitrogen, Hydrogen, Oxygen, and Air to 30,000°K, *Tech. Mem. RAD-TM-63-7*, Avco Corporation, Wilmington, Massachusetts, March 22, 1963. (Available from Defense Documentation Center as AD 435053.)

Appendix A: Some Suggestions for
Future Lightning Research

Most of the measurements made on the lightning discharge can be assigned to one of five general categories: (1) Photographic measurements (Chap. 2), (2) electric and magnetic field measurements (Chap. 3), (3) electric current measurements (Chap. 4), (4) spectroscopic measurements (Chap. 5), and (5) acoustic measurements (Chap. 6).   In each of these areas of research more and better data are needed.   In this appendix specific suggestions for future lightning research are presented.   References to the literature given in this appendix are to be found in the bibliographies at the ends of the appropriate chapters.

1. *Photographic measurements:* Lightning photography with photographic film should be supplemented by "photography" by photocell and by electronic image-converter and intensifier cameras, thus providing an increase in sensitivity and time resolution from previous work.   It would be worthwhile to obtain data on the absolute value of light output from the channels of the stepped leader, dart leader, and return stroke over the total time duration of these discharges.   Measurements of the luminous formation of stepped-leader steps and of the propagation of luminosity associated with M components would be most valuable. More photographic data are needed regarding downward-moving leaders carrying positive charge as well as upward-moving leaders carrying both charge signs.

2. *Electric and magnetic field measurements:* More electric field measurements of stepped-leader field changes and appropriate analyses of these

measurements are needed. A good starting point for these studies would be a detailed examination of the prestroke field changes which Clarence and Malan (1957) have labeled B, I, and L (Sec. 3.7.2).

There are some differences among investigators as to the field changes observed between strokes and the interpretation of those field changes (Sec. 3.7.5). For example, Malan (1965) reports that a slow, negative interstroke field change is observed in South Africa at distances of 25 to 100 km. Apparently no other investigator has reported a slow, negative field change between strokes for strokes observed at considerable distances. More measurements are needed.

Electric field measurements have been applied to the study of the intracloud discharge with the result that various investigators have obtained conflicting results (Sec. 3.8). The original work of Workman et al. (1942) indicated that the cloud discharges were essentially horizontal. Later Reynolds and Neill (1955) reported that the cloud discharges were usually inclined at about a 45° angle from the vertical and involved a moment change of the order of 10 coul-km. Similar moment changes have been found by Tamura et al. (1958). On the other hand, the majority of cloud-discharge field-change studies indicate a moment change of the order of 100 coul-km (Wilson, 1920; Wormell, 1939; Pierce, 1955a,b). Most researchers agree that in the cloud discharge usually the positive charge center is located above the negative charge center. Smith (1957) found that usually negative charge was raised during the discharge. Takagi (1961) and Ogawa and Brook (1964) found that positive charge was usually lowered. There are numerous other conflicting data regarding the cloud discharge and little doubt that more information on cloud discharges is needed.

Magnetic field measurements have been made by Hatakeyama (1936), by Meese and Evans (1962), and by Williams and Brook (1963) (Sec. 3.9). Williams and Brook (1963) report data on continuing currents and on stepped-leader currents obtained from magnetic field measurements. Additional measurements of this type would be very valuable. From magnetometer measurements, Meese and Evans (1962) derived values of charge transferred by a flash which were about an order of magnitude greater than those values usually determined from electric-field-change data. A resolution of this apparent discrepancy would be desirable.

In calculating the electric and magnetic energy radiated due to various lightning processes, the dipole approximation (Sec. 3.4) is almost always used. It would be of some value to investigate the validity and effects of this approximation, particularly in its application to charge motion within the cloud (K changes) (Arnold and Pierce, 1964) and to leader-step phenomenon (Schonland, 1953; Hodges, 1954).

More radar studies of intracloud lightning processes of the type carried out by Hewitt (1957) (Sec. 3.10) could provide valuable information about cloud discharges and about processes going on in the cloud during the time between cloud-to-ground strokes.

The radiation fields from lightning should be analyzed so as to determine stroke properties.

3. *Electric current measurements:* The very fine oscillographic data obtained recently by Berger and his co-workers (Berger and Vogelsanger, 1965) in Switzerland have added greatly to our knowledge of lightning currents. More detailed and corroborating data on stroke currents and their rise times are needed.

More statistical data on interstroke continuing currents are needed. It is important not to confuse interstroke continuing currents with the currents due to upward-initiated discharges. The latter currents were originally termed continuing currents by Hagenguth and Anderson (1952) who present statistical data for them.

In the absence of continuing currents, do small currents flow in the lightning channel in the time between strokes? McCann (1944) has reported that in most of the multiple-stroke flashes studied by him the channel current fell below 0.1 amp between strokes. Berger (1967) has reported that the current between strokes was measured to be less than 1 amp. Corroborating data would be valuable.

Leaders currents are generally estimated from charge motion and charge-motion times derived from electric-field-change data. Currents due to upward-moving leaders have been recorded by Berger and Vogelsanger (1965); a few currents due to downward-moving stepped leaders have been determined from magnetic field measurements by Williams and Brook (1963). More data are needed on currents in upward-moving and downward-moving stepped leaders and in dart leaders.

Almost all measurements of the time variation of lightning current, with the exception of the work of Norinder and co-workers, have been made near ground level and represent the current flowing at the base of the lightning channel. It is not apparent that the current in the channel an arbitrary height above the ground should be identical or even very similar to the current measured at ground level (Sec. 4.1). It would certainly be worthwhile to obtain lightning current waveshapes for positions in the channel above the ground contact. Perhaps an adequate theory of the return stroke will enable one to calculate the current in the channel.

There appear to be no studies of the currents associated with M components. Such studies would be valuable.

4. *Spectroscopic measurements:* Spectroscopic studies of the cloud-to-ground lightning return stroke with about 10-Å wavelength resolution have recently yielded values for the stroke temperature and other properties on a 2 to 5 μsec time scale (Orville, 1966, 1968). It would be very desirable to obtain similar information on a shorter time scale and with

better wavelength resolution. Better time-resolution would allow a more detailed determination of the temperature variation and a more accurate determination of the peak temperature. Better wavelength resolution would make possible an accurate determination of channel opacity as well as allowing an electron density determination to be made from measured Stark broadening of certain nitrogen ion lines. Data to corroborate that obtained from the visible spectrum should be obtained by extending the spectral range of the lightning spectroscopic studies as far into the ultraviolet and infrared as is possible. It would be most valuable to obtain spectroscopic data correlated with data on electrical current and electric field changes. With these data, thermodynamic properties of the channel such as temperature could be correlated with electrical properties such as current or charge transfer.

It is important to obtain spectra of the stepped and dart leaders. Such data are needed in the formulation of realistic theories of the leaders. See Appendix D for a report on the first stepped-leader spectrum obtained.

In order to obtain time-resolved spectral data, the spectrometer must be coupled to a recording device. In most of the previous time-resolved lightning studies the recording was done on photographic film which was moved at high velocity. Better time-resolution and sensitivity could be obtained by recording the spectrum with electronic image converter or image intensifier cameras. Further, photomultiplier systems could be used to examine certain spectral features with good time-resolution over many orders of magnitude of luminosity.

5. *Acoustic measurements:* The theory of thunder given by Few et al. (1967) should be verified by additional experiments. In particular, acoustic measurements should be made very close to the lightning channel, and new studies of the frequency spectrum of distant thunder should be undertaken. A check on the infrasonic spectrum reported by Bhartendu (1964) (Fig. 6.7) is needed. It would be worthwhile to determine if there are differences between the acoustic spectra of cloud strokes and of ground strokes. Such information would apparently make possible a calculation of the ratio of energy input of a cloud discharge to that of a ground discharge.

Correlated measurements of thunder duration and channel length should be made to determine if the thunder duration is due to the time difference of the sound reaching the observer from opposite ends of the channel as Latham's (1964) data would suggest or if the thunder duration is longer than would be expected from the usual channel length as argued by Remillard (1960).

Three-microphone acoustic measurements and correlated stereophotography should be used to identify those features of the lightning channel which are responsible for the claps and peals; and hence to verify or deny the hypothesis presented in Sec. 6.3.2 regarding their origin.

Appendix B: Bead Lightning

Bead or chain lightning (*éclair en chapelet; Perlschnurblitz*) is a visually well-documented phenomenon in which the lightning channel to ground breaks up, or appears to break up, into luminous fragments generally reported to be some tens of meters in length.   These beads appear to persist for a longer time than does the usual cloud-to-ground discharge channel.   Bead lightning occurs relatively infrequently, and apparently no reliable photograph of the phenomenon has been published in the literature.   Further, it is not clear that one would expect to record a beaded effect using a still camera.   If the channel is first continuous and then breaks up into beads, the image recorded on the film will be a super-position of the continuous channel and the beads.   The beads will only be apparent on the photograph if the energy output of the beads to which the film is sensitive is at least a small fraction of the light output of the channel prior to the formation of beads and if the film is not overexposed. Still-camera photographs of what may be bead lightning have been published by Scheminzky and Wolf (1948) and by Matthias and Buchsbaum (1962).   In neither case was the phenomenon in question seen by the observer.   In both cases the regions between the beads are dark.   The photograph of Scheminzky and Wolf shows 73 beads each of about 10 m length centered along what appears to be a black channel.   The origin of the black channel is unknown.   The photograph of Matthias and Buchs-baum shows three beads each of a different length at the top of the photo-graph with no visible channel beneath to ground.   A high-speed motion picture, taken by the U.S. Navy, of lightning striking a plume of water rising from a depth charge (Brook et al., 1961; Young, 1962) shows in the final frames of each of three strokes a channel which has broken up into light and dark sections. The sections are a few meters in length. Unfortunately, neither Brook et al. nor Young gives a photograph showing the beaded effect.

Various theories of bead lightning have been set forth and some of these are now considered:

1. The visual appearance of bead lightning is caused by viewing portions of the lightning channel end-on. That is, if the lightning channel is coming toward or going away from an observer, the observer will see a greater length of the channel within a given viewing angle than would be the case were the channel perpendicular to his viewing direction. The greater length of channel appears to the eye as a normal channel of greater-than-normal brightness. To account for the beaded effect, the channel must periodically be slanted toward or away from the observer.

2. Bead lightning is due to the periodic masking of a normal cloud-to-ground channel by clouds or rain.

3. Bead lightning is due to a pinch-effect instability by which the current-carrying channel is distorted into a "string of sausages" with the strong light emission coming from the "necked-off" regions of high-current density and high-particle density (Uman, 1962). As noted in Sec. 7.6, it is possible for the magnetic-pinch effect to occur if the lightning current is high enough while the channel radius is small. For example, for a channel radius of 1 cm, a current of 8×10^4 amp must flow before the magnetic pressure at the surface of the channel exceeds 10 atm. For a channel radius of 0.1 cm, the current must exceed 8×10^3 amp. Whether conditions conducive to a magnetic pinch ever do occur in lightning is not known. To obtain a series of beads by the pinch-effect mechanism, the channel must somehow be conditioned periodically as a function of height. Possibly, the periodicity is just an accident of nature.

4. Bead lightning is a series of spherical arcs formed out of a defunct lightning channel by a resurgence of channel current (Uman and Helstrom, 1966). Calculations describing one of these spherical arcs (ball lightning?) are presented by Uman and Helstrom (1966). Each spherical arc is formed by the constriction of a diffuse cloud-to-ground current in a region of high temperature and high conductivity, a sort of thermal pinch effect operating in space. The constricted current and the energy input associated with it serve to maintain the arc.

5. Bead lightning is due to the long luminous lifetimes of sections of the lightning channel of exceptionally large radius (Uman and Voshall, 1968). As discussed in Sec. 7.7 (see also Fig. 7.2), channels of large radius take longer to cool than channels of small radius. To a first approximation the cooling time is proportional to the square of the radius.

If the lightning channel radius were somehow periodically modulated as a function of height, then as the channel luminosity decayed, the channel would take on the appearance of a string of beads. Perhaps this modulation occurs accidentally when the channel consists of a large number of kinks or bends.

In order to account for the observed long persistence of the beads, those theories which require current flow may be invoked in conjunction with long continuing currents. Perhaps the observed long persistence is due to the eye, rather than to the channel.

References

1. Brook, M., G. Armstrong, R. P. H. Winder, B. Vonnegut, and C. B. Moore: Artificial Initiation of Lightning Discharges: *J. Geophys. Res.*, **66**:3967–3969 (1961).

2. Matthias, B. T., and S. J. Buchsbaum: Pinched Lightning, *Nature*, **194**:327 (1962).

3. Scheminzky, F., and F. Wolf: Photographie eines Perlschnurblitzes, *Sitzber. Akad. Wiss. Wien Abt. IIA*, **156**(1–2):1–8 (1948).

4. Uman, M. A.: Bead Lightning and the Pinch Effect, *J. Atmospheric Terrest. Phys.*, **24**:43–45 (1962).

5. Uman, M. A., and C. W. Helstrom: A Theory of Ball Lightning, *J. Geophys. Res.*, **71**:1975–1984 (1966).

6. Uman, M. A., and R. E. Voshall: Time Interval between Lightning Strokes and the Initiation of Dart Leaders, *J. Geophys. Res.*, **73**:497–506 (1968).

7. Young, G. A.: A Lightning Strike of an Underwater Explosion Plume, *U.S. Naval Ordinance Lab., NOLTR 61-43*, March, 1962.

Appendix C: Ball Lightning

Ball lightning (*boules de feu* or *foudre sphérique; Kugelblitz*) is the name given to the mobile luminous spheres which have been observed during thunderstorms. A typical ball lightning is about the size of an orange or a grapefruit and has a lifetime of a few seconds. Compilations of eye-witness reports of ball lightning have been published by Brand (1923), Rodewald (1954), Dewan (1964), Silberg (1965), McNally (1966), and Rayle (1966), among others. Visual sightings are often accompanied by sound, odor, and permanent material damage, and hence it would appear difficult to deny the reality of the phenomenon [as Humphreys (1936) has done]. In a letter to the editor of the London *Daily Mail*, Morris (1936) described an unusual incident in which a ball lightning caused a tub of water to boil:

> *During a thunderstorm I saw a large, red hot ball come down from the sky. It struck our house, cut the telephone wire, burnt the window frame, and then buried itself in a tub of water which was underneath.*
>
> *The water boiled for some minutes afterwards, but when it was cool enough for me to search I could find nothing in it.*

Photographs purported to be of ball lightning have been published by Jensen (1933), Kuhn (1951), Wolf (1956), Davidov (1958), Jennings (1962), and Müller-Hillebrand (1963).

A phenomenon very similar to, if not identical with, ball lightning has been reported to occur in submarines due to the discharge of a high current (about 150,000 amp direct current from a 260-volt source) across a circuit breaker (Silberg, 1962). In addition, the writer has received a number of reports of ball-lightning-like phenomena being initiated accidentally in high-power electrical equipment.

It is generally thought that ball lightning is a rare phenomenon. However, Rayle (1966) states that

> *Surveys of NASA Lewis Research Center personnel were conducted to obtain information about ball lightning occurrences. A comparison of the frequency of observation of ball lightning with that of ordinary lightning impact points reveals that ball lightning is not a particularly rare phenomenon. Contrary to widely accepted ideas, the occurrence of ball lightning may be nearly as frequent as that of ordinary cloud-to-ground strokes.*

Ball lightning and St. Elmo's fire are sometimes confused. St. Elmo's fire is a corona discharge from a pointed conducting object in a strong electric field. Like ball lightning, St. Elmo's fire may assume a spherical shape. Unlike ball lightning, St. Elmo's fire must remain attached to a conductor, although it may exhibit some motion along the conductor. Further, St. Elmo's fire can have a lifetime much greater than the lifetime of the usual ball lightning.

From the many published ball lightning observations, it is possible to compile a list of ball lightning characteristics:

1. *Occurrence:* Most observations of ball lightning are made during thunderstorm activity. Most, but not all, of thunderstorm-related ball lightnings appear almost simultaneously with a cloud-to-ground lightning discharge. These ball lightnings appear within a few meters of the ground. Sometimes ball lightnings are reported to appear near ground in the absence of a lightning discharge. Ball lightnings have also been observed to hang in mid-air far above the ground and have been observed falling from a cloud toward the ground.

2. *Appearance:* Ball lightnings are generally spherical, although other shapes have been reported. They are usually 10 to 20 cm in diameter, with reported diameters ranging from 1 to over 100 cm. Ball lightnings come in various colors, the most common colors being red, orange, and yellow. Ball lightnings are generally not exceptionally bright, but can be seen clearly in daylight. They are usually reported to maintain a relatively constant brightness and size during their lifetimes, although ball lightnings which change in brightness and size are not uncommon.

3. *Lifetime:* Ball lightnings generally have a lifetime of less than 5 sec. A small fraction of ball-lightning reports indicate a lifetime of over a minute.

4. *Motion:* Ball lightnings usually move horizontally at a velocity of a few meters per second. They may also remain motionless in mid-air or may descend from a cloud toward the ground. They do not often rise,

as would be the case if they were spheres of hot air at atmospheric pressure in the presence of only a gravitational force. Many reports describe ball lightnings which appear to spin or rotate as they move. Ball lightnings are sometimes reported to bounce off solid objects, typically the ground.

5. *Heat, sound, and odor:* Rarely do observers of ball lightning report the sensation of heat. However, accounts of ball lightnings which burned barns and melted wires do exist. One report found in McNally (1966) describes a ball lightning which hit a pond of water with a sound "as if putting a red hot piece of iron into the water." Sometimes ball lightnings are reported to emit a hissing sound. Many observers report a distinctive odor accompanying ball lightning. The odor is usually described as sharp and repugnant, resembling ozone, burning sulphur, or nitric oxide.

6. *Attraction to objects and enclosures:* Ball lightnings are often reported to be attached to metallic objects such as wire fences or telephone lines. When attached to metallic objects, they generally move along those objects. Some or all these observations may refer to a type of St. Elmo's fire. Ball lightnings often enter houses through screens or chimneys. Sometimes they are reported to enter houses through glass window panes. They are also reported to originate within buildings, on occasion from telephones. Ball lightnings can exist in an all-metal enclosure such as the interior of an airplane (Uman, 1968).

7. *Demise:* Ball lightnings decay in one of two modes, either silently or explosively. The explosive decay takes place rapidly and is accompanied by a loud noise. The silent decay can take place either rapidly or slowly. The majority of ball lightnings apparently exhibit a rapid decay. After the ball has decayed, it is sometimes reported that a mist or residue remains. Occasionally a ball lightning has been observed to break up into two or more smaller ball lightnings.

8. *Types:* There may be more than one type of ball lightning. For example, the ball lightning that attaches to conductors may be different from the free-floating ball lightning; and the ball lightning that appears near ground level may be different from the ball lightning that hangs high in the air or the ball lightning that falls out of a cloud.

We look now at the theories which have been proposed to "explain" ball lightning. Unfortunately, no theory of ball lightning exists which can account for both the degree of mobility that the ball exhibits and for the fact that it does not rise. Thus, despite the numerous theoretical models proposed for the phenomenon, the mechanisms which cause the ball lightning remain unknown. All ball lightning theories fall into one of two general classes: Those whose energy source for the ball is outside the ball (externally powered ball lightning) and those whose energy source for the ball is stored within the ball itself (internally powered ball lightning).

We look first at the internally powered models. Within this class of models there are essentially six subclasses: (1) The ball lightning is gas or air behaving in an "unusual" way. It has been suggested that the ball lightning is slowly burning gas, is the radiation from the slow recombination of unspecified ions existing in the ball, is the radiation from long-lived metastable states of air particles or from particles which absorb energy from the metastables, is due to chemical reactions involving dust, soot, etc., and so on. (2) Ball lightning is a sphere of heated air at atmospheric pressure. Uman and Lowke (1968), using an extension of the theory discussed in Sec. 7.7, have calculated the temporal and spatial characteristics of an isolated sphere of hot air. It was found that for a sphere of about 20 cm diameter, the cooling rate was about $1000°K/sec$ in the temperature range near $3000°K$ and that the sphere maintained an essentially constant radius during the cooling process. Unfortunately, the relatively slow cooling rate apparently does not lead to a relatively constant ball brightness. (3) Ball lightning is a very high-density plasma (electron density of 10^{25} m^{-3}) which exhibits quantum mechanical properties characteristic of the solid state (Neugebauer, 1937). (4) Ball lightning is due to one of several suggested configurations of closed-loop current flow contained by its own magnetic field. Finkelstein and Rubinstein (1964) have shown that plasma containment of this type is not possible under normal conditions in air. (5) Ball lightning is due to some sort of air vortex (like a smoke ring) providing containment for luminous gases. (6) Ball lightning is a microwave radiation field contained within a thin spherical shell of plasma (Dawson and Jones, 1968).

We look now at those ball lightning theories which provide the ball with an external power source. Three types of power sources have been suggested: (1) A high-frequency (hundreds of megacycles) electromagnetic field, (2) a steady current flow from cloud to ground, and (3) focused cosmic-ray particles. (1) Cerrillo (1943) and Kapitza (1955) proposed that focused rf energy from the thundercloud could create and maintain a ball lightning. The high electric fields necessary to effect this mechanism have never been observed in thunderstorms. (2) Finkelstein and Rubinstein (1964) and Uman and Helstrom (1966) have suggested that a steady current flowing from cloud-to-ground would contract in cross section in a region of high conductivity (the ball) and that the increased energy input due to the constriction of current could maintain the ball. This type of theory cannot account for the existence of ball lightning inside structures, particularly inside metal structures. (3) Arabadzhi (1957) has suggested that radioactive cosmic-ray particles could be focused by the electric fields of the thundercloud so that they would create an air discharge at one point in space.

The ball lightning literature is voluminous and to do justice to the subject would require a separate book.   No attempt at completeness has been made in this appendix.   For the reader who wishes to investigate the subject further, the references cited in this appendix should provide a satisfactory starting point.

References

1. Arabadzhi, V. I.: The Theory of Atmospheric Electricity Phenomena, *Uch. Zap. Minsk. Gos. Univ., im A. M. Gor'kogo, Ser. Fiz.-Mat.*, no. 5, 1957.   (Translation available as RJ-1314 from Associated Technical Services, Inc., Glen Ridge, New Jersey.)
2. Brand, W.: "Der Kugelblitz," Grand, Hamburg, Germany, 1923.
3. Cerrillo, M.: Sombre las posibles interpretaciones electromagneticas del fenomena de las centellas, *Comision Impulsora Coordinadora Invest. Cient., Mexico, Ann.*, **1**:151–178 (1943).
4. Davidov, B., Rare Photograph of Ball Lightning, *Priroda*, **47**:96–97 (1958).   In Russian.
5. Dawson, G. A., and R. C. Jones: Ball Lightning as a Radiation Bubble, Fourth International Conference on the Universal Aspects of Atmospheric Electricity, Tokyo, Japan, May, 1968.
6. Dewan, E. M.: Eyewitness Accounts of Kugelblitz, Microwave Physics Laboratory, *Air Force Cambridge Res. Lab., CRD-125*, March, 1964.
7. Finkelstein, D., and J. Rubinstein: Ball Lightning, *Phys. Rev.*, **135**:A390–A396 (1964).
8. Humphreys, W. J.: Ball Lightning, *Proc. Am. Phil. Soc.*, **76**:613–626 (1936).
9. Jennings, R. C., Path of a Thunderbolt, *New Scientist*, **13** (no. 270):156, January 18, 1962.
10. Jensen, J. C.: Ball Lightning, *Physics* (now *J. Appl. Phys.*), **4**:372–374 (1933).
11. Kapitza, P.: The Nature of Ball Lightning, *Dokl. Akad. Nauk SSSR*, **101**:245–248 (1955).   In Russian.
12. Kuhn, E.: Ein Kugelblitz auf einer Moment-Aufnahme? *Naturwissenshaften*, **38**:518–519 (1951).
13. McNally, J. Rand, Jr.: Preliminary Report on Ball Lightning, *Oak Ridge Natl. Lab., ORNL-3938, UC-34-Phys.*, May, 1966.
14. Morris, W.: A Thunderstorm Mystery, letters to the editor of *Daily Mail* of London, Nov. 5, 1936.
15. Müller-Hillebrand, D.: Zur Frage des Kugelblitzes, *Elektrie*, **17**:211–214 (1963).
16. Neugebauer, T.: Zu dem Problem des Kugelblitzes, *Z. Physik*, **106**:474–484 (1937).
17. Rayle, W. D.: Ball Lightning Characteristics, *NASA Technical Note D-3188*, January, 1966.
18. Rodewald, M.: Kugelblitzbeobachtungen, *Z. Meteorol.*, **8**:27–29 (1954).
19. Silberg, P. A., Ball Lightning and Plasmoids, *J. Geophys. Res.*, **67**:4941–4942 (1962).
20. Silberg, P. A.: A Review of Ball Lightning, in S. C. Coronti (ed.), "Problems of

Atmospheric and Space Electricity," pp. 436–454, American Elsevier Publishing Company, New York, 1965.

21. Uman, M. A.: Some Comments on Ball Lightning, *J. Atmospheric Terrest. Phys.*, **30**:1245–1246 (1968).

22. Uman, M. A., and C. W. Helstrom: A Theory of Ball Lightning, *J. Geophys. Res.*, **71**:1975–1984 (1966).

23. Uman, M. A., and J. J. Lowke: in the paper by M. A. Uman: Decaying Lightning Channels, Bead Lightning and Ball Lightning, Fourth International Conference on the Universal Aspects of Atmospheric Electricity, Tokyo, Japan, May, 1968.

24. Wolf, F.: Interessante Aufnahme eines Kugelblitzes, *Naturwissenschaften,* **43**: 415–417 (1956).

Appendix D: Recent Developments

The following material was received too late for inclusion in the body of the book:

Mackerras (1968), working in Australia, has compared the characteristics of the electric field changes due to cloud-to-ground with those due to intracloud flashes. He observed that impulsive discharges (R-changes for ground strokes; K-changes for cloud strokes) occurred more frequently and caused larger field changes in ground flashes than in cloud flashes. Impulsive discharges were found to occur in about 88 percent of ground flashes and about 35 percent of cloud flashes. Nonimpulsive discharges (slowly changing electric fields) were found to occur in about 58 percent of ground flashes and 96 percent of cloud flashes. Cloud flashes in the upper part of the thundercloud were found to involve fewer impulsive discharges than cloud flashes near cloud base level. Cloud-to-ground flashes lowering positive charge to earth were observed.

Orville (1968) has obtained the first stepped-leader spectrum. A 2-m section of the leader channel was isolated. The spectrometer viewed a fixed height above ground. During the progression of the stepped leader toward ground, seven more or less discrete light pulses were recorded before the upward-moving return stroke occurred. The discrete spectral emissions from the leader channel were recorded at intervals of about 30 to 40 μsec. The first two discrete pulses recorded are apparently due to sections of two successive leader tips (steps). The spectra of these pulses are characterized by strong NII emission and moderate H_α emission. The leader-tip temperature was calculated to be between 20,000 and 35,000°K. Above the leader tip, the NII radiation became weaker and the H_α radiation stronger during the light pulses, and a weak H_α spectrum was recorded between light pulses. From the structure of the NII spectral lines observed with the slitless spectrometer, it was deduced that the diameter of the region emitting the NII radiation was less than about 0.35 m. How much less was indeterminate.

Conner (1967) has reported spectral measurements of the energy radiated in the 3900 to 6900 Å range by seven lightning strokes; he has also reported simultaneous electric field measurements of the total energy input to the same strokes. Conner's values for energy radiated were between 59 and 2,000 joules/m with five of the seven measurements falling between 220 and 570 joules/m. These values are in good agreement with the value of 870 joules/m for the wavelength range 4000 to 11,000 Å obtained by Krider et al. (1968) for a single-stroke lightning flash (Secs. 5.5.1 and 7.3.3), despite the fact that Conner's data had to be strongly corrected for transmission through rain. Conner found an average value of the ratio of energy radiated to energy input of about 0.007, the values for seven strokes varying between 0.011 and 0.0026. The comparable value obtained by Krider et al. was 0.004.

Hill (1968) has studied the tortuosity of the channels of 13 lightning flashes. He concludes that the changes in direction of sections of lightning channels are distributed randomly (Gaussian distribution) and that the mean absolute value of direction change is approximately constant from flash to flash. For the segment lengths studied (between about 5 and 70 m) and for total channel lengths analyzed (between 1 and 4.3 km), the mean absolute value of the channel direction change was 16°.

Jones et al. (1968) have presented a theoretical model of the shock wave from a lightning discharge. Their results are essentially similar to those of Few et al. (1967) (Sec. 6.3.1). However, Jones et al. have derived a different functional form for the shock overpressure in the weak-shock region from that used by Few et al. The two approaches and the resultant numerical values are compared by D. Jones (1968). For the case that Few et al. calculated a minimum thunder frequency of 57 Hz, Jones, using the theory of Jones et al., finds a value of 33 Hz. The discrepancy between the two values does not change the conclusion reached in Sec. 6.3.2 that the dominant frequencies in thunder are not infrasonic.

A Ph.D. thesis entitled "Thunder" was completed in 1968 by A. A. Few of Rice University, Houston, Texas.

Bhartendu (1968) has published a concise version of his Ph.D. thesis on thunder (see Chap. 6).

R. Jones (1968) has assumed that the magnetic forces due to the lightning current balance the kinetic overpressure tending to make the channel expand (see Sec. 7.6) and has determined an expression for the channel radius at which this equilibrium occurs (see Uman, 1964, for a more detailed derivation of this relation). For an assumed current of 12 ka and an assumed channel electron density of 10^{25} m^{-3} at 23,000°K (a pressure of the order of 100 atm), Jones finds a radius of 0.15 cm. It is not clear, as noted in Sec. 7.6, that the channel will be sufficiently small at sufficiently high currents to allow a magnetic pinch to occur. Jones has related his calculated radius to the size of burn marks on conductors struck by lightning. As noted in Sec. 2.5.2, any deduction of channel radius from an observed electrode effect must be considered unreliable.

Oetzel (1968) has determined theoretically the return-stroke diameter during the time around the peak current. He has used two models for the return stroke: (1) a lumped resonant circuit model in which the inductive and resistive elements are due to the return stroke and are functions of the return-stroke conductivity, diameter, and length, and (2) a charged transmission line model. In both cases, a damping constant determined primarily by the rise time of the return stroke current can be used to compute the ratio of the channel resistance and inductance per unit length. This damping constant is to first approximation a function of only channel diameter for a given channel conductivity. Using reasonable values for the lightning conductivity, Oetzel finds that the diameters of first return strokes lie in the range 1 to 4 cm, while subsequent return strokes have diameters in the range 0.2 to 0.5 cm. Subsequent stroke diameters are found to be smaller because their experimentally observed currents rise more rapidly [Sec. 4.3, Eq. (4.1) and discussion after]; the observed rapid rise necessitates a smaller diameter for a given conductivity.

For lightning to set fire to an object, the lightning current must cause the object to be subjected to a sufficiently high temperature for a sufficient length of time to cause ignition. It has often been contended that it is the long-continuing current which sets fire to flammable objects such as trees. Many laboratory tests have shown that currents of the order of 10 ka or higher, lasting for times of the order of 10 μsec, will not set fire to wood. The first detailed evidence that it is indeed the continuing currents that cause forest fires has been presented by Fuquay et al. (1967). From electric field measurements, Fuquay et al. determined that seven lightning discharges which caused forest fires all contained at least one long continuing current phase exceeding 40 msec duration. Fuquay et al. report that of 856 cloud-to-ground discharges observed during 1965 and 1966, about half contained a long continuing current phase. This finding is in excellent agreement with similar data published by Kitagawa et al. (1962) and Brook et al. (1962) (Sec. 2.5.3 and Sec. 3.7.5).

The December, 1967, issue of the *Monthly Weather Review* (Vol. 95, no. 12) is devoted to atmospheric electrical phenomena. Of particular interest are the article by L. B. Loeb, "Contributions to the Mechanisms of the Lightning Stroke," the article by D. R. Fitzgerald, "Probable Aircraft 'Triggering' of Lightning in Certain Thunderstorms," and the several articles on thundercloud processes.

The Fourth International Conference on the Universal Aspects of Atmospheric Electricity was held in Tokyo, Japan, May 12–18, 1968. The proceedings of this meeting will be published as a book with S. C. Coroniti, Avco Corporation Space Systems Division, Lowell, Massachusetts, as editor. A number of papers on lightning were delivered. Of particular interest are the review paper, "The Radio Emission from Close Lightning," by G. N. Oetzel and E. T. Pierce, the review paper, "The Time-Resolved Characteristics of the Lightning Return Stroke," by R. E. Orville, and the paper, "Some Results on the Lightning Stroke Current Measurements in Japan," by S. Tsurumi, G. Ikeda, and K. Kinoshita.

References

1. Bhartendu: A Study of Atmospheric Pressure Variations from Lightning Discharges, *Can. J. Phys.*, **46**:269–281 (1968).
2. Conner, T. R.: The 1965 ARPA-AEC Joint Lightning Study at Los Alamos, vol. 1, The Lightning Spectrum, Charge Transfer in Lightning, Efficiency of Conversion of Electrical Energy into Visible Radiation, Los Alamos Scientific Laboratory Report LA-3754, Dec. 5, 1967.
3. Fuquay, D. M., R. G. Baughman, A. R. Taylor, and R. G. Hawe: Characteristics of Seven Lightning Discharges that Caused Forest Fires, *J. Geophys. Res.*, **72**:6371–6373 (1967).
4. Hill, R. D.: Analysis of Irregular Paths of Lightning Channels, *J. Geophys. Res.*, **73**:1897–1906 (1968).
5. Jones, D. L.: Comments on 'A Dominant 200-Hertz Peak in the Acoustic Spectrum of Thunder,' *J. Geophys. Res.*, **73**:4776–4777 (1968).
6. Jones, D. L., G. Goyer, and M. Plooster: Shock Wave from a Lightning Discharge, *J. Geophys. Res.*, **73**:3121–3127 (1968).
7. Jones, R. C.: Return Stroke Core Diameter, *J. Geophys. Res.*, **73**:809–814 (1968).
8. Mackerras, D.: A Comparison of Discharge Processes in Cloud and Ground Lightning Flashes, *J. Geophys. Res.*, **73**:1175–1183 (1968).
9. Oetzel, G. N.: Computation of the Diameter of a Lightning Return Stroke, *J. Geophys. Res.*, **73**:1889–1896 (1968).
10. Orville, R. E.: Spectrum of the Lightning Stepped-Leader, *J. Geophys. Res.*, **73**:6999–7008 (1968).
11. Uman, M. A.: "Introduction to Plasma Physics," pp. 184–189, McGraw-Hill Book Company, New York, 1964.

Appendix E: A Review of Natural Lightning: Experimental Data and Modeling

I. OVERVIEW OF LIGHTNING

A. Introduction

IN the last few years, it has become evident that the electric and magnetic fields of all important lightning processes can vary on a submicrosecond time scale (e.g., Weidman and Krider, 1978, 1980; Baum *et al.*, 1980; Weidman *et al.*, 1981). The existence of these fast fields, in turn, implies that the lightning currents which produce them contain large submicrosecond components (e.g., Weidman and Krider, 1978; Clifford *et al.*, 1979; Weidman and Krider, 1980). Recent airborne measurements have directly confirmed the existence of submicrosecond current rise times for some cloud discharge processes (e.g., Pitts and Thomas, 1981). Submicrosecond fields and currents represent a hazard to aircraft because they can efficiently excite aircraft resonances. This hazard is particularly serious for the newer aircraft in which the increased use of nonmetallic structural materials has reduced the electromagnetic shielding of the interior of the aircraft making flight-critical low-voltage digital electronics susceptible to upset or damage. In the present paper, we attempt to review all of the lightning literature pertinent to this hazard.

Lightning is a transient, high-current electric discharge whose path length is measured in kilometers. The most common source of lightning is the electric charge separated in ordinary thunderstorm clouds (cumulonimbus). Well over half of all lightning discharges occur within the cloud (intracloud discharges). Cloud-to-ground lightning (sometimes called streaked or forked lightning) has been studied more extensively than other forms because of its practical interest (e.g., as the cause of human injuries and death, disturbances in power and communication systems, and the ignition of forest fires) and because the channels below cloud level are more easily photographed and studied with optical instruments. Cloud-to-cloud and cloud-to-air discharges are less common than intracloud or cloud-to-ground lightning.

The definition and terminology of discharge components for cloud-to-ground flashes and intracloud discharges which follows is adapted from Schonland (1956) and Uman (1969). While our primary interest is in the individual lightning flash, it is worth noting that many basic lightning parameters, such as the area density of flashes, the fraction of discharges which are to ground versus the storm phase, the number of lightnings versus storm duration, and the maximum and average flashing rates, are subjects of current research interest (see, for example, Livingston and Krider, 1978); and the questions of whether the characteristics of individual flashes depend on the meteorological environment, the geographic location, or other factors are also being studied. Thomson (1980), for example, reports no significant correlation between the average number of strokes per flash or the average interstroke time interval and the geographic latitude of the measurement. Although the average lightning characteristics at any latitude may be similar, there are certainly differences within a given region: frontal storms produce a higher flashing rate and more strokes per flash than local convective storms (e.g., Holzer, 1953; Schonland, 1956; Kitterman, 1980); topography affects the channel lengths to ground and other properties (e.g., McEachron, 1939; Winn *et al.*, 1973); and there are seasonal effects such as winter thunderstorms producing positive discharges to ground (Takeuti *et al.*, 1973, 1976, 1977, 1978, 1980; Brook *et al.*, 1982).

B. Cloud-to-Ground Lightning

A typical discharge between cloud and ground starts in the cloud and eventually neutralizes tens of coulombs of negative cloud charge. The total discharge is called a *flash* and lasts about 0.5 s. A flash is made up of various discharge components, among which are 3 or 4 high-current pulses called *strokes*. Each stroke lasts about 1 ms and the separation time between strokes is typically 40 to 80 ms. Lightning often appears to "flicker" because the human eye can just resolve the individual light pulses associated with each stroke.

In the idealized model of the cloud charges shown in Fig. 1(a) the main charge regions P and N are of the order of many

Appendix E first appeared in *IEEE Transactions on Electromagnetic Compatibility* in May 1982 and was written in conjunction with E. Philip Krider of the Institute of Atmospheric Physics, University of Arizona at Tucson.

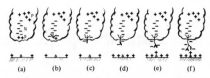

Fig. 1.   Stepped-leader initiation and propagation. (a) cloud charge distribution prior to lightning. (b) discharge called "preliminary breakdown" in lower cloud. (c)–(f) stepped-leader progression toward ground. Scale of drawing is distorted for illustrative purposes. Adapted from Uman (1971).

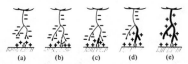

Fig. 2.   Return-stroke initiation and propagation. (a) final stage of stepped-leader descent. (b) initiation of upward-moving discharges. (c)–(e) return-stroke propagation from ground to cloud. Scale of drawing is distorted for illustrative purposes. Adapted from Uman (1971).

tens of coulombs of positive and negative charge, respectively, and p is a smaller positive charge. The *stepped leader* initiates the first stroke in a flash by moving from cloud to ground as sketched in Figs. 1 and 2. The stepped leader is itself initiated by a *preliminary breakdown* within the cloud, although there is still some disagreement about the exact form and location of this process. In Fig. 1, the preliminary breakdown is shown in the lower part of the cloud between the N and p regions. The preliminary breakdown sets the stage for negative charge (electrons) to be channeled toward ground in a series of short luminous steps (hence the name stepped leader). Photographically observed leader steps are typically 1 μs in duration and tens of meters in length, with a pause time between steps of about 50 μs (Fig. 1(c–f), Fig. 2(a)). A fully developed stepped leader lowers about 5 C of negative cloud charge toward ground in tens of milliseconds, and the average downward velocity is about 2×10^5 m/s. The steps have pulse currents of at least 1 kA, and associated with these currents are electric- and magnetic-field pulses with widths of about 1 μs or less and risetimes of about 0.1 μs or less. The average leader current is of the order of 100 A. The stepped-leader channel branches in a downward direction during its development toward ground. The preliminary breakdown, the subsequent lowering of negative charge by the stepped leader, and the resultant depletion of negative charge in the cloud combine to produce a total electric-field change with a duration between a few and a few hundred milliseconds.

The electric potential of the leader channel with respect to ground is about -1×10^8 V. As the leader tip nears ground, the electric field beneath it becomes very large and causes one or more upward-moving discharges to be initiated at the ground (Fig. 2), which starts the *attachment process*. The attachment process to the ground has, as we shall see, many elements in common with the attachment to an aircraft in flight. When one of the upward-moving discharges from the

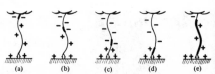

Fig. 3.   Dart-leader and subsequent return stroke. (a)–(c) dart leader deposits negative charge on defunct first-stroke channel. (d)–(e) return-stroke propagates from ground to cloud. Scale of drawing is distorted for illustrative purposes. Adapted from Uman (1971).

ground contacts the downward-moving leader some tens of meters above the ground, the leader tip is connected to ground potential. The leader channel is then discharged when a ground potential wave, the *return stroke*, propagates up the previously ionized leader path. The upward velocity of a return stroke is typically one-third the speed of light (Fig. 2), and the total transit time from ground to the top of the channel is typically about 100 μs. The return stroke, at least in its lower portion, produces a peak current of typically 30 kA, with a time from zero to peak of a few microseconds. Currents measured at the ground fall to half of the peak value in about 50 μs, and currents of the order of hundreds of amperes may flow for milliseconds or longer. The rapid release of return-stroke energy heats the leader channel to a temperature near 30 000 K and generates a high-pressure channel which expands and creates the shock waves which eventually become thunder. The return stoke lowers the charge originally deposited on the stepped leader channel to ground and, in so doing, produces an electric-field change with time variations which range from a submicrosecond scale to many milliseconds.

After the return-stroke current has ceased to flow, the flash may end; on the other hand, if additional charge is made available to the top of the channel by discharges within the cloud known as *J-* and *K-processes,* a continuous *dart leader* (Fig. 3) may propagate down the residual first-stroke channel at a velocity of about 3×10^6 m/s. The dart leader lowers a charge of the order of 1 C and the average current is typically 500 A. The dart leader then initiates the second (or any subsequent) return stroke. Dart leaders and strokes subsequent to the first are usually not branched.

Some leaders begin as dart leaders but toward the end of their trips toward ground they become stepped leaders. These processes are known as *dart-stepped* leaders. Dart-leader electric-field changes usually have a duration of about 1 ms, and the subsequent-stroke field changes are similar to, but usually a factor of two or so smaller than, first-stroke field changes. Subsequent stroke currents have faster zero-to-peak rise times than first strokes but a similar maximum rate-of-change. The largest field impulses in a flash are often produced by return strokes that are preceded by dart-stepped leaders.

The time between successive return strokes in a flash is usually 40 to 80 ms, but can be tenths of a second if a *continuing current* flows in the channel after the stroke. Continuing currents are on the order of 100 A and represent a direct transfer of charge from cloud to ground. The typical electric-field change produced by a continuing current is linear for roughly 0.1 s and is consistent with the lowering

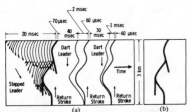

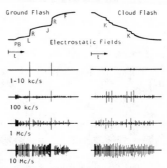

Fig. 4. (a) The luminous features of a typical lightning flash as would be recorded by a camera with relative motion (horizontal and continuous) between lens and film, a so-called streak camera. Scale of drawing is distorted for illustrative purposes. (b) The same lightning flash as recorded by an ordinary camera. Adapted from Uman (1969).

Fig. 5. Electric-field changes in the time domain and the corresponding fields at different frequencies for a typical cloud-to-ground flash and a typical intracloud flash, both at a distance of about 10 to 15 km and both having a duration of about 0.5 s. PB = preliminary breakdown; L = leader; R = return stroke; J = J-process; F = Final process which may be a continuing current or a cloud discharge; K = K-change. Amplitude scales for different frequencies are not the same. Adapted from Malan (1963).

of about 10 C of cloud charge to ground. Between one-quarter and one-half of all cloud-to-ground flashes contain a continuing current component.

A drawing of a streak photograph and a still photograph of a typical lightning flash is shown in Fig. 4. In addition to the usual downward-moving negatively-charged stepped leader shown in Figs. 1, 2, and 4, lightning may also be initiated by a positive downward-moving stepped leader. Positive discharges are rather rare in most thunderstorms, but the peak currents and total charge transfers can be extremely large (Berger and Vogelsanger, 1966; Berger 1967, 1972; Takeuti et al., 1973, 1976, 1977, 1978, 1980; Brook et. al., 1982). Furthermore, lightning can be initiated at the ground, usually from tall structures or mountains, by upward-going stepped leaders which can be either positively or negatively charged (Berger and Vogelsanger, 1966; Berger, 1967, 1972). The upward-going leaders branch in an upward direction. In this study, we will concentrate on the most-common form of cloud-to-ground lightning, namely that which lowers negative charge from cloud to ground and is initiated by a downward-moving, negative-charged stepped leader.

C. Cloud Discharges

Intracloud and intercloud lightning discharges occur between positive and negative cloud charges and have total durations about equal to those of ground discharges, 0.5 s. A typical cloud discharge neutralizes 10 to 30 C of charge over a total path length of 5 to 10 km. The discharge is thought to consist of a continuously propagating leader which generates 5 or 6 weak return strokes called recoil streamers or K-changes when the leader contacts pockets of space charge opposite to its own. The cloud K-changes are very similar to the K-changes which occur in the intervals between return strokes in ground discharges. Cloud discharges have not been studied nearly as extensively as discharges to ground, and hence much less is known about their detailed physical characteristics. The charge motion produces electric fields whose frequency spectra have roughly the same amplitude distribution as those of ground discharges for frequencies below about 1 kHz and above 100 kHz. Between 1 kHz and 100 kHz, the ground discharge is a more efficient radiator because of the very energetic return strokes.

Fig. 5. shows a comparison of typical ground-flash and cloud-flash electric fields in both the time and frequency domains. The time-domain field at close range shows a ramp-like change with superimposed steps for both types of discharges, but the steps in ground discharges, which are due to the return strokes, are much larger than the cloud-flash steps due to K-changes. The outputs of narrow-band receivers centered around the indicated frequencies are drawn below the time-domain fields of each flash. For instance, the output from a 100 kHz receiver during a ground flash first shows a series of pulses corresponding to preliminary breakdown and leader activity and then a major pulse at the time of the first and each subsequent return stroke. The pulses above about 10 MHz are of roughly equal magnitude for both ground and cloud flashes and are probably produced by relatively small-scale discharges that are primarily within the cloud. The return-stroke and K-change frequency spectra both have maxima in the 5-kHz range, but the return-stroke spectrum has a larger amplitude.

The term "cloud discharge" refers to all discharges which do not contact the ground. It is important to keep in mind that all cloud-to-ground lightning flashes have significant in-cloud components and hence share many common features with cloud discharges.

II. DETAILED REVIEW OF LIGHTNING FIELDS AND CURRENTS

A. Introduction

The brief overview of lightning phenomenology given in Section I provides the background which is necessary for a detailed discussion of the fields and currents which are produced by a discharge. In this review, we will first consider cloud fields and charges, followed by the basic elements of cloud-to-ground flashes; that is, the preliminary breakdown, the stepped leader, the attachment process, the first return

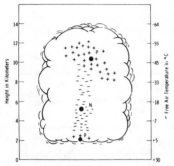

Fig. 6. Probable distribution of typical thunderstorm-cloud charge distribution in South Africa according to Malan (1952). Solid black circles indicate locations of effective point charges to give measured electric-field strength at the ground.

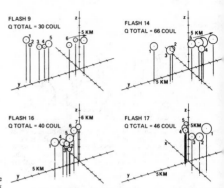

Fig. 7. Charge location for strokes and continuing current in four multiple-stroke flashes in New Mexico. Each charge location is numbered sequentially. The numbers which contain two or more locations (i.e., flash 9, charge 6; flash 14, charge 5; flash 17, charge 4) represent sequential continuing-current locations. The size of the charge spheres is determined by measuring the charge and assuming the charge density to be 20 C/km³. Adapted from Krehbiel et al. (1979).

stroke, the dart leader, subsequent return strokes, the continuing current, and J- and K-processes. For return strokes, we will discuss the current, the electric and magnetic fields, the frequency spectra, and the velocity and luminosity. We will then discuss various aspects of cloud discharges. Finally, we consider frequency spectra for both cloud and ground discharges. In the discussions to follow, we will give only the most pertinent literature references; however, the Bibliography contains a relatively complete list of references for each of the individual lightning processes.

B. Cloud Charges and Static Electric Fields

Recent reviews of the electrification processes which are active in thunderstorms have been given by Magono (1980) and by Latham (1981). Although some measurements have been made of the electrical properties of nonthunderstorm clouds (Simpson, 1949; Imyanitov and Chubarina, 1967; Imyanitov et al., 1972), the cumulonimbus is the most common lightning generator and is the cloud type which has received the most study.

By the early 1930's, a simple model for the charge structure of a thunderstorm had evolved from ground-based measurements of the cloud electric fields and lightning field changes (Wilson, 1916, 1920; Appleton et al., 1920; Schonland and Craib, 1927). In this model, the cloud charges form a positive electric dipole; that is, a large positive charge, P, is located above a large negative charge, N, as shown in Fig. 6. Simpson and Scrase (1937) were able to verify this basic dipole structure with sounding balloons, and they also found a small concentration of positive charge, p, at the base of the cloud as is also shown in Fig. 6. More recent measurements have confirmed the general validity of this overall charge structure (e.g., Simpson and Robinson, 1941; Malan, 1952; Huzita and Ogawa, 1976; Winn et al., 1981), although it is now recognized that there can be large horizontal displacements between the positive and the negative charges and that the positive charge may be highly diffuse.

Because of the difficulty in interpreting electric-field measurements made outside the cloud in the presence of spatial-

and time-varying conductivities, the magnitudes and heights of the actual cloud charges are uncertain (see Kasemir, 1965; Moore and Vonnegut, 1977). The space-charge layers on the surface of the cloud, for example, can lead to a substantial underestimation of the cloud charge magnitudes (Brown et al., 1971; Hoppel and Phillips, 1971; Klett, 1972). Values of p, N, and P inferred from remote measurements and their altitudes above ground level are roughly $+10$ C at 2 km, -40 C at 5 km, and $+40$ C at 10 km in South Africa, ground level being about 1.8 mk above sea level (Malan, 1952); and $+24$ C at 3 km, -120 C at 6 km, and $+120$ C at 8.5 km in Japan, ground level being about 1 km above sea level (Huzita and Ogawa, 1976). Jacobson and Krider (1970) have summarized most of the available data for the location and size of the N charge neutralized by cloud-to-ground lightning, and their values in Florida (at sea level) are -10 to -40 C at an altitude of 6 to 9.5 km. Measurements of the electric fields inside clouds using rockets, aircraft and balloons can provide relatively accurate measurements of the charge altitudes (e.g., Simpson and Scrase, 1937; Winn et al., 1981) but cannot determine the charge magnitudes without making questionable assumptions as to the size and shape of the charge distributions. Simpson and Robinson (1941) reported $+4$ C at 1.5 km, -24 C at 3 km, and $+24$ C at 6 km above ground in England, the ground being about 1 km above sea level. Winn et al. (1981) found about $+1$C on rain at about 4 km above sea level and negative charge of density 5×10^{-9} C/m³ between 4.8 and 5.8 km, the -2 to $-5°$C temperature range. Thus despite various uncertainties and inaccuracies, both internal and external measurements lead to a general model such as that shown in Fig. 6. The overall charge associated with each major region is probably not uniformly distributed, but is concentrated in localized pockets of high space-charge density. Evidence for these localized charge distributions is provided by the observation that in

TABLE I
THUNDERSTORM ELECTRIC FIELDS MEASURED IN
AIRBORN EXPERIMENTS

Investigation	Typical (V/m)	High Values Occasionally Observed	Measurement Type
Winn et al. (1974)	5-8 × 10⁴	2 × 10⁵	Rockets
Winn et al. (1981)	-	1.4 × 10⁵	Balloons
Rust, Kasemir (private comm.)	1.5 × 10⁵	3.0 × 10⁵	Aircraft
Kasemir and Perkins (1978)	1 × 10⁵	2.8 × 10⁵	Aircraft
Imyanitov et al. (1972)	1 × 10⁵	2.5 × 10⁵	Aircraft
Evans (1969)	-	2 × 10⁵	Parachuted Sonde
Fitzgerald (1976)	2-4 × 10⁵	8 × 10⁵	Aircraft

dividual return strokes in multiple-stroke ground flashes tap different regions of cloud charge in succession (Krehbiel *et al.*, 1979), where these regions are displaced horizontally from one another by distances of 1 to 3 km. Examples of these charge locations are given in Fig. 7.

Since the threshold electric field for breakdown of dry air between plane, parallel electrodes at sea level is about 3×10^6 V/m, one can assume that, in a cloud of water drops and ice crystals, at the reduced pressures characteristic of 4- to 5-km altitude, the fields will probably not exceed 5×10^5 to 1×10^6 V/m. A summary of the electric fields measured inside thunderstorms is given in Table I. Winn *et al.* (1974) report a peak horizontal field of 1×10^6 V/m on one rocket flight, although they express reservations about the validity of that measurement. Fitzgerald (1976) reports measuring a peak field of 1.2×10^6 V/m with an aircraft. Gunn (1947) found an electric field of 3.4×10^5 V/m on the underbelly of an aircraft just before it was struck by lightning and considers this to be an underestimate of the true field.

C. Preliminary Breakdown

The electric field change just before the first return stroke in a cloud-to-ground flash has a duration from a few milliseconds to a few hundred milliseconds with a typical value of some tens of milliseconds (e.g., Clarence and Malan, 1957; Kitagawa and Brook, 1960; Takeuti *et al.*, 1960; Harris and Salman, 1972; Thomson, 1980; Beasley *et al.*, 1982). Since the duration of the stepped leader (see Section D) is usually shorter than the duration of the overall pre-stroke field change, some of the prestroke field behavior is attributed to a so-called "preliminary breakdown" process within the cloud. Photographic evidence for cloud discharges preceding the stepped leader is given by Malan (1952, 1955) who showed that clouds often produce luminosity for a hundred or more milliseconds before the emergence of the stepped leader from cloud base. Stepped leaders propagate at a typical average velocity of about 2×10^5 m/s and thus should take about 25 ms to travel 5 km from cloud to ground. It is a matter of some dispute as to whether the relatively long prestroke field changes, say over 100 ms, should be associated with processes which initiate stepped leaders, as argued by Clarence and Malan (1957),

or whether the long prestroke field changes should be treated as relatively independent cloud discharges, as argued by Kitagawa and Brook (1960) and by Thomson (1980). Here we define the preliminary breakdown to be those discharge processes which lead directly to the initiation of a stepped leader. In the absence of a causal relationship, the prestroke processes will be assumed to have the properties of a cloud discharge as discussed in Section II-J.

Here our main interest is the location of the preliminary breakdown within the cloud, the nature of the electromagnetic fields generated by it, and the currents which are necessary to produce these fields. The location of the preliminary breakdown has been determined in three ways: 1) from the average variation of the initial part of the electric-field change with distance for a number of single-station measurements (Clarence and Malan, 1957), 2) from field changes measured simultaneously at eight ground stations (Krehbiel *et al.*, 1979), and 3) from the location of the sources of the initial VHF (30 to 50 MHz) radiation produced within the cloud (Rustan, 1979; Rustan *et al.*, 1980; Beasley *et al.*, 1982).

Clarence and Malan (1957) assumed that the preliminary breakdown channels were vertical and concluded that the initial breakdown began between the N charge center and the lower p as shown in Fig. 1(b). That the discharge was just as likely to start from p and go upward as to start from N and go downward. The location of this breakdown was between 1.4 km and 3.6 km above the ground in South Africa where ground is about 1.8 km above sea level. Clarence and Malan also suggest that, after the preliminary breakdown which lasts 2 to 10 ms, there is an intermediate stage which lasts from zero to hundreds of milliseconds. During the intermediate stage, the breakdown channels become negatively charged to the point where they can generate a downward-propagating stepped leader. During the preliminary breakdown, Clarence and Malan (1957) report large VLF pulses, but their detailed shapes were not resolved.

Unfortunately, investigators since Clarence and Malan (1957) have not been able to verify the general validity of the Clarence and Malan model (Kitagawa and Brook, 1960; Thomson, 1980; Krehbiel *et al.*, 1979; Beasley *et al.*, 1982). Krehbiel *et al.* (1979), for example, in a detailed study of two flashes

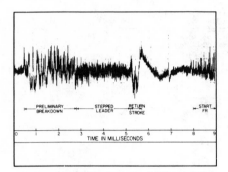

Fig. 8. Log-amplitude VHF radiation, 40 to 60 MHz, at the beginning of the 165959 flash on July 19, 1976 which stuck the 150-m weather tower at the Kennedy Space Center. Adapted from Rustan (1979).

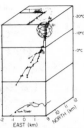

Fig. 9. VHF source locations for the preliminary breakdown (AB) and stepped leader (below B) for the lightning to ground whose VHF radiation is shown in Fig. 8. Locations are 94-μs averages determined by finding time delays between VHF pulses arriving at four ground stations from cross-correlating the recorded signals for that time period. The first stroke charge Q_1 is reported in Uman et al. (1978) and determined using the technique of Krehbiel et al. (1979). Adapted from Rustan et al. (1980).

with long-duration preliminary field changes, found that the initial charge motion was vertical but soon became horizontal. Negative charge moved downward and horizontally away from the cloud volume that was neutralized by the first return stroke, and eventually one of the succession of these breakdown events initiated a leader to ground. Krehbiel et al. (1979) suggest that the field changes observed by Clarence and Malan (1957) can be interpreted more simply in terms of horizontal channels, and hence there is no necessity for an intermediate stage.

Rustan (1979), Rustan et al. (1980), and Beasley et al. (1982) have located the sources of individual radio impulses produced by lightning in Florida using the difference in the time-of-arrival of the impulses at an array of VHF receivers. The durations of the preliminary breakdown in the flashes that were analyzed were on the order of several milliseconds. These investigators found results contrary to those of Clarence and Malan (1957) but in agreement with Krehbiel et al. (1979); that is, the initial sources of VHF tended to be vertical between 4 and 10 km, but the appearance of these sources preceded any significant change in the electrostatic field. The stepped leader emerged from the bottom of the preliminary breakdown column, was characterized by a significant change in the electrostatic field, and generated VHF pulses of much lower amplitude and shorter duration than those of the preliminary breakdown. Both the preliminary breakdown and the stepped leader lowered negative charge toward ground. The transition from the preliminary breakdown to the stepped leader is marked by large VLF pulses which will be discussed in the next paragraph. Fig. 8 shows an example of the VHF signals which are detected during the preliminary breakdown, stepped leader, return stroke, and FR (following first return stroke) phases of a cloud-to-ground flash, and Fig. 9 shows an example of the source locations determined using the VHF time-of-arrival technique.

Many investigators (e.g., Kitagawa, 1957; Clarence and Malan, 1957; Kitagawa and Kobayashi, 1959; Kitagawa and Brook, 1960; Krider and Radda, 1975; Weidman and Krider, 1979; Beasley et al., 1982) have published data which suggest

that the beginning of the stepped leader or perhaps the end of the preliminary breakdown is characterized by a train of relatively large bipolar pulses which can be detected by systems operating at frequencies ranging from VLF to VHF. In much of the literature (e.g., Clarence and Malan, 1957) these pulses are identified as the beginning of a β-leader, which will be discussed in Section II-D, but Beasley et al. (1982) have recently determined that these pulses are produced at the location where the preliminary breakdown ends and where the stepped leader begins. Examples of the waveforms of these pulses are shown in Fig. 10(a), and (b) and a frequency spectrum is shown in Fig. 11.

Weidman and Krider (1979) have characterized the shapes of the characteristic VLF pulses as follows: The initial polarity tends to be the same as the return-stroke electric field which follows. The overall shape is bipolar with about a 10-μs rise to peak and an overall duration of about 50 μs. Two or three microsecond-width pulses are superimposed on the initial half-cycle, but the negative half-cycle is smooth. The peak amplitude of a pulse often approaches that of the following return stroke, and there are usually several pulses which occur at intervals of about 100 μs. It is not known whether the β-pulses of Clarence and Malan were similar to those described by Weidman and Krider (1979), but it is likely in view of the fact that the shapes of the pulses in cloud discharges which do not precede return strokes are similar to the shapes of the pulses during the preliminary breakdown, although the initial polarities of the two types of pulses tend to be opposite (Weidman and Krider, 1979). Since the cloud pulses and the preliminary breakdown pulses are apparently generated by an intermittent or stepped process, it is natural to expect that the literature regarding the differences between these pulses and pulses associated with the photographed stepped leader would be confused, as is apparently the case.

Prior to the occurrence of the characteristic VLF pulses shown in Fig. 10(a) and (b), there is appreciable VHF radiation from the cloud (Malan, 1958; Brook and Kitagawa, 1964; Iwata and Kenada, 1967; Rustan, 1979) which is probably produced by localized breakdown processes that involve rela-

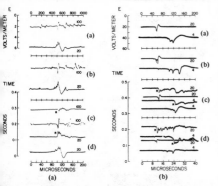

Fig. 10. (a) Large bipolar electric fields radiated by lightning discharges at end of preliminary breakdown or beginning of stepped leader at distances of 100 to 200 km. Each waveform is shown on both a slow (100 μs/div) and a fast (20 μs/div) time scale, the latter inverted with respect to the former. The fast trace has been positioned so that the centers of each trace coincide in time. A positive field is shown as a downward deflection on the slow trace. The time interval between the return stroke R and the discharge preceding it in (c) and (d) is shown on the scale at left. Adapted from Weidman and Krider (1979). (b) Large bipolar electric fields radiated by lightning discharges at end of preliminary breakdown or beginning of stepped leader at distances of 30 to 50 km. Each waveform is shown on both a slow (20 μs/div) and a fast (4 μs/div) time scale, the former indented with respect to the latter. A positive field is shown as a downward deflection on all records. The time interval between the return stroke R and the discharge preceding it in (c) and (d) is shown on the scale at left. Adapted from Weidman and Krider (1979).

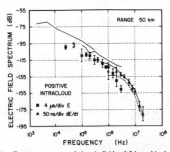

Fig. 11. Frequency spectra of electric fields of 5 large bipolar pulses between 15 and 45 km, normalized to 50 km. The decibel values given are 20 times the logarithm (base 10) of the Fourier transform of the electric-field intensity. Squares and triangles represent mean values; vertical bars, standard deviations. Solid lines represent the first-return-stroke frequency spectra given in Fig. 19. Time-domain waveforms of bipolar pulses are shown in Fig. 10(a) and (b). Adapted from Weidman et al. (1981).

tively small charge transfers. The detailed physics of the preliminary breakdown is not understood; however, there have been a number of suggestions as to how breakdown could start in the cloud, continue to grow, and eventually produce a stepped leader (e.g., Loeb, 1966; 1970; Dawson and Duff,

1970; Phelps, 1974; Griffiths and Phelps, 1976, a, b; Diachuk and Muchnik, 1979).

D. Stepped Leader

A significant fraction of what is known today about stepped leaders was determined photographically by Schonland and his coworkers in South Africa using streak cameras (Schonland et al., 1938 a, b; Schonland, 1956). These photographic measurements were also supplemented by slow (millisecond scale) electric-field measurements at close range (e.g., Schonland et al., 1938; Malan and Schonland, 1947; Schonland, 1956). More recently, Krider and Radda (1975), Krider et al. (1977), Weidman and Krider (1980), Baum et al. (1980), and Weidman et al. (1981) have studied the electromagnetic fields radiated by individual leader steps with microsecond and submicrosecond resolution.

We now list some of the more important characteristics of stepped leaders.

1) On the basis of step length and average earthward velocity, Schonland (1938) and Schonland et al. (1938 a, b) have divided stepped leaders into two categories, α and β. The type α leaders have a uniform earthward velocity, of the order of 10^5 m/s, have steps that are shorter and much less luminous than the β steps, and do not vary appreciably in length or brightness. Type β leaders begin with long, bright steps and a high average earthward velocity, of the order of 10^6 m/s, exhibit extensive branching near the cloud base, and, as they approach the earth, they assume the characteristics of α leaders. Schonland (1956) states that the majority of photographed leaders are type α, whereas the majority of electric-field measurements suggest type β. This fact and the fact that the non-α characteristics of β's are photographed at high altitude suggest that the initial β processes are probably associated with the end of the preliminary breakdown. As noted in the previous section, we will adopt this interpretation and consider all stepped leaders to have only α characteristics.

The step lengths of type-α leaders are typically 50 m when the leader is relatively far above the ground, with a pause time between steps ranging from 40 to 100 μs (Schonland, 1956). Longer pause times are followed by longer step lengths, and vice versa. From time-resolved photographs, Schonland (1956) states that the average two-dimensional velocity of a stepped leader is between 0.08 and 2.4 × 10^6 m/s, with the most probable value being close to 2 × 10^5 m/s. From electric-field records, Kitagawa (1957) observed a mean pause time of 50 μs for steps far above the ground, decreasing to 13 μs as the leader tip approached the ground. Recent work has verified that leader pulses on electric-field records just before the return stroke occur at about 15-μs intervals (Krider and Radda, 1975; Krider et al., 1977); however, the shorter intervals may, in part, be due to the fact that there may be steps in different branches developing at the same time (Berger, 1967).

2) The luminosity of a step rises to peak in about 1 μs or less and falls to half this value in roughly the same time (Schonland et al., 1935; Schonland, 1956; Orville, 1968; Krider, 1974). Thus for a 50-m step the velocity of propagation of the light along the step must be at least 5 × 10^7 m/s. Negatively

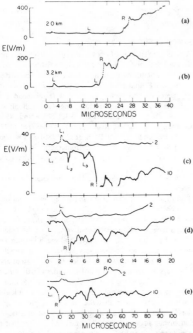

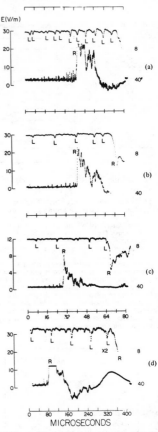

Fig. 12. Electric fields produced by close leader steps and return strokes in Florida. (a), (b) Oscilloscope records from discharges at 2 and 3.2 km (c)-(e) Records from discharges at 15-30 km over water. The initial portion of the lower (10 μs/div) trace in (c)-(e) is shown inverted with respect to (a), (b) and on a 2 μs/division time base on the upper trace. Adapted from Krider *et al.* (1977).

charged leaders are photographically dark between steps, but positively charged leaders emit some light and have less distinct steps (Schonland, 1956; Berger and Vogelsanger, 1966).

3) Photographs show a faint corona discharge extending for about one step length in front of the bright leader step (Schonland *et al.*, 1935; Berger and Vogelsanger, 1966). The luminosity of this advance corona does not appear to develop between steps but rather occurs simultaneously with the creation of the bright step behind it. Luminous stepped-leader diameters have been measured photographically to be between 1 and 10 m with no apparent central core (Schonland, 1953). The expectation that there is a central current-carrying core follows from the spectral measurements of Orville (1968) and the fact that laboratory arcs of several hundred amperes, the average current needed to lower 5 C in about 10 ms, have channel diameters of some centimeters.

4) The millisecond-scale electrostatic field of the stepped leader is reasonably well understood and is adequately modeled by the lowering of negative charge down a vertical column (Malan and Schonland, 1947; Beasley *et al.*, 1982). The measured field change is relatively smooth, which implies that the

Fig. 13. Electric-field waveforms produced by four lightning discharges in Florida at distances of 20 to 100 km. Each record contains an abrupt return-stroke transition R preceded by small pulses characteristic of leader steps. The polarities of all waveforms reproduced are in the sense of negative charge being lowered to ground. The same waveform is shown on both a slow (40-μs/div) and an inverted fast (8-μs/div) time scale. The vertical gain of the fast trace for d has been magnified by a factor of 2 in relation to the lower trace. Adapted from Krider *et al.* (1977).

leader lowers charge continuously between steps and that the step process itself does not lower appreciable charge (Schonland, 1953; Krider *et al.*, 1977). Typical electrostatic fields measured on the ground at various distances from the stepped leader are given by Beasley *et al.* (1982).

5) Measurements of the fields radiated by leader steps near the ground by Krider *et al.* (1977) and Weidman and Krider (1980) are given in Figs. 12 and 13. These data and the data of Baum *et al.* (1980) suggest that the step currents are in the kiloamperes range or larger, with rise times of the order of 0.1 μs

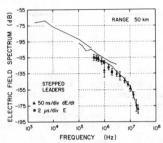

Fig. 14. Frequency spectra of the electric fields of 9 leader steps between 20 and 50 km normalized to 50 km. The decibel values given are 20 times the logarithm (base 10) of the magnitude of the Fourier transform of the electric field intensity. Squares and triangles represent mean values; vertical bars, standard deviations. Solid lines represent the first return stroke frequency spectra given in Fig. 19. Adapted from Weidman et al. (1981).

or less. The average frequency spectrum of the field pulses from individual steps obtained by Weidman et al. (1981) is shown in Fig. 14.

There is considerable theory, primarily qualitative, on the formation and propagation of the stepped leader (e.g., Schonland, 1956, 1953, 1938; Bruce, 1941, 1944; Loeb, 1966, 1968; Klingbeil and Tidman, 1974 a, b; Szpor, 1970, 1977; Wagner and Hileman, 1958, 1961; Krider et al., 1977).

C. Attachment Process

When the stepped leader approaches any conducting object, such as an aircraft or a transmission-line tower, the electric field produced by the charge on the leader can be enhanced by the object to the point where upward discharges (called leaders, connecting leaders, connecting discharges, or, sometimes, streamers) are emitted from the object. The characteristics of these discharges are not well understood, but have been the subject of considerable discussion in the context of modeling lightning strikes to power lines where the attachment process plays a significant role in the design of overhead ground-wire protection.

An important parameter in lightning protection is the "striking distance": the distance between the object to be struck and the tip of the downward-moving leader at the instant that the connecting leader is initiated from the object. It is assumed that, at this instant of time, the point of strike is determined. It follows that the actual junction point is somewhere between the object and the tip of the last leader step, and it is often assumed to be halfway between.

We now examine the attachment process as it relates to lightning strikes to ground or to objects attached to the ground. General reviews of this phenomenon have been given by Golde (1967, 1977) who outlines the following analytical approach: a reasonable charge distribution is assumed for the leader channel, and the resultant fields on the ground or nearby objects are calculated. The leader is assumed to be at the striking distance when the field at some point exceeds a critical breakdown value that is determined by laboratory tests. Various authors have derived relations between the

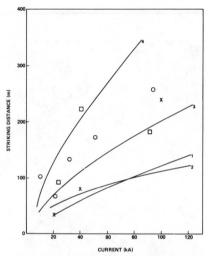

Fig. 15. Striking distance versus return-stroke peak current for objects attached to ground. [curve 1: Golde (1945); curve 2: Wagner (1963); curve 3: Love (1973); curve 4: Ruhling (1972); ×: Davis (1962); ○: estimates from 2-D photographs by Eriksson (1978); □: estimates from 3-D photography by Eriksson (1978)]. Adapted from Golde (1977) and Eriksson (1978).

striking distance and the leader charge (e.g., Golde, 1945); however, the relationship of more practical value in power line design is that of striking distance to the peak current of the following return stroke. To make this connection, the peak current must be related to the leader charge distribution. It has not been proven that these two quantities are actually related, since the leader charge may be spread over a rather large volume and in various branches, whereas the peak stroke current is determined in a few microseconds in the short channel section that is attached to ground. On the other hand, Berger (1972) shows that there is a correlation between the measured peak currents in return strokes and the total charge transfer to ground in the first 1 ms or so. According to Berger, the best fit relating peak current I to charge transfer Q for 89 negative strokes is

$$I = 10.6Q^{0.7}$$

with I measured in kiloamperes and Q in coulombs. According to this expression, a typical peak current of 25 kA corresponds to a total leader charge of 3.3 C. When this expression is combined with the relation between charge and breakdown field, a relation for the striking distance d_s can be found in terms of peak current. One example of several theoretical analyses reviewed by Golde (1977) yields

$$d_s = 10I^{0.65}$$

where d_s is in meters and I in kiloamperes. In Fig. 15, several

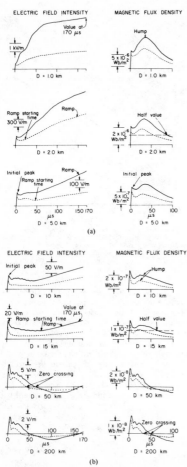

ELECTRIC FIELD INTENSITY   MAGNETIC FLUX DENSITY

(a)

ELECTRIC FIELD INTENSITY   MAGNETIC FLUX DENSITY

(b)

Fig. 16.   (a) Typical electric-field strength (left column) and magnetic flux density (right column) for first (solid line) and subsequent (dotted line) return strokes at distances of 1, 2, and 5 km. The following characteristic features of the waveforms are identified: for electric field, initial peak, ramp-starting time, ramp, and 170-μs value; for magnetic field, initial peak, hump, and half-value. (b) Typical fields as described in Fig. 16(a) for distances of 10, 15, 50, and 200 km. Characteristic waveform features identified in addition to those noted in Fig. 16(a) are electric- and magnetic-field zero crossings. Adapted from Lin *et al.* (1979).

theoretical curves discussed by Golde (1977) are shown together with some experimental data. From the available experimental data and theory, it is possible to conclude that striking dis-

tances are generally between a few tens and a few hundred meters.

F. Return Strokes

The return stroke is the most studied and hence best understood lightning process. This research has been motivated both by practical considerations (e.g., the need to reduce lightning injuries and deaths and lightning damage) and by the fact that, of all the phases of lightning, the return stoke is easiest to measure.

Several types of experimental data are available on return strokes: 1) Wide band (dc to some megahertz) electric and magnetic fields at ground level (e.g., Tiller *et al.*, 1976; Weidman and Krider, 1978; Lin *et al.*, 1979; Weidman and Krider, 1980) and, to a very limited extent, above ground level (e.g., Baum, 1980; Nanevicz and Vance, 1980; Pitts, 1981; Pitts and Thomas, 1981); 2) measured electric-field frequency spectra (e.g., Taylor, 1963, Serhan *et al.*, 1980; Weidman *et al.*, 1981); 3) current waveforms at ground level (e.g., Berger *et al.*, 1975; Garbagnati *et al.*, 1975; Eriksson, 1978) and, to a very limited extent, above ground (e.g., Petterson and Wood, 1968; Pitts, 1981; Pitts and Thomas, 1981); and 4) return-stroke velocities and luminosities (e.g., Schonland *et al.*, 1935, Boyle and Orville, 1976; Orville *et al.*, 1978; Saint-Privat D'Allier Research Group, 1979; Hubert and Mouget, 1981; Idone and Orville, 1982). In order for any model of the return stroke to be valid, it must be capable of describing in a self-consistent way the above independently measured fields, currents, velocities, and luminosities. We now consider the above four types of experimental data in detail.

1) The most complete description of the overall return-stroke electric- and magnetic-field waveforms is given by Lin *et al.* (1979). The bandwidths of the measuring systems extended from near dc to over 1 MHz. Measurements were made simultaneously at two Florida stations separated by either 50 or 200 km, and fields were recorded from discharges at distances from 0.2 to 200 km. Typical first- and subsequent-stroke waveforms from Lin *et al.* (1979) are sketched in Fig. 16(a), and (b). Actual waveforms are found in Figs. 10, 12, 13, and 18.

Since Lin *et al.* (1979) and previous studies in the same experimental program (e.g., Fisher and Uman, 1972; Tiller *et al.*, 1976; Uman *et al.*, 1976a) concentrated primarily on the overall shape of the waveforms, Weidman and Krider (1978, 1980) have recently examined the microsecond and submicrosecond structure of electric-field, E, and field-derivative, dE/dt, signatures. These authors find that the first return stroke in a cloud-to-ground flash produces an electric-field "front" which rises in 2 to 8 μs to about half of the peak field amplitude and is followed by a fast transition to peak whose mean 10 to 90 percent risetime is about 90 ns. Subsequent-stroke fields have fast transitions very similar to first strokes, but fronts which last only 0.5 to 1 μs and which rise to only about 20 percent of the peak field. This fine structure in return-stroke fields is sketched in Fig. 17 and illustrated by the waveforms in Figs 18, as well as in Figs. 10 and 12.

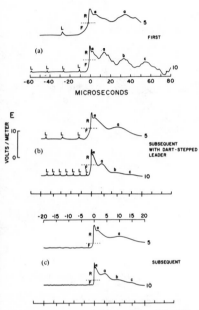

Fig. 17. Sketches of the detailed shapes of the radiation fields produced by (a) the first return stroke, (b) a subsequent return stroke preceded by a dart-stepped leader, and (c) a subsequent stroke preceded by a dart leader in lightning discharges to ground. The field amplitudes are normalized to a distance of 100 km. The small pulses characteristic of leader steps L are followed by a slow front F and an abrupt, fast transition to peak R. Following the fast transition, there is a small secondary peak or shoulder α and large subsidiary peaks, a, b, c, etc. The lower trace in each case shows the field on a time scale of 20 μs/div, and the upper trace is at 5 μs/div. The origin of the time axis is chosen at the peak field in each trace. Adapted from Weidman and Krider (1978).

Unfortunately, the origin of the fronts in return-stroke fields is still not well understood, particularly for first strokes (Weidman and Krider, 1978). If fronts are produced by upward-connecting discharges, then these discharges must have lengths in excess of 100 m and peak currents of 10 kA or more. A front may be produced by a slow surge of current in the leader channel prior to the fast transition, but then this surge must contain 10 kA or more and the associated channel length must be at least 1 km. To date, the available optical data are not adequate to determine whether either of these processes (or both) does actually occur.

The fast rise times that are found in the initial return-stroke fields suggest that return-stroke currents contain large submicrosecond variations (Weidman and Krider, 1978, 1980 a, b). Submicrosecond rise times have been observed in currents measured during strikes to towers, but very few of these are as

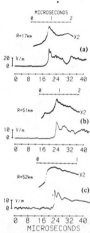

Fig. 18. Electric-field waveforms produced by three lightning return strokes. The same waveform is shown on both a slow (4-μs/div) and a fast (200-ns/div in (a) and (b), 100-ns/div in (c) time scale. The field amplitude is shown to the left of the lower record. The vertical gain of the upper trace is twice (×2) that of the lower trace. The range R to the lightning discharge is shown for each flash. Adapted from Weidman and Krider (1980).

fast as 90 ns (Berger et al., 1975; Table II). Thus the origin of the fast transition is still not clear, although it is possible that an upward discharge or the tower itself alters the current rise-times measured at the ground from the values actually present in the channel above ground.

To measure distant dE/dt signatures on a 10-ns time scale, it is essential that the field propagation from the lightning source to the receiving antenna be over salt water; otherwise, there can be a significant degradation in the high-frequency content of the fields due to propagation over the relatively poorly conducting earth (e.g., Uman et al., 1976b; Weidman and Krider, 1980 a, b). It is possible that lightning striking salt water could produce inherently faster rise times than lightning striking ground, but Weidman and Krider (1978) argue that this is probably not the case. Electric and magnetic fields from close lightning where the source is sufficiently elevated can be directly measured with 10-ns resolution (Baum et al., 1980).

2) Data on return-stroke frequency spectra below 1 MHz are given by Serhan et al. (1980). These spectra were obtained by Fourier-analyzing the time-domain electric-field waveforms of Lin et al. (1979) and Tiller et al. (1976) and extend over a frequency range from 1 kHz to 700 kHz for lightning at distances over land between 1 km and 200 km. First-stroke spectra at 50 km are shown in Fig. 19. To obtain the spectra of strokes within 10 km, Serhan et al. (1980) truncated the electric-field records (see Fig. 16) and this artifically introduced a high-frequency content which does not exist in nature. The

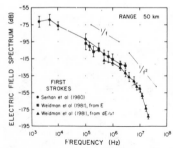

Fig. 19. Frequency spectra for 13 first strokes over land (Serhan *et al.*, 1980) and 5 first return strokes over salt water (Weidman *et al.*, 1981) at a distance of about 50 km. The decibel values given are 20 times the logarithm (base 10) of the magnitude of the Fourier transform of the electric-field strength. Solid symbols indicate mean values; vertical bars, standard deviations.

extent of this artificial high-frequency enhancement is presently under study.

Recently, Weidman *et al.* (1981) have extended the first-stroke spectra of Serhan *et al.* (1980) to 20 MHz, as shown in Fig. 19. The spectra of Weidman *et al.* (1981) were derived from E and dE/dt measurements of lightning at distances of about 50 km over salt water. Since the lightning and the propagation paths were entirely over salt water, there should be negligible ground-wave attenuation below about 10 MHz. Such attenuation is clearly evident in the 50- and 200-km spectra of Serhan *et al.* (1980) and, in fact, was used by them to estimate the average conductivity of Florida soil.

LeVine and Krider (1977) have made narrow-band RF measurements at 3, 30, 139, and 295 MHz correlated with wideband electric-field measurements. They show that first strokes have strong radiation at all of these frequencies, but that the radiation does not peak until 10 to 30 μs after the start of the wide band waveform, as is evident in Figs. 20(a), (b). This evidence suggests that much of the RF emission may be due to the effects of branches in first strokes and cloud processes. Supporting this suggestion is the observation that subsequent strokes, which usually do not have branches, generate little HF or VHF radiation as shown in Figs. 21(a), (b), and (c). There is, however, considerable radiation preceding a subsequent return stroke and due to the dart leader (see Section II-G). The observations of LeVine and Krider (1977) are in reasonable agreement with those of Takagi (1969) at 60, 150, and 420 MHz, but not with those of Brook and Kitagawa (1964) at 420 and 850 MHz. Brook and Kitagawa (1964) report that the RF was always delayed from 60 to 100 μs in the 50 percent of return stroke for which RF was detected and argue that the RF is due to breakdown at the top of the return-stroke channel.

3) Return-stroke currents measured in Switzerland are reviewed by Berger *et al.* (1975); in Italy by Garbagnati *et al.* (1975, 1978); and in South Africa by Eriksson (1978). Berger's and Garbagnati's measurements were obtained at the tops of towers on mountains, and because of this, their published current waveforms may be different from those of strikes to low

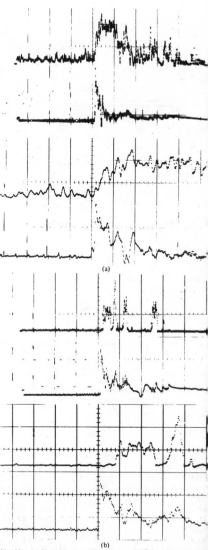

Fig. 20. (a) Simultaneous oscilloscope records of the 3-MHz RF signal (top trace) and the electric field (lower trace) due to a first return stroke. The time base is 100 μs/large division for the top two traces and 20 μs/large division for the bottom two traces. Adapted from LeVine and Krider (1977). (b) Simultaneous records of the 139-MHz RF with horizontal polarization (top trace) and the electric field (lower trace) due to a first return stroke. The time base is 100 μs/large division for the top two traces and 20 μs/large division for the bottom two traces. Adapted from LeVine and Krider (1977).

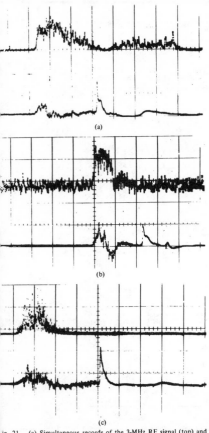

ig. 21. (a) Simultaneous records of the 3-MHz RF signal (top) and the electric field (bottom) due to a subsequent return stroke. The time base is 100 μs/large division. Adapted from LeVine and Krider (1977). (b) Simultaneous records of the 295-MHz RF with horizontal polarization (top) and the electric field (bottom) due to a subsequent return stroke. The continuous pulses on the RF channel prior to any electric-field activity are due to receiver noise. The time base is 100 μs/large division. Adapted from LeVine and Krider (1977). (c) Simultaneous records of the 139-MHz RF with vertical polarization (top) and the electric field (bottom) due to a subsequent return stroke. The time base is 100 μs/large division. Adapted from Levine and Krider (1977).

objects or the ground. Of particular interest is the early portion of the first-stroke waveform, since this must partially be due to an upward-going leader (see Section II-E) and thus might be different for strikes to a tall structure than for strikes to ground, low objects, or an aircraft. Also, first-stroke currents striking tall objects are expected to be larger, on the average, than those to normal ground (Sargent, 1972). In any event, the extensive measurements of Berger et al. (1975) are sum-

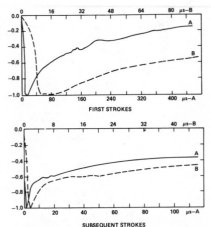

Fig. 22. Average return-stroke currents measured by Berger and co-workers as reported by Berger et al. (1975).

marized in Fig. 22 and Table II. The measurements of Garbagnati et al. (1975, 1978) and Eriksson (1978) are, in general, consistent with those of Berger.

The long first-stroke rise times shown in Fig. 22 must include a contribution due to an upward-going leader. Subsequent strokes have rise times for which the median value from 2 kA to peak is reported to be 1 μs. Berger et al. (1975) state that 5 percent of the 120 front times measured were less than 0.2 μs and that 5 percent of the maximum rates-of-rise of current exceeded 120 kA/μs. Fieux et al. (1978) report that their 10- to 90-percent subsequent strokes rise times were less than 1 μs in 70 percent of 63 measurements. Weidman and Krider (1980) have derived maximum rates of rise of current from first- and subsequent-stroke fields and find a mean of about 180 kA/μs with maximum values about twice the mean. (The mean value of 90 kA/μs found in Weidman and Krider (1980) is in error by a factor of 2 due to a similar error in the equation they use to relate the rate of change of field to rate of change of current.) Peak currents for first strokes are generally thought to be in the 20- to 40-kA range with 200 kA occurring at about the 1-percent level, although there is some disagreement about the exact statistics (Szpor, 1969; Sargent, 1972).

Lightning currents which pass through an aircraft in flight have been measured by Petterson and Wood (1968), Pitts (1981), and Pitts and Thomas (1981). The aircraft were apparently involved with few, if any, return strokes, but rather with other portions of ground discharges or with cloud discharges.

4) Return-stroke velocity measurements may be, to some extent, arbitrary because the luminosity of the return-stroke wavefront has a shape which varies with height and, therefore, it is often difficult to identify the same luminous feature at different heights (Hubert and Mouget, 1981; Jordan and Uman, 1980; Weidman and Krider, 1980). Recent velocity

TABLE II
LIGHTNING CURRENT PARAMETERS, ADAPTED FROM BERGER
et al. (1975)

Number of Events	Parameters	Unit	Percent of cases exceeding tabulated value		
			95%	50%	5%
	Peak current (minimum 2 kA)				
101	negative first strokes	kA	14	30	80
135	negative subsequent strokes	kA	4.6	12	30
26	positive first strokes (no positive subsequent strokes recorded)	kA	4.6	35	250
	Charge				
93	negative first strokes	C	1.1	5.2	24
122	negative subsequent strokes	C	0.2	1.4	11
94	negative flashes	C	1.3	7.5	40
26 *	positive flashes	C	20	80	350
	Impulse charge				
90	negative first strokes	C	1.1	4.5	20
117	negative subsequent strokes	C	0.22	0.95	4.0
25	positive first strokes	C	2.0	16	150
	Front duration				
89	negative first strokes	μs	1.8	5.5	18
118	negative subsequent strokes	μs	0.22	1.1	4.5
19	positive first strokes	μs	3.5	22	200
	Maximum di/dt				
92	negative first strokes	kA/μs	5.5	12	32
122	negative subsequent strokes	kA/μs	12	40	120
21	positive first strokes	kA/μs	0.20	2.4	32
	Stroke duration				
90	negative first strokes	μs	30	75	200
115	negative subsequent strokes	μs	6.5	32	140
16	positive first strokes	μs	25	230	2000
	Integral ($i^2 dt$)				
91	negative first strokes	A^2 s	6.0×10^3	5.5×10^4	5.5×10^5
88	negative subsequent strokes	A^2 s	5.5×10^2	6.0×10^3	5.2×10^4
26	positive first strokes	A^2 s	2.5×10^4	6.5×10^5	1.5×10^7
	Time				
133	between negative strokes	ms	7	33	150
	Flash duration				
94	negative (including single stroke flashes)	ms	0.15	13	1100
39	negative (excluding single stroke flashes)	ms	31	180	900
24	positive (only single stroke flashes)	ms	14	85	500

data have been published by Boyle and Orville (1976), Orville *et al.* (1978), the Saint-Privat D'Allier Research Group (1979), Hubert and Mouget (1981), and Idone and Orville (1982). The last paper is the most comprehensive to date on the subject. Seventeen first-stroke and forty-six subsequent-stroke velocities are given. The first-stroke mean velocity within about 1 km of ground was 9.6×10^7 m/s; the subsequent-stroke mean velocity was 1.2×10^8 m/s. All velocities decreased with height. The fact that these velocities are generally higher than the earlier results (Schonland and Collens, 1934; Schonland *et al.*, 1935; McEachron, 1947) is attributed to the fact that the recent measurements are made near ground where return-stroke velocity is a maximum.

G. Dart Leaders

Return strokes subsequent to the first in a flash to ground are usually initiated by dart leaders. Dart leaders are so named because they appear on streak camera photographs as a 50-m long dart of light propagating toward earth. Dart leaders carry cloud potential earthward via an ionizing wave of potential gradient (Loeb, 1966) and lower about 1 C of negative charge (Brook *et al.*, 1962) in about 2 ms. It follows that these leaders must have a current of the order of 500 A. Dart leader velocities range from about 1 to 27×10^6 m/s, with the higher velocities being related to shorter interstroke intervals and the lower velocities to longer intervals (Winn, 1965; Schonland *et al.*, 1956; Schonland, 1956).

The electric-field changes produced by dart leaders have been described by Malan and Schonland (1951). At a range of 5 to 8 km, the first dart leaders in multistroke flashes produce positive field changes and later ones have hook-shaped field changes which begin with a negative polarity. Malan and Schonland (1951) suggest that succeeding leaders originate from charge volumes higher in the cloud, the typical increase in height between successive leaders being about 0.7 km according to Malan and Schonland (1951) and about 0.3 km according to Brook et al. (1962). Apparent leader heights varied between 2 and 13 km (Brook et al., 1962). Krehbiel et al. (1979) have pointed out that single-station field observations could also be explained in terms of a horizontal rather than a vertical lengthening of the channel and have presented evidence to show that this is the case. Schonland et al. (1938) have inferred from the field ratios of 46 dart leaders and return-strokes that the dart-leader channels tend to be uniformly charged, although, as we will argue in Section IV-C, the charge removed by subsequent return strokes is apparently not uniform.

If the time interval between strokes is long, the dart leader may make a transition from a continuously moving leader to a stepped leader, a so-called "dart-stepped leader." The stepped portion has a relatively high downward velocity (about 10^6 m/s), short step lengths (about 10 m), and short time intervals between steps (about 10 μs) (Schonland, 1956; Krider et al., 1977).

The in-cloud portion of the dart leader produces considerable radiation at VHF (Takagi, 1969 a, b; LeVine and Krider, 1977; Rustan, 1979; Rustan et al., 1980) and in the microwave region from 400 to 1000 MHz (Brook and Kitagawa, 1964). This radiation emanates primarily from within the cloud rather than from the channel to ground (Proctor, 1971, 1976; Rustan, 1979; Rustan et al., 1980). The RF associated with the dart leader starts about 250 μs before the return stroke (LeVine and Krider, 1977) and, at frequencies above about 100 MHz, ceases about 100 μs prior to the return stroke (LeVine and Krider, 1977; Brook and Kitagawa, 1964). At 3 MHz, the dart-leader radiation often continues up to and during the return stroke (LeVine and Krider, 1977). These effects are illustrated in Fig. 21(a), (b), and (c).

H. Continuing Current

Return-stroke currents can last for about a millisecond with the "intermediate" current during the final stage decreasing from a few kiloamperes to zero (Hagenguth and Anderson, 1952). Following the current associated with the return stroke proper, there may also be a low-level "continuing current" to ground whose charge source is distributed horizontally in the cloud as illustrated in Fig. 7 (Krehbiel et al., 1979). Continuing currents have been measured directly in strikes to towers (Berger and Vogelsanger, 1965) and have been inferred from remote measurements of interstroke electric and magnetic fields. Brook et al. (1962) and Kitagawa et al. (1962) have made correlated photographic and electric field observations in New Mexico. So-called "long" continuing-current durations, defined by the New Mexico group as those lasting longer than a typical 40-ms interstroke interval, were found to have durations up to 500 ms, the average being 150 ms. The charges lowered by continuing currents were between 3.4 and 29.2 C, the average being about 12 C. Current values were between 38 and 130 A. Williams and Brook (1963), also in New Mexico, used a magnetometer to measure magnetic field from which continuing current was derived. They found an average current of 184 A, an average charge transfer of 31 C, and an average duration of 184 ms. Krehbiel et al. (1979), using multiple electric-field measuring stations in New Mexico,

found continuing currents between 50 and 580 A in three discharges. The initial current values were 580, 185, and 150 A, and all decreased with time. Livingston and Krider (1978) have reviewed the statistics on the occurrence of continuing current and have presented their own finding that 29 to 46 percent of the flashes in Florida had a continuing current. About half of the 200 flashes in New Mexico measured by Brook et al. (1962) and Kitagawa et al. (1962) had long continuing currents and about one-quarter of all the interstroke intervals had one. Schonland (1956) states that about 20 percent of all flashes in South Africa contain a continuing current; Thomson (1980) gives a figure of 48 percent for New Guinea.

The continuing currents measured by Berger and Vogelsanger (1965) in Switzerland following normal negative strikes to a tower were found to be of the order of 100 to 300 A. Half the flashes containing continuing currents lowered over 25 C, and the maximum charge lowered was 80 C. These values can be compared to the stroke currents and charges given in Table II.

The "long" continuing currents referred to in the preceding paragraphs are apparently the cause of burning effects (e.g., forest fires (Fuqua et al., 1967), burned-through overhead ground wires on power systems, metal damage on airplanes). Brook et al. (1962) and Kitagawa et al. (1962) also discuss "short" continuing currents, i.e., those that last less than 40 ms. In either case, after the continuing current ends, a normal interstroke interval, presumably containing a J-change, follows if there is to be another return stroke to ground.

An unusual class of continuing currents has been observed in Japanese winter thunderstorms by Brook et al. (1982). These lower positive charge and are characterized by larger currents and charge transfer than normal negative lightning. Brook et al. (1982) studied a continuing current which peaked at 10^5 A and which remained above 10^4 A for more than 3 ms. In two of eight positive flashes, the charge transfer to ground was 200 C and 300 C.

I. J- and K-Processes in Discharges to Ground

During the time interval between successive return strokes, there is usually a slow, relatively steady change in the electric field due to charge motion within the cloud. This change is called a J-change, the J standing for "Junction." Impulsive discharges termed K-changes are usually superimposed on the J-change at intervals of 5 to 10 ms (Kitagawa and Kobayashi, 1959; Kitagawa and Brook, 1960; Kitagawa et al., 1958). Kitagawa and Brook (1960) and Kitagawa (1965) suggest that the slow J-process is actually the instrumentally-smoothed sum of the field changes due to the rapid K-processes, each of which lasts less than 1 ms. Kitagawa and Brook (1960) show that the distribution of time intervals between K-changes in the interstroke intervals of discharges to ground is essentially the same as the distribution of K-change intervals in the final portion, the so-called J-portion, of an intracloud discharge, although the polarity of the K-changes may be different in cloud and ground discharges. Ogawa and Brook (1964) state that K-changes in cloud discharges are an order of magnitude larger than K-changes in ground discharges while others (e.g., Ishikawa, 1961; Wadhera and Tantry, 1967 a, b) claim they are of the same magnitude. We will discuss the J- and K-changes of cloud discharges in Section II-J.

The J-field change in ground discharges is almost always negative for flashes within a few kilometers and can be positive or negative for discharges beyond about 5 km (Malan and Schonland, 1951 a, b; Malan, 1955, 1963). During the J-change, there is no appreciable luminosity in the channel between cloud base and ground. Further, Malan and Schonland (1951b) report that, for flashes in the 5- to 12-km range, J-field changes occuring early in the flash are positive, while later ones are negative. However, Malan (1965) states that 44 percent of interstroke field changes beyond 25 km are negative, a result similar to the authors' observations in Florida. If all distant J-changes were positive instead of a mixture of positive and negative, that would be clear evidence for J-processes being due to a vertical motion of either negative charge downward or positive charge upward. It is likely that all distant J-changes are not positive because 1) many of the J-processes are more horizontal than vertical, and 2) some of the apparent negative field change that occurs during the interstroke period is due to the rearrangement of charge in the atmosphere between the measurement point and the source, rather than at the source (Illingworth, 1971). Evidence for nonvertical J-processes has recently been reported by Ogawa and Brook (1969) and Krehbiel et al. (1979). Rustan (1979) found vertical J-processes in some flashes and horizontal ones in others. The exact orientation of the J-process is undoubtedly related to the electrical structure of the cloud.

Schonland (1938) suggests that, during the time interval between strokes, J-discharges progress downward from previously untapped negative charge centers in the cloud to the top of the previous return-stroke channel. Apparent confirmation of this view of the J-process is given by Rustan et al. (1980) for one flash. On the other hand, Bruce and Golde (1941) and Malan and Schonland (1951b) argue that the J-change represents a raising of positive charge from the top of the previous return-stroke channel into a new region of negative charge. Further, visual observations (Brook and Vonnegut, 1960) tend to confirm this view. Krehbiel et al. (1979) have deduced from multiple-station electric-field measurements that J-processes in New Mexico usually move negative charge horizontally toward the top of a previous stroke, but this is not necessarily the same negative charge that is involved in the next stroke to ground.

The K-processes are generally thought to be "recoil streamers" or small return strokes which occur when a propagating channel encounters a pocket of opposite charge within the cloud. In this view, the J-changes are not necessarily the sum of their K-changes. The detailed characteristics of the K-change currents are very much in doubt, although there is evidence that there may be one or more fast pulses over an interval that is between about 500 μs and 750 μs (Kitagawa and Kobayashi, 1959; Brook and Kitagawa, 1964; Rustan, 1979; Rustan et al., 1980). Rustan (1979) and Rustan et al. (1980) measured the location of VHF sources termed "solitary pulses," which are probably associated with K-changes, and found them to propagate upward for a few kilometers at velocities between 1 and 4 $\times$ 10^7 m/s. Arnold and Pierce (1964) give a median value of 0.1 for the ratio of the maxi-

mum K-electric-field amplitude to the return-stroke electric-field peak, which implies that K-processes are associated with peak currents of the order of several thousand amperes, since the K-processes have velocities several times lower than return strokes (Rustan, 1979, Rustan et al., 1980, Arnold and Pierce, 1964). Steptoe (1958) and Muller–Hillebrand (1968) have proposed expressions for K-process currents. Both investigators suggest that the current rises to peak in 9 μs, falls to half value in tens of microseconds, and then remains at a low level until a significant fraction of a coulomb of charge is transferred.

If an intermediate or a continuing current is flowing to ground when a K-change occurs, the K-change will brighten the channel below cloud and produce a luminous event which is called an "M-component" (Malan and Schonland, 1947 Kitagawa et al., 1962). The implication of this observation is that the continuing-current discharge is propagating away from the top of the return-stroke channel when it contacts a region of negative charge, and that the resulting luminosity propagates downward to ground, perhaps in a manner similar to that of a dart leader.

J. Cloud Discharges

The term "cloud discharge" may well refer to a variety of different phenomena. The cloud discharge which sometimes precedes a flash to ground by more than 100 ms (see Section II-C) may be different from an isolated cloud flash; and there may be several types of isolated cloud flashes, the distinction between intracloud lightning, intercloud lightning, and cloud-to-air lightning being hardly discussed in the literature. Ogawa and Brook (1964) and Brook and Ogawa (1977) argue that these three types of cloud flashes are probably similar. Since relatively little is known about cloud discharges, we shall concentrate on the type about which the most is known, the typical isolated intracloud flash.

The average durations of cloud discharges have been given by a number of investigators, for example, 245 ms by Pierce (1955), 490 ms by Mackerras (1968), 300 ms by Takagi (1961), 420 ms by Ishikawa (1961), and 500 ms by Ogawa and Brook (1964). The average change in the total electric moment produced by cloud discharges has been found to be of the order of 100 C·km (Pierce, 1955; Mackerras, 1968 Takagi, 1960; Wang, 1963). Since the distance between charge centers is roughly 5 km, the charge transfers are of the order to 10 C, values which are similar to ground flashes. The bulk of the available ground-based electric-field data suggests that a typical intracloud flash contains an initial downward-moving discharge which propagates at a velocity of about 10^4 m/s and which lowers positive charge from the upper cloud region toward the lower negative charge (see Figs. 6 and 7). When contact occurs, several upward-moving K-changes are initiated Electrostatic-field measurements by Ogawa and Brook (1964) Takagi (1961), Takeuti (1965), and Ishikawa and Takeuch (1966) show that 60 to 75 percent of the cloud-discharge field changes are consistent with the downward movement of a positive charge. On the other hand, Smith (1957), working in Florida, found that 56 percent of the cloud discharges were initiated by negative discharges moving upward from the lowe part of the thundercloud and only 17 percent were of the downward positive type. Smith described the remaining cloud ischarges as the neutralization of a dipole cloud charge where he negative charge was above the positive. It is worth noting hat the interpretations of charge motion discussed above are ased on the assumption of vertical discharges and measure- ments at one or two observing stations, and hence the existence of nonvertical discharges could lead to errors in interpretation Huzita and Ogawa, 1976). Proctor (1981), using a VHF source-location technique in conjunction with a ground-based electric-field measurement, found that most cloud flashes in South Africa were horizontal and were initiated by several leaders moving negative charge away from a common origin. Detailed data are given on three cloud discharges which moved negative charge horizontally at altitudes of 2.9, 4.5, and 6.5 km above ground level, one which moved negative charge ver- tically between 4.8 and 11.5 km, and one which moved posi- tive charge horizontally at an altitude of 4 km, 1 km above the freezing level. None of these discharges discussed by Proctor 1981) are consistent with previously accepted views of a typical cloud discharge based only on electric-field measure- ments. Additional information is found in Proctor (1976).

To find the location of the cloud charges from ground-based electric-field measurements, a minimum of five ground stations are needed if the discharge is a vertical dipole with the ap- propriate symmetry and a minimum of seven stations if it is nonvertical. Seven stations provide enough information to solve for the charge and the x, y, z coordinates of each end- point (assuming the charge distribution to be spherically sym- metric), and five stations provide enough data to determine the charge, the x, y coordinates, and altitude of each end of the dipole. Analyses of multiple-station electric-field records by Workman et al. (1942) and by Reynolds and Neill (1955) show that cloud discharges in New Mexico are primarily hori- zontal: vertical charge separations are typically 0.6 km, with the positive charge generally above the negative, and the hori- zontal charge separations are typically 1 to 10 km. The mo- ment changes are of the order of 10 C·km, an order of mag- nitude smaller than the more recent measurements referenced earlier, probably due to the small vertical charge separation, but the charge transfers are similar, typically 10 to 20 C. Smith (1957) used only two ground stations in Florida, and hence could not find accurate charge locations. Takeuti (1965) used only three stations in Japan, but was able to estimate total path lengths of 2 to 8 km. It should be noted that si- multaneous multiple measurements of electric fields are diffi- cult to make and that data from different sets of stations in the same experiment can give quite different results, pre- sumably because of errors in measurement, difficulty in identifying a unique portion of the discharge at all stations, and lack of simple dipole behavior (Krehbiel et al., 1979).

The luminosity variation of an intracloud event is usually composed of a continuous, slowly-varying background on which is superimposed a series of rapid pulses (Kitagawa and Kobayashi, 1959; Takagi, 1961; Malan, 1955; Brook and Kitagawa, 1960).

The electric-field change produced by a typical intracloud discharge on a millisecond scale is shown in Fig. 23. Kitagawa and Brook (1960) have divided this overall field change into three phases: 1) an initial portion having a duration of 50 to 300 ms and characterized by small pulses with a mean pulse

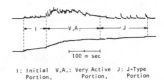

Fig. 23.   Diagram of the typical field change of a cloud discharge. The upper and lower traces are recorded simultaneously by the fast and slow antennas, respectively. Adapted from Kitagawa and Brook (1960).

interval of 680 μs. This initial process is quite different from the more-rapid pulsations that characterize the beginning of a discharge to ground; 2) a very active portion lasting of the order of 100 ms and characterized by large VLF pulses and a relatively rapid electrostatic-field change; and 3) the final or J-type portion which is similar to the J-change in ground dis- charge and which contains impulsive K-changes at intervals ranging from 2 to 20 ms. Kitagawa and Brook report that 50 percent of 1400 cloud-discharge field changes contained all three of the above phases, 40 percent were missing the initial portion, and 10 percent contained just the initial portion or the very active portion or both.

The larger fast pulses superimposed on the slow electro- static-field change are usually called K-changes; the smaller ones do not have a name. Ogawa and Brook (1964) state that K-changes do not occur during the initial or very active por- tions and interpret K changes as being due to upward-propa- gating negative recoil streamers that are initiated by a down- ward-moving positive discharge. Apparently, the cloud-dis- charge K-changes are similar to those which occur during the period between return strokes in flashes to ground (see Sec- tion I) and hence the discussion in Section I is also valid for the present case. On the other hand, the K-changes in cloud and ground discharge usually have opposite polarity and, according to Ogawa and Brook (1964), the mean moment change of cloud K-changes, 8 C·km, is considerably larger than the largest moment change, 2 C·km, found for ground K-changes. Ogawa and Brook (1964) suggest that the negative cloud K-changes (or recoil streamers) have the following prop- erties: time duration 1 to 3 ms, channel length 1 to 3 km, veloc- ity 2×10^6 m/s, charge neutralized 3.5 C, average current 1 to 4 kA. They also report that during the J-period there are most frequently six of these changes. Rustan (1979), Rustan et al. (1980), and Proctor (1981) have used a VHF source-location technique to follow the propagation of noise sources apparently associated with K-changes. Five studied by Rustan (1979) and Rustan et al. (1980) in one discharge had a typical velocity of 2×10^7 m/s and propagated upward from a height of 5 to 8 km to a height of about 13 km. The cloud discharge involved the raising of negative charge or the lowering of positive. Proctor (1981) studied 19 events having VHF noise charac- teristic of K-changes, but sometimes occurring without a K- electric-field change and found velocities from 2.5×10^6 m/s to 4.4×10^7 m/s for path lengths from 136 m to 4.4 km. All but one of these events transferred positive charge, consistent with the observed cloud discharges initially moving negative charge out of the main negative region of the cloud. Although

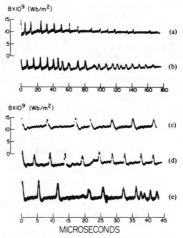

Fig. 24. Trains of unipolar magnetic-field pulses produced by five different intracloud lightning discharges within 50 km in Arizona. Adapted from Krider *et al.* (1975).

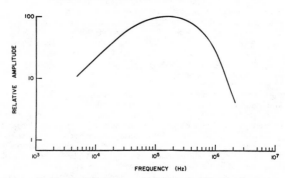

Fig. 25. The relative Fourier amplitude as a function of frequency for a typical intracloud pulse sequence as shown in Fig. 24. Adapted from Krider *et al.* (1975).

Ogawa and Brook (1964) report that *K*-changes occur only in the final portion of a cloud discharge, it is difficult to see how their occurrence could be ruled out in the very active portion, particularly in view of the multitude of electric-field pulses that are observed during that time.

Krider *et al.* (1975), Weidman and Krider (1979), and Krider *et al.* (1979) have examined the detailed shapes of individual field pulses generated by cloud discharges. Two types of pulses were found: 1) regular sequences of primarily unipolar pulses which occur at 5-μs intervals and which have a total sequence duration of 100 to 400 μs. 2) large-amplitude,

bipolar pulses similar to those we have associated with the pre liminary breakdown (see Section II-C) but with opposite po larity and larger pulse intervals.

The unipolar pulses have risetimes of 0.2 μs or less and a full width at half maximum of about 0.75 μs. Examples o these pulses are given in Fig. 24 and the frequency spectrum o one pulse sequence is given in Fig. 25. Krider *et al.* (1975 noted the similarity of these pulse sequences to those radiate by the dart-stepped leaders which sometimes preceded subse quent strokes in ground discharges (see Section II-G) and sug gest that a similar dart-stepping process may occur in the clou

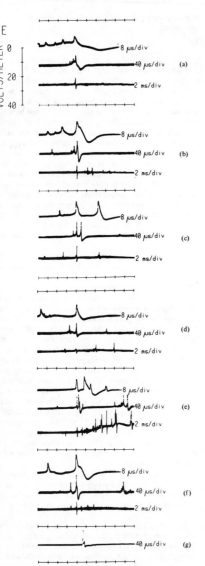

VOLTS/METER

E

0

20

40

8 μs/div
40 μs/div   (a)
2 ms/div

8 μs/div
40 μs/div   (b)
2 ms/div

8 μs/div
40 μs/div   (c)
2 ms/div

8 μs/div
40 μs/div   (d)
2 ms/div

8 μs/div
40 μs/div   (e)
2 ms/div

8 μs/div
40 μs/div   (f)
2 ms/div

40 μs/div   (g)

Fig. 26.   Large bipolar electric fields radiated by intracloud discharges at distances of 15 to 30 km. Each event is shown on time scales of 2 ms/div, 40 μs/div, and 8 μs/div. A positive field is shown as a downward deflection on all records. Adapted from Weidman and Krider (1979).

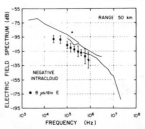

Fig. 27.   Frequency spectra of 6 large bipolar electric fields from intracloud discharges as shown in Fig. 26. The decibel values given are 20 times the logarithm (base 10) of the magnitude of the Fourier transform of the electric-field intensity. Solid symbols indicate mean values; vertical bars, standard deviations. Lightning between 15 and 30 km, normalized to 50 km. The solid lines represent the first-return-stroke frequency spectra given in Fig. 19.

discharge. If each of the 20 to 80 pulses occurring at 5-μs intervals has a length of 10 m (a value similar to dart-stepped leaders), the total channel length would be 200 to 800 m and the propagation velocity would be 2×10^6 m/s. These values are similar to the Ogawa and Brook (1964) description of a K-change. Krider et al. (1975) find that the sequences of regular pulses occur throughout the cloud discharge, but state that there is a tendency for the sequences to occur toward the end of the discharge. This situation implies that some may well be associated with K-changes.

Examples of the large bipolar waveforms are shown in Fig. 26 and the frequency spectrum of individual pulses is shown in Fig. 27. Weidman and Krider (1979) have found that the large pulses often have several fast unipolar pulses superimposed on the initial rise to peak. The shape of the fast impulses is similar to individual pulses in the regular sequences just discussed. The large pulses have a mean full width of 63 μs and a mean pulse interval of 780 μs, a value which is similar to the 680 μs mean pulse interval reported by Kitagawa and Brook (1960) for the initial portion of the discharge. On the other hand, the bipolar pulses are among the largest events in the cloud discharge and these perhaps should be associated with the very active portion of the discharge. Weidman and Krider (1979) suggest that the fast pulses on the initial rise are due to a step-like breakdown current and that the subsequent large bipolar field is due to a slower current surge flowing in the channel established by the steps. Krider et al. (1979) have shown that RF radiation at frequencies between 3 and 295 MHz is coincident with each bipolar pulse, as illustrated in Fig. 28, in turn implying that the process which generates the pulse probably involves the breakdown of virgin air rather than propagation along an already-existing channel.

K. Flash Frequency Spectra

A number of investigators have used narrow-band receivers at a variety of frequencies to sample portions of the lightning frequency spectrum. A composite view of the flash integrated frequency spectrum which incorporates essentially all pub-

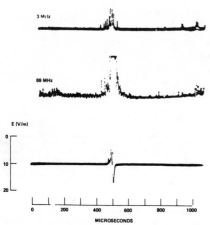

Fig. 28. A large bipolar electric field and associated RF emissions at 3 and 69 MHz produced by an intracloud lightning discharge. The polarity of the electric field is such that a negative potential gradient is displaced upward. Adapted from Krider *et al.* (1979).

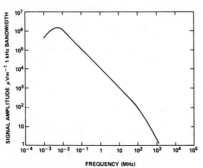

Fig. 29. Spectrum of total lightning flash normalized to a bandwidth of 1 kHz and a distance of 10 km.

lished measurements is shown in Fig. 29. Similar composite spectra have been published by Kimpara (1965), Oh (1969), Oetzel and Pierce (1969), Pierce (1977), and others.

Since most measurements have been made for lightning at distances greater than 10 km, the frequency content above about 1 MHz in Fig. 27 must be considered a lower limit to the actual value since the higher frequencies can be strongly attenuated by ground-wave propagation (see Section II-F and Fig. 18(a), (b)). Unfortunately, from an integrated spectrum, it is not possible to determine which lightning process produces a particular spectral emission.

Spectra below about 20 MHz are best determined by a Fourier analysis of time-domain waveforms, but the effects of ground-wave propagation must be considered at the higher frequencies.

III. AIRCRAFT MEASUREMENTS

An aircraft in flight can become attached to a lightning channel by initiating discharges which connect to the stepped leader or to any other phase of the discharge in which there is a propagating channel (e.g., the *J*-process discussed in Section II-I). Further, an aircraft can "trigger" lightning which otherwise would not have occurred had the aircraft not been there (Fitzgerald, 1967). Triggering is made possible by the field enhancement caused by the presence of the aircraft in a region where the field is already high, and the resultant generation and propagation of leaders in two directions away from the aircraft toward regions of opposite charge. Any charge on the aircraft due to precipitation interactions may increase the probability of a lightning strike. Further, the charge deposited in the wake of the aircraft (which is opposite in polarity to that on the plane) due to precipitation charging may serve either to guide the lightning to the plane or to shield the plane from it. A discussion of these and other effects is given by Vonnegut (1965) and in the paper by Clifford and Kasemir in this issue.

The attachment of a typical lightning to an aircraft in flight probably takes place as follows: A leader with a typical charge density of 10^{-3} C/m approaches the aircraft at a velocity of 10^5 to 10^6 m/s. When the fields on the wings, nose, tail, and other extremities exceed a critical value, corona will begin and outward-going discharges will be initiated. These discharges will lower the electric field at the surfaces of the plane because of shielding. The "striking distance" is determined by the distance to the incoming leader at the time when the aircraft initiates the outward-going connecting discharges. The time for the leader and connecting discharge to join is probably of the order of 10 μs, and the striking distance is probably of the order of the plane length. When contact takes place, the aircraft will be raised to the 10^8- to 10^9-V potential of the lightning channel in a characteristic charging time determined by the channel surge impedance, roughly 1000 Ω, and the aircraft capacitance, 10^{-9} to 10^{-10} F. Thus the charging time will be 10^{-7} to 10^{-6} s, and the rate of change of the aircraft voltage will be 10^8 to 10^{10} V·μs. Contact will also cause the electric field at the surface of the plane to reverse direction, and the aircraft may produce additional corona and leaders. In any event, the leader current will propagate through the aircraft in a relatively steady fashion, and then there may be superimposed fast current pulses due to leader steps, *K*-changes, return strokes, or any other impulsive lightning process.

Currents and photographs of 52 lightning events measured with an instrumented F-100F aircraft are given by Petterson and Wood (1968). It is not known in which phases of the various lightning discharges the plane was involved. Since positive charge was usually transferred into the nose boom, Petterson and Wood (1968) infer that most of the strikes were intracloud discharges between the upper positive and lower negative cloud charge regions (Figs. 6 and 7). Most peak currents were several thousand amperes with millisecond rise times. There were also fast current pulses with zero to peak risetimes of about 1 μs. The maximum current was 22 kA for which the risetime was not recorded. One point of interest is that, when a lightning channel became attached to the F-100F, the motion of the plane did not detach the channel, but rather stretched it.

Data on electric and magnetic fields and currents during strikes to aircraft have been reported by Pitts (1981) and by Pitts and Thomas (1981). Ten strikes to an F-106B were recorded during 1980 as part of a research program at the NASA Langley Research Center. Some of the electric- and magnetic-field pulses have widths of less than 1 μs and rise times in the tens of nanoseconds range. The peak currents, however, are only between 100 and 200 A with rise times between about 0.1 and 1.0 μs. No apparent relation exists between the simultaneous electric field, magnetic field, and current, although the data still have not been completely analyzed. Information on the instrumentation and sensors used in the NASA study is given by Baum *et al.* in this issue.

Baum (1980) has presented the initial results of airborne electric- and magnetic-field and current measurements made on a WC-130 aircraft as part of a U.S. Air Force study. Many waveforms similar to those of return strokes measured at ground level were observed from lightning in the 10-km range, a result which is expected from theory (see Fig. 33). In addition, series of pulses were recorded with rise times as short as 40 ns or less and with widths of about 1 μs or less, similar to those observed in the NASA study. Information on the instrumentation and sensors used in the Air Force study is given by Baum *et al.* in this issue.

Some data from recent French airborn measurements using a Transall C-160 aircraft are reported by Clifford and Kasemir in this issue. Current pulses between 1 and 70 kA were found to be superimposed on a steady current of the order of 1 kA or less, the pulses having widths between 8 and 200 μs.

IV. MODELING AND THE RELATION OF LIGHTNING CURRENTS TO REMOTE ELECTRIC AND MAGNETIC FIELDS

A. Modeling

A model can be defined as a physical or mathematical construct which approximates to some degree certain aspects of natural or man-made phenomena. Perhaps the most important aspect of a model is that it can be used for prediction. We will be concerned with the mathematical modeling of lightning processes which result in the prediction of the electric and magnetic fields produced by those processes.

There are basically three levels of sophistication in the mathematical modeling of lightning.

1) The least sophisticated approach, but to date the one which provides the closest approximation to nature, simply assumes a temporal and spatial form for the channel current and then uses this current to calculate the remote fields. The form of assumed current is constrained to agree with the properties of lightning currents measured at the ground and

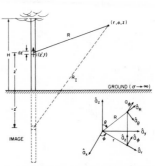

Fig. 30. A drawing defining all geometrical parameters needed in the calculation of electric and magnetic fields.

by the available data on the electric and magnetic fields. Lin *et al.* (1980) have reviewed the literature on this type of model for return strokes and have presented a new and superior variation. Master *et al.* (1981) have further improved the model of Lin *et al.* (1980) so as to provide more realistic fields above ground level.

2) A more sophisticated approach involves mathematically describing the lightning channel as an *RLC* transmission line with circuit elements that may vary with height and time. The intent is to predict a channel current as a function of height and time, and to use this current to calculate the fields. Price and Pierce (1977) and Little (1978, 1979) have used this approach to model return strokes with reasonable success.

3) The most sophisticated model attempts to describe the detailed physics of the lightning channel using equations of conservation of mass, momentum, and energy, equations of state, and Maxwell's equations. Unfortunately, this type of model requires a detailed knowledge of the physical parameters such as the ionization and recombination coefficients and of thermodynamic properties such as the thermal and electrical conductivities; but one can still attempt to predict the channel current as a function of height and time. A model of this type has recently been developed for lightning return strokes by Strawe (1979) and holds considerable promise for providing a better understanding of the discharge. At present, however, this model still does not produce realistic results.

Here, we only consider models of type (1) and judge their validity by how well the assumed currents and predicted fields agree with measurements.

B. The Relation of Lightning Currents to Remote Electric and Magnetic Fields

For the present discussion, we assume all lightning currents are contained in straight, vertical channels of negligible cross-section above a perfectly conducting ground plane as shown is Fig. 30. The extension of this approach to tortuous channels with horizontal components will be discussed later. The differential electric and magnetic fields at altitude z and range r from a short length of channel dz' at height z' carrying a time-varying current $i(z', t)$ are

$$
d\vec{E}(x, y, z, t)
$$
$$
= \frac{dz'}{4\pi\epsilon_0 R} \left[\left\{ \frac{3r(z-z')}{R^4} \int_0^t i(z', \tau - R/c)\, d\tau \right. \right.
$$
$$
+ \frac{3r(z-z')}{cR^3} \cdot i(z', t - R/c) + \frac{r(z-z')}{c^2 R^2}
$$
$$
\left. \cdot \frac{\partial i(z', t - R/c)}{\partial t} \right\} \vec{a}_r + \left\{ \frac{2(z-z')^2 - r^2}{R^4} \right.
$$
$$
\cdot \int_0^t i(z', \tau - R/c)\, d\tau + \frac{2(z-z')^2 - r^2}{cR^3}
$$
$$
\left. \left. \cdot i(z', t - R/c) - \frac{r^2}{c^2 R^2} \cdot \frac{\partial i(z', t - R/c)}{\partial t} \right\} \vec{a}_z \right] \quad (1)
$$

$$
d\vec{B}(x, y, z, t) = \frac{\mu_0 dz'}{4\pi R} \left\{ \frac{r}{R^2} \cdot i(z', t - R/c) + \frac{r}{cR} \right.
$$
$$
\left. \cdot \frac{\partial i(z', t - R/c)}{\partial t} \right\} \vec{a}_\phi \quad (2)
$$

where (1) and (2) are in cylindrical coordinates, ϵ_0 is the permittivity and μ_0 the permeability of vacuum, and all geometrical factors are defined in Fig. 30 (Uman *et al.*, 1975; Master *et al.*, 1981; Master and Uman, 1982). The effects of the perfectly conducting ground plane on the radiation from the source at z' are included by replacing the plane by an image current as shown in Fig. 30 (Stratton, 1941). The electric and magnetic fields of the image are obtained by substituting R_I for R and z_I for z' in (1) and (2). Once the expressions for the fields of a short channel section are formulated, the fields for the total channel can be obtained by integration over the channel.

An important special case is that of the fields at the ground. For a vertical channel between the heights of H_B and H_T, these are (Uman *et al.*, 1975)

$$
E_z(x, y, 0, t)
$$
$$
= \frac{1}{2\pi\epsilon_0 R} \left[\int_{H_B}^{H_T} \frac{(2z'^2 - r^2)}{R^4} \int_0^t i(z', \tau - R/c)\, d\tau\, dz' \right.
$$
$$
+ \int_{H_B}^{H_T} \frac{(2z'^2 - r^2)}{cR^3} i(z', t - R/c)\, dz'
$$
$$
\left. - \int_{H_B}^{H_T} \frac{r^2}{c^2 R^2} \frac{\partial i(z', t - R/c)}{\partial t}\, dz' \right] \quad (3)
$$

$$
B_\phi(x, y, z, t) = \frac{\mu_0}{2\pi R} \left[\int_{H_B}^{H_T} \frac{r}{R^2} i(z', t - R/c)\, dz' \right.
$$
$$
\left. + \int_{H_B}^{H_T} \frac{r}{cR} \frac{\partial i(z', t - R/c)}{\partial t}\, dt' \right] \quad (4)
$$

C. Calculations

Equations (1) and (2) have been used to calculate the electric and magnetic fields as a function of time at any specified point in space with the channel currents as input parameters. From the time-domain fields, the field frequency spectra are obtained. As an example of the technique, we will present the fields and frequency spectra for a typical subsequent return stroke using the model of Lin *et al.* (1980) and a modification of that model due to Master *et al.* (1981). Other important lightning processes (e.g., preliminary breakdown, stepped leader, first return stroke) have been modeled using similar techniques (Uman and Krider, 1981).

The model of Lin *et al.* (1980) postulates that the return stroke current is composed of three components: a) a short-duration upward-propagating pulse of *constant* magnitude and waveshape that is associated with the electrical breakdown at the return-stroke wavefront and that produces the fast peak current. This pulse is assumed to propagate at a constant velocity; b) a uniform current which is already flowing in the leader channel (or which may start to flow soon after the commencement of the return stroke); and c) a "corona" current which is caused by a flow of charge radially inward and then downward which removes the charge initially stored in the corona sheath around the leader channel after the passage of the return-stroke wavefront. These three current components are illustrated in Fig. 31.

Two observations form the basis for a modification of the Lin *et al.* (1980) model by Master *et al.* (1981). i) Formerly, subsequent strokes were thought to have both luminosities (and, therefore, currents) and velocities that were constant with height (Schonland, 1956). However, Jordan and Uman (1980) have recently found that the peak luminosity of subsequent strokes decreases to half the initial peak in less than one kilometer above ground. The implication of this observation is that the breakdown pulse current [(a) in previous paragraph] must also decrease with height. ii) When the breakdown pulse reaches the top of the channel, the model of Lin *et al.* (1980) predicts a field pulse of opposite polarity to that of the initial field with a waveshape that is a "mirror image" of the initial field. A detailed discussion of the "mirror image" effect is given by Uman *et al.* (1975). Mirror images are observed occasionally in the fields from first-return strokes, but almost never in the fields from subsequent-return strokes (Lin *et al.*, 1979). This observation is reasonable if the breakdown pulse current decays with height so that it has a negligible effect when it reaches the end of the channel. In view of i) and ii) above, Master *et al.* (1981) have proposed the following modification to the model of Lin *et al.* (1980): the breakdown pulse current is allowed to decrease with height above ground, but all other features of the original model remain unchanged. As we shall see, the fields at ground level produced by this new model are essentially the same as those of Lin *et al.* except for the absence of the "mirror image." The fields in the air, however, differ considerably, especially at close ranges.

We first calculate the electric and magnetic fields of a typical subsequent return stroke using the model of Lin *et al.* (1980), and then we repeat the calculation using the improved model. The current components at ground level are shown in

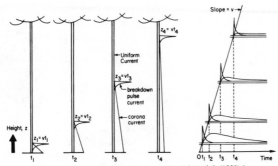

Fig. 31. Current distribution for the model of Lin et al. (1980) in which the breakdown pulse current is constant with height. The constant velocity of the breakdown pulse current is v. Current profiles are shown at four different times t_1 through t_4, when the return-stroke wavefront and the breakdown pulse current are at four different heights z_1 through z_4, respectively.

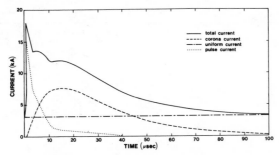

Fig. 32. Return-stroke current components at ground for a typical subsequent stroke calculated from measured electric and magnetic fields from Lin et al. (1980).

Fig. 32. The current parameters used in the field calculations are as follows.

a) Breakdown current: increases from 0 to 3 kA in 1.0 μs, rises abruptly to a peak of 14.9 kA at 1.1 μs, falls to half peak value at 3.8 μs, and zero at 40 μs. The fast transition risetime given in the previous sentence has been altered from the value used by Lin et al. (1980) so as to be consistent with the measurements of Weidman and Krider (1978). The breakdown current pulse propagates upward with an assumed constant velocity of 1×10^8 m/s.

b) Uniform current: 3100 A.

c) Corona current per meter of channel: $I_0 e^{-z'/\lambda}(e^{-\alpha t} - e^{-\beta t})$, with $I_0 = 21$ A/m, $\lambda = 1500$ m, $\alpha = 10^5$ s^{-1}, and $\beta = 3 \times 10^6$ s^{-1}. The total charge initially stored on the leader and lowered by return stroke is 0.3 C. The channel length is 7.5 km. In the improved model, the amplitude of the breakdown pulse current decreases with height as $e^{-z'/\lambda}$, with $\lambda = 1500$ m; that is, the breakdown pulse current decreases with the same scale height as the corona current. All other parameters in the improved model are the same as those for the original model.

Calculated vertical and horizontal electric fields are shown in Fig. 33(a), (b), respectively, and calculated magnetic fields in Fig. 33(c). Solid lines represent the original model of Lin et al. (1980), dashed lines the modified version of the model in which the breakdown pulse current decreases with height. All zero times on the figures indicate the time at which the return-stroke current originates at ground level. The waveforms at the field points begin after the appropriate propagation-time delay. The intersections of the slanted solid lines with the horizontal dotted lines at various heights indicate the times at which the return-stroke wavefront passes those heights. Salient features of these waveforms are discussed by Master et al. (1981). Fig. 34(a) and (b) give the amplitude spectra of the vertical electric field at ground level at distances of 50 km and 200 m, respectively. The data in Fig. 34(a) compare favorably with the measurements at 50 km given in Fig. 18(b).

The type of lightning model just discussed can easily be

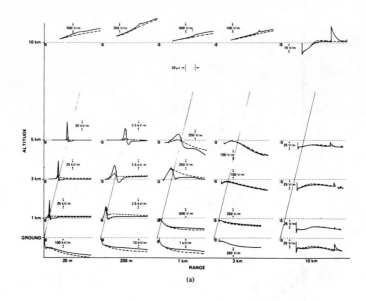

(a)

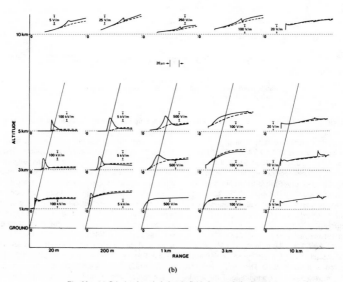

(b)

Fig. 33.   (a) Calculated vertical electric fields for a typical subsequent
return stroke from Master *et al.* (1982). (b) Calculated horizontal
electric fields for a typical subsequent return stroke from Master
et al. (1982).

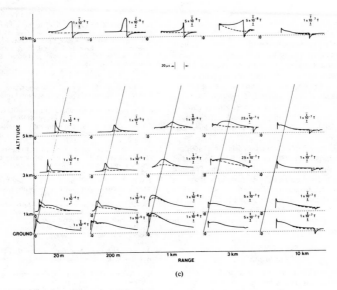

(c)

Fig. 33.   (c) Calculated magnetic fields for a typical subsequent return stroke from Master *et al.* (1982).

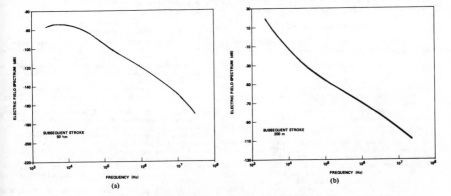

Fig. 34.   (a) Calculated frequency spectra for the vertical electric field
of a typical subsequent return stroke at 50 km on the ground. The
dB values given are 20 times the logarithm (base 10) of the magnitude of the Fourier transform of the electric-field strength. (b) Calculated frequency spectra for the vertical electric field of a typical
subsequent return stroke at 200 m on the ground. the decibel values
given are 20 times the logarithm (base 10) of the magnitude of the
Fourier transform of the electric-field strength.

generalized to include the effects of nonvertical channels, channel tortuosity, and branching. It can also be applied to any other lightning process as long as the electric and magnetic fields produced by that process have been adequately measured as a function of distance. Uman and Krider (1981) have listed the best present estimates of the input current parameters for other important lightning processes that can be used in the modeling approach discussed above. Unfortunately, the fields, and hence the currents, for many of these processes are not adequately known, and, therefore, more and better measurements are needed before the modeling can be considered completely satisfactory.

Using simple return-stroke current models, Hill (1969) and LeVine and Meneghini (1978) have investigated the effect of channel tortuosity on distant radiation fields and Pearlman (1979) has studied the effect on very close fields. It appears that tortuosity adds frequency content above about 100 kHz to the distant radiation fields, but does not have a significant effect on the close fields. Additional work is needed to combine a realistic model for tortuosity and branching with an adequate return-stroke current model.

SECTION V
CONCLUSIONS

Although lightning has received serious study for most of the 20th century, it is only within the last decade that the electric and magnetic fields have been adequately measured and that realistic models to predict those fields have been proposed. The discovery of large submicrosecond variations in fields and currents is, in part, responsible for the major reexamination of aircraft lightning protection and hazard testing now in progress. Research currently taking place throughout the world should soon fill in many of the present deficiencies in our knowledge evident from this review.

ACKNOWLEDGMENT

The authors would like to express their appreciation to M. J. Master for his many helpful comments and suggestions.

BIBLIOGRAPHY

SECTION I

C. E. Baum, E. L. Breen, J. P. O'Neill, C. B. Moore, and D. L. Hall, "Measurement of electromagnetic properties of lightning with 10 nanosecond resolution in Lightning Technology," NASA Conf. Publ. 2128, FAA-RD-80-30, 1980, pp. 39–82.

K. Berger, "Novel observations on lightning discharges: results of research on Mount San Salvatore," J. Franklin Inst., 283, pp. 478–528, 1967.

——, "Methoden und resultate der blitzforschung auf dem Monte San Salvatore bei Lugano in den jahre 1963-1971," Bull. Schweiz. Elektrotech. Ver., vol. 63, pp. 1403–1422, 1972.

K. Berger and E. Vogelsanger, "Photographische blitzuntersuchugen der jahre 1955-1963 auf dem Monte San Salvatore," Bull. Schweiz. Elektrotech. Ver., vol. 57, pp. 1–22, 1966.

M. Brook, M. Nakano, P. Krehbiel, and P. Takeuti, "The electrical structure of the Hokuriku winter thunderstorms," J. Geophys. Res., vol. 87, 1201–1215, 1982.

D. W. Clifford, E. P. Krider, and M. A. Uman, "A case of submicrosecond risetime current pulses for use in aircraft induced coupling studies," presented at 1979 IEEE-EMC Conf., IEEE Rep. 79CH-1383-9 EMC, Library of Congress Catalog No. 78-15514.

R. E. Holzer, "Simultaneous measurement of sferics signals and thunderstorm activity," in Thunderstorm Electricity, H. R. Byers, Ed. Chicago, IL.: University of Chicago Press, 1953, pp. 267–275.

C. G. Kitterman, "Characteristics of lightning from frontal system thunderstorms," J. Geophys. Res., vol. 85, pp. 5503–5505, 1980.

J. M. Livingston, and E. P. Krider, "Electric fields produced by Florida thunderstorms," J. Geophys. Res., vol. 83, pp. 385–401, 1978.

D. J. Malan, The Physics of Lightning. London, England: English Universities Press, 1963.

K. B. McEachron, "Lightning to the Empire State Building," J. Franklin Inst., vol. 227, pp. 149–217, 1939.

F. L. Pitts, and M. E. Thomas, "1980 direct strike lightning data," NASA Tech. Memo. 81946, Langley Research Center, Hampton, VA, Feb. 1981.

B. F. J. Schonland, "The lightning discharge," in Handbuch der Physik, vol. 22, S. Flugge Ed. New York: pp. 576–628, 1956.

T. Takeuti, M. Nakano, M. Nagatani, and H. Nakada, "On lightning discharges in winter thunderstorms," J. Met. Soc. Japan, vol. 51, pp. 494–496, 1973.

T. Takeuti, M. Nakano, and Y. Yamamoto, "Remarkable characteristics of cloud-to-ground discharges observed in winter thunderstorms in Hokuriku area, Japan," J. Met. Soc. Jap., vol. 54, pp. 436–439, 1976.

T. Takeuti, M. Nakano, M. Brook, D. J. Raymond, and P. Krehbiel, "The anamalous winter thunderstorms of the Hokuriku coast," J. Geophys. Res., vol. 83, pp. 2385–239, 1978.

T. Takeuti, S. Israelsson, M. Nakano, D. Lundquist, and E. Astrom, "On thunderstorms producing positive ground flakes," Proc. Res. Inst. of Atmospherics, Nagoya University, Japan, 27-A, pp. 1–17, 1980.

E. M. Thomson, "The dependence of lightning return stroke characteristics on latitude," J. Geophys. Res., vol. 85, pp. 1050–1056, 1980.

M. A. Uman, Lightning, New York: McGraw-Hill, 1969.

M. A. Uman, Understanding Lightning, BEK Industries, Inc., Pittsburgh Pa., 1971.

C. D. Weidman and E. P. Krider, "The fine structure of lightning return stroke waveforms," J. Geophys. Res., vol. 83, pp. 6239–6247, 1978.

C. D. Weidman and E. P. Krider, "Submicrosecond risetimes in lightning return stroke fields," Geophys. Res. Lett., vol. 7, pp. 955–958, 1980a.

C. D. Weidman, and E. P. Krider, "Submicrosecond risetimes in lightning radiation fields, in Lightning Technology," NASA Conf. Publ. 2128, FAA-RD-80-30, 1980b, pp. 29–38.

C. D. Weidman, E. P. Krider, and M. A. Uman, "Lightning amplitude spectra in the interval from 100 kHz to 20 MHz," Geophys. Res. Lett., vol. 8, pp. 931–934, 1981.

W. P. Winn, T. V. Aldridge, and C. B. Moore, "Video tape recording of lightning flashes," J. Geophys. Res., vol. 78, pp. 4515–4519, 1973.

SECTION II

A. Cloud Charges and Static Electric Fields

E. V. Appleton, R. A. Watson-Watt, and J. F. Herd, "Investigations on lightning discharges and on the electric fields of thunderstorms," in Proc. Roy. Soc. London, A221, pp. 73–115, 1920.

V. I. Arabadzhi, "The measurement of electric field intensity in thunderclouds by means of a radiosonde. Doklady Akad. Nauk. SSSR 111, pp. 85–88, 1956.

V. Barnard, "The approximate mean height of the thundercloud charge taking part in a flash to ground," J. Geophys. Res., vol. 56, pp. 33–35, 1951.

K. A. Brown, P. R. Krehbiel, C. B. Moore, and G. N. Sargent "Electrical screening layers around charged clouds," J. Geophys. Res., vol. 76, pp. 2825–2836, 1971.

W. H. Evans, "Electric fields and conductivity in thunderclouds," J. Geophys. Res., vol. 74, pp. 939–948, 1969.

W. H. Evans, "The measurement of electric fields in clouds," Rev. Pure Appl. Geophys., vol. 62, pp. 191–197, 1965.

D. R. Fitzgerald, "Experimental studies of thunderstorm electrification," Air Force Geophys. Lab., AFGL-TR-76-0128, AD-A032374, June 1976.

O. P. Gish, and G. R. Wait, "Thunderstorms and the earth's general electrification," J. Geophys. Res., vol. 55, pp. 473, 1950.

R. Gunn, "The electric field intensity and its systematic changes under an active thunderstorm," J. Atmos. Sci., vol. 22, pp. 498–50, 1965.

R. Gunn, "Electric field intensity at the ground under active thunderstorms and tornadoes," J. Meteor., vol. 13, pp. 269–273, 1956.

R. Gunn, "The electrification of precipitation and thunderstorms," Proc. IRE, vol. 45, pp. 1131–1358, 1957.

R. Gunn, "Electric field intensity inside of natural clouds," J. Appl. Phys., vol. 19, pp. 481–484, 1957.

R. E. Holzer, and D. S. Saxon, "Distribution of electrical conduction currents in the vicinity of thunderstorms," *J. Geophys. Res.*, vol. 57, pp. 207–216, 1952.

W. A. Hoppel, and B. B. Phillips, "The electrical shielding layer around charged clouds and its roll in thunderstorm electricity," *J. Atmos. Sci.*, vol. 28, pp. 1258–1271, 1971.

A. Huzita, and T. Ogawa, "Charge distribution in the average thunderstorm cloud," *J. Meteor. Soc. Japan*, vol. 54, pp. 284–288, 1976.

I. M. Imyanitov, Y. V. Chubarina, and Y. M. Shvarts, "Electricity in clouds," NASA Tech. Trans. from Russia, NASA TT-F-718, 1972.

I. M. Imyanitov, and Y. V. Chubarina, "Electricity of the free atmosphere, translated from Russian by Israel Program for Scientific Translations," Jerusalem, NASA TT-F-425, TT 67-51374, Clearinghouse for Federal Scientific and Technical Information, 1967.

E. A. Jacobson, and E. P. Krider, "Electrostatic field changes produced by Florida lightning," *J. Atmos. Sci.*, vol. 33, pp. 113–117, 1976.

H. W. Kasemir, *The Thundercloud, Problems of Atmospheric and Space Electricity*, S. C. Coroniti, Ed. New York: American Elsevier Publishing Company, pp. 215–235, 1965.

H. W. Kasemir, and F. Perkins, "Lightning trigger field of the orbiter," *Final Rep.*, Kennedy Space Center Contract CC 69694A, 1978.

J. D. Klett, "Charge screening layers around electrified clouds," *J. Geophys. Res.*, vol. 77, pp. 3187–3195, 1972.

P. R. Krehbiel, M. Brook, and R. A. McCrory, "An analysis of the charge structure of lightning discharges to the ground," *J. Geophys. Res.*, vol. 84, pp. 2432–2456, 1979.

J. Latham, "The electrification of thunderstorms," *Quart. J. Roy. Meteorol. Soc.*, vol. 107, pp. 277–298, 1981.

J. M. Livingston, and E. P. Krider, "Electric fields produced by Florida thunderstorms," *J. Geophys. Res.*, vol. 83, pp. 385–401, 1978.

C. Magono, *Thunderstorms*. New York: Elsevier, 1980.

D. J. Malan, "Les decharges dans l'air et la charge inferieure positive d'un nuage orageux," *Ann. Geophys.*, vol. 8, pp. 385–401, 1952.

D. J. Malan, and B. F. J. Schonland, "The distribution of electricity in thunderclouds," *Proc. Roy. Soc.*, vol. 209, pp. 158–177, 1951.

C. B. Moore, and B. Vonnegut, "The thundercloud," in *Lightning, 1, Physics of Lightning*, R. H. Golde, Ed. New York: Academic Press, pp. 51–98, 1977.

T. Ogawa, and M. Brook, "Charge distribution in thunderstorm clouds," *Quart. J. Roy. Meteorol. Soc.*, vol. 95 pp. 513–525, 1969.

B. B. Phillips, "Charge distribution in a quasistatic thundercloud model," *Monthly Weather Rev.*, vol. 95, pp. 847–853, 1967.

B. B. Phillips, "Convected cloud charge in thunderstorms," *Monthly Weather Rev.*, vol. 95, pp. 863–870, 1967.

B. J. F. Schonland, and J. Craib, "The electric fields of South African thunderstorms," *Proc. Roy. Soc.*, A114, pp. 229–243, 1927.

G. C. Simpson, "Atmospheric electricity during disturbed weather," *Geophys. Mem. London*, vol. 84, pp. 1–51, 1949.

G. C. Simpson, and G. D. Robinson, "The distribution of electricity in the thunderclouds," Pt. 11, *Proc. Roy. Soc. London*, A117, pp. 281–329, 1941.

G. C. Simpson, and F. J. Scrase, "The distribution of electricity in thunderclouds," *Proc. Roy. Soc. London*, A161, pp. 309–352, 1937.

C. T. R. Wilson, "On some determinations of the sign and magnitude of electric discharges in lightning flashes," *Proc. Roy. Soc. London*, A82, pp. 555–574, 1916.

C. T. R. Wilson, "Investigations on lightning discharges and on the lightning field of thunderstorms," *Phil. Trans. Roy. Soc. London*, A92, pp. 73–115, 1920.

W. P. Winn, and C. B. Moore, "Electric field structure in an active part of a small isolated thundercloud," *J. Geophys. Res.*, vol. 86, pp. 1187–1193, 1981.

W. P. Winn, and L. G. Byerley, "Electric field growth in thunderclouds," *Quart. J. Roy. Meteorol. Soc.*, vol. 101, pp. 979–994, 1975.

W. P. Winn, and C. B. Moore, "Electric field measurements in thunderclouds using instrumented rockets," *J. Geophys. Res.*, vol. 76, pp. 5003–5018, 1971.

W. P. Winn, G. W. Schwede, and C. B. Moore, "Measurements of electric fields in thunderclouds," *J. Geophys. Res.*, vol. 79, pp. 1761–1767, 1974.

B. Preliminary Breakdown

W. H. Beasley, M. A. Uman, and P. L. Rustan, "Electric fields preceding cloud-to-ground lightning flashes," *J. Geophys. Res.*, vol. 87, 1982, to be published.

N. D. Clarence, and D. J. Malan, "Preliminary discharge processes in lightning flashes to ground," *Quart. J. Roy. Meteorol. Soc.*, vol. 83, pp. 161–172, 1957.

J. A. Crabb, and J. Latham, "Corona from colliding drops as a possible mechanism for the triggering of lightning," *Quart. J. Roy. Meteorol. Soc.*, vol. 100, pp. 191–202, 1974.

G. A. Dawson, and D. G. Duff, "Initiation of cloud-to-ground lightning strokes," *J. Geophys. Res.*, vol. 75, pp. 5858–5867, 1970.

V. A. Diachuk, and V. M. Muchnik, "Corona discharge of watered hailstones as a basic mechanism of lightning initiation," *Doklady Akad. Nauk SSSR*, vol. 248, pp. 60–63, 1979.

N. D. Gladshteyn, "Statistical properties of noise from the predischarge part of a lightning discharge," *Geomagn. Aeron.*, vol. 5, pp. 741–742, 1967.

R. F. Griffiths, and C. T. Phelps, "A model of lightning initiation arising from positive corona streamer development," *J. Geophys. Res.*, vol. 31, pp. 3671–3676, 1976a.

R. F. Griffiths, and C. T. Phelps, "The effects of air pressure and water vapor content on the propagation of positive corona streamers, and their implications to lightning initiation," *Quart. J. Roy. Meteorol. Soc.*, vol. 102, pp. 412–426, 1976b.

D. J. Harris, and Y. E. Salman, "The measurement of lightning characteristics in northern Nigeria," *J. Atmospheric Terrest. Phys.*, vol. 34, pp. 775–786, 1972.

N. Kitagawa, and M. Kobayashi, "Field changes and variations of luminosity due to lightning flashes," in *Rec. Adv. Atmos. Electricity*. London: Pergamon, 1959, pp. 485–501.

N. Kitagawa, and M. Brook, "A comparison of intracloud and cloud-to-ground lightning discharges," *J. Geophys. Res.*, vol. 65, pp. 1189–1201, 1960.

P. R. Krehbiel, M. Brook, and R. A. McCrory, "An analysis of the charge structure of lightning discharges to ground," *J. Geophys. Res.*, vol. 84, pp. 2432–2456, 1979.

E. P. Krider, C. D. Weidman, and D. M. LeVine, "The temporal structure of the HF and VHF radiation produced by intracloud lightning discharges," *J. Geophys. Res.*, vol. 74, pp. 5760–5762, 1979.

L. Latham, and J. M. Stromberg, *Point Discharges in Lightning*, vol. 1, R. H. Golde, Ed. New York: Academic Press, 1977, pp. 99–117.

L. B. Loeb, "The mechanism of stepped and dart leaders in cloud-to-ground lightning strokes," *J. Geophys. Res.*, vol. 71, pp. 4711–4721, 1966.

L. B. Loeb, "Mechanism of charge drainage from thunderstorm clouds," *J. Geophys. Res.*, vol. 75, pp. 5882–5889, 1970.

D. J. Malan, "Les decharges dans l'air et la charge inferieure positive d'un nuage orageux," *Ann. Geophys.*, vol. 8, pp. 385–401, 1952.

D. J. Malan, "Les decharges lumineuses dans les nuages orageux," *Ann. Geophys.*, vol. 11, pp. 427–434, 1955.

D. J. Malan, and B. F. J. Schonland, "The distribution of electricity in thunderclouds," *Proc. Roy. Soc. London*, A209, pp. 158–177, 1951.

C. T. Phelps, "Positive streamer system identification and its possible role in lightning initiation," *J. Atmospheric Terrest. Phys.* vol. 26, pp. 103–111, 1974.

P. L. Rustan, Jr., *Properties of Lightning Derived from Time Series Analysis of VHF Radiation Data*, Ph.D. Thesis, University of Florida, 1979.

P. L. Rustan, M. A. Uman, D. G. Childers, W. H. Beasley, and C. L. Lennon, "Lightning source locations from VHF radiation data for a flash at Kennedy Space Center," *J. Geophys. Res.*, vol. 85, pp. 4893–4903, 1980.

T. Takeuti, H. Ishikawa, and M. Takagi, "On the cloud discharge preceding the first ground stroke," *Proc. Res. Inst. Atmospherics*, Nagoya Univ., vol. 7, pp. 1–6, 1960.

E. M. Thomson, "Characteristics of Port Moresby ground flashes," *J. Geophys. Res.*, vol. 85, pp. 1027–1036, 1980.

M. A. Uman, W. H. Beasley, J. A. Tiller, Y. T. Lin, E. P. Krider, C. D. Weidman, P. R. Krehbiel, M. Brook, A. A. Few, J. L. Bohannon, C. L. Lennon, H. A. Poehler, W. Jafferis, J. R. Gulick, and J. R. Nicholson, "An unusual lightning flash at Kennedy Space Center," *Science*, vol. 201, pp. 9–16, 1978.

C. D. Weidman, and E. P. Krider, "The radiation field waveforms produced by intracloud lightning discharge processes," *J. Geophys. Res.*, vol. 84, pp. 3157–3164, 1979.

C. D. Weidman, E. P. Krider, and M. A. Uman, "Lightning amplitude spectra in the interval from 100 kHz to 20 MHz, *Geophys. Res. Lett.*, vol. 8, pp. 931–934, 1981.

K. L. Zonge, and W. H. Evans, "Prestroke radiation from thunde clouds," *J. Geophys. Res.*, vol. 71, pp. 1519–1523, 1966.

C. Stepped Leader

H. R. Arnold, and E. T. Pierce, "Leader and junction processes in the lightning discharge as a source of VLF atmospherics," *Radio Sci.*, vol. 68D, pp. 771–776, 1964.

C. E. Baum, E. L. Breen, J. P. O'Neill, C. B. Moore, and D. L. Hall, "Measurement of electromagnetic properties of lightning with 10 nanosecond resolution," in *Lightning Technology, NASA Conf. Publ.*, 2128, FAA-RD-80-30, Apr. 1980, pp. 39–84.

W. H. Beasley, M. A. Uman, and P. L. Rustan, "Electric fields preceding cloud-to-ground lightning flashes," *J. Geophys. Res.*, vol. 87, 1982, to be published.

W. D. Bensema, "Pulse spacing in high-frequency atmospheric noise burst," *J. Geophys. Res.*, vol. 74, pp. 2780–2782, 1969.

K. Berger, and E. Vogelsanger, "Messungen und resultate der blitz-forschung der jahre 1955-1963 auf dem Monte Sans Salvatore," *Bull. Schweiz. Elektrotech. Ver.*, vol. 56, pp. 2–22, 1965.

K. Berger, and E. Vogelsanger, "Photographische blitzuntersuchungen der jahre 1955-1965 auf dem Monte San Salvatore," *Bull. Schweiz. Elektrotech. Ver.*, vol. 57, pp. 1–22, 1966.

C. Berger, "Novel observations on lightning disharges," *J. Franklin Inst.*, vol. 283, pp. 478–525, 1967.

C. E. R. Bruce, "The lightning and spark discharges," *Nature*, vol. 147, pp. 805–806, 1941.

C. E. R. Bruce, "The initiation of long electrical discharges," *Proc. Roy. Soc.*, A183, pp. 228–242, 1944.

S. P. Gupta, M. Rao, and B. A. P. Tantry, "VLF spectra radiated by stepped leaders," *J. Geophys. Res.*, vol. 77, pp. 3924–3927, 1972.

J. H. Hagenguth, and J. G. Anderson, "Lightning to the Empire State Building, Pt. 111," *AIEE Trans.*, 71 (pt. 3), pp. 641–749, 1952.

D. B. Hodges, "A comparison of the rates of change of current in the step and return processes of lightning flashes," *Proc. Phys. Soc. London*, B67, pp. 582–587, 1954.

H. Ishikawa, M. Takgai, and T. Takeuti, "On the leader waveforms of atmospherics near the origin," *Proc. Res. Inst. Atmospherics*, Nagoya Univ., vol. 5, pp. 1–11, 1958.

S. R. Khastgir, and D. Ghosh, "Theory of stepped-leader in cloud-to-ground electrical discharges," *J. Atmos. Terr. Phys.*, vol. 34, pp. 109–113, 1972.

S. R. Khastgir, "Leader stroke current in a lightning discharge according to the streamer theory," *Phys. Rev.*, vol. 106, pp. 616–617, 1957.

N. Kitagawa, "On the electric field change due to the leader process and some of their discharge mechanism," *Papers Meterol. Geophys. Tokyo*, vol. 7, pp. 400–414, 1957.

N. Kitagawa, and M. Brook, "A comparison of intracloud and cloud-to-ground lightning discharges," *J. Geophys. Res.*, vol. 65, pp. 1189–1201, 1960.

N. Kitagawa, and M. Kobayashi, "Distribution of negative charge in the cloud taking part in a flash to ground," *Papers meteorol. Geophys. Tokyo*, vol. 9, pp. 99–105, 1958.

R. Klingbeil, and D. A. Tidman, "Theory and computer model of the lightning stepped leader," *J. Geophys. Res.*, vol. 79, pp. 865–869, 1974a.

R. Klingbeil, and D. A. Tidman, "Reply to Comments on the brief report theory and computer model of the lightning stepped leader by C. T. Phelps," *J. Geophys. Res.*, vol. 79, pp. 5669–5670, 1974b.

E. P. Krider, "The relative light intensity produced by a lightning stepped leader," *J. Geophys, Res.*, vol. 79, pp. 4542–4544, 1974.

E. P. Krider, and G. J. Radda, "Radiation field wave forms produced by lightning stepped leaders," *J. Geophys. Res.*, vol. 80, pp. 2635–2657, 1975.

E. P. Krider, C. D. Weidman, and R. C. Noggle, "The electric fields produced by lightning stepped leaders," *J. Geophys. Res.*, vol. 82, pp. 951–960, 1977.

L. J. Loeb, "The mechanism of stepped and dart leaders in cloud-to-ground lightning strokes," *J. Geophys. Res.*, vol. 71, pp. 4711–4721, 1966.

L. J. Loeb, "Confirmation and extension of a proposed mechanism of the stepped leader lightning stroke," *J. Geophys. Res.*, vol. 73, pp. 5813–5817, 1968.

D. J. Malan, and B. J. F. Schonland, "Progressive lightning, pt. 7, directly correlated photographic and electrical studies of lightning near thunderstorms," *Proc. Roy. Soc. London*, A191, pp. 485–503, 1947.

R. E. Orville, "Spectrum of the lightning stepped leader," *J. Geophys. Res.*, vol. 73, pp. 6999–7008, 1968.

C. T. Phelps, Comments on brief report by R. Klingbeil, and D. A. Tidman, "Theory and computer model of the lightning stepped leader," *J. Geophys. Res.*, vol. 79, pp. 5669, 1974.

B. F. J. Schonland, Progressive lightning, Pt. 4, *Proc. Roy. Soc. London*, A164, pp. 132–150, 1938.

B. F. J. Schonland, "The pilot streamer in lightning and the long spark," *Proc. Roy. Soc. London*, A220, pp. 25–38, 1953.

B. F. J. Schonland, "The lightning discharge," *Handbuch der Physik*, vol. 22, pp. 576–628, Berlin, Germany: Springer-Verlag, 1956.

B. F. J. Schonland, "Lightning and the long electric spark," *Advan. Science*, pp. 306–313, 1962.

B. F. J. Schonland, D. B. Hodges, and H. Collens, "Progressive lightning, Pt. 5, A comparison of photographic and electrical studies of the discharge process," *Proc. Roy. Soc. London*, A166, pp. 56–75, 1938a.

B. F. J. Schonland, D. J. Malan, and H. Collens, Progressive lightning, Pt. 2, *Proc. Roy. Soc. London*, A152, pp. 595–625, 1935.

B. F. J. Schonland, D. J. Malan, and H. Collens, "Progressive lightning, Pt. 6," *Proc. Roy. Soc. London*, A168, pp. 455–469, 1938b.

C. M. Srivastava, and S. R. Khasgir, "On the maintenance of current in the stepped leader stroke of lightning discharge," *J. Science Ind. Res.*, vol. 14B, pp. 34–35, 1955.

B. J. Steptoe, "Some observations on the spectrum and propagation of atmospherics," Ph.D. Thesis, University of London, England, 1958.

S. Szpor, "Review of the relaxation theory of the lightning stepped leader," *Acta Geophysica Polonica*, vol. 18, pp. 73–77, 1970.

S. Szpor, "Critical comparison of theories of stepped leaders," *Archiwum elektrotechniki*, vol. 26, pp. 291–299, 1977.

E. M. Thomson, "Characteristics of Port Moresby ground flashes," *J. Geophys. Res.*, vol. 85, pp. 1027–1036, 1980.

M. A. Uman, "Lightning breakdown, proc. of the U.S. -Japan seminar on gas breakdown and its fundamental processes," S. Takeda, Ed., *Japan Soc. for the Promotion of Science*, 1973.

M. A. Uman, and K. D. McLain, "Radiation field and current of the lightning stepped leader," *J. Geophys. Res.*, vol. 75, pp. 1058–1066, 1970.

C. F. Wagner, and A. R. Hileman, "The lightning stroke," *AIEE Trans.*, 77 (pt. 3), pp. 229–242, 1958.

C. F. Wagner, and A. R. Hileman, "The lightning stroke," (2), *AIEE Trans.*, vol. 80 (pt. 3), pp. 622–642, 1961.

C. D. Weidman, and E. P. Krider, "Submicrosecond risetimes in lightning radiation fields, in Lightning Technology," *NASA Conf. Publ.*, 2128, FAA-RD-80-30, Apr. 1980, pp. 29–38.

C. D. Weidman, E. P. Krider, and M. A. Uman, "Lightning amplitude spectra in the interval from 100 kHz to 20 MHz," *Geophys. Res. Lett.*, vol. 8, pp. 931–934, 1981.

D. Attachment Process

K. Berger, "Methoden und resultate der blitzforschung auf dem Monte San Salvatore bei Lugano in dem jahren 1963-1971," *Bull Schweiz. Elektrotech. Ver.*, vol. 63, pp. 1403–1422, 1972.

K. Berger, and E. Vogelsanger, "Messungen und resultate der blitz-forschung der jahre 1955–1963 auf dem Monte San Salvatore," *Bull. Schweiz. Elektech. Ver.*, vol. 56, pp. 2–22, 1965.

A. J. Eriksson, "A discussion on lightning and tall structures," *CSIR Spec. Rep. ELEK 152*, National Electrical Engineering Research Institute, Pretoria, S. A., July 1978; Lightning and tall structures, Trans. South African Institute of Electrical Engineers, vol. 69, pp. 2–16. Aug. 1978.

R. P. Fieux, C. H. Gary, B. P. Hutzler, A. R. Eybert-Berard, P. L. Hubert, A. C. Meesters, P. H. Perroud, J. H. Hamelin, and J. M. Person, "Research on artificially triggered lightning in France," *IEEE Trans. Power App. Syst.*, PAS-97, pp. 725–733, 1978.

R. H. Golde, "The frequency of occurence and the distribution of lightning flashes to transmission lines," *AIEE Trans.*, vol. 64, pp. 902–910, 1945.

R. H. Golde, "Occurence of upward streamers in lightning discharges," *Nature*, vol. 160, pp. 395–396, 1947.

R. H. Golde, "The attractive effects of a lightning conductor," *J. Inst. of Elec. Eng. London*, vol. 9, pp. 212–213, 1963.

R. H. Golde, "The lightning conductor," *J. Franklin Inst.*, vol. 283, pp. 451–477, 1967.

R. H. Golde, "The lightning conductor," in *Lightning, vol. II, Lightning Protection*, R. H. Golde, Ed. New York: Academic Press, 1977, pp. 545–576.

R. H. Golde, "Lightning and tall structures," *IEE Proc. Inst. Elec. Eng. London*, vol. 125, pp. 347–351, 1978.

J. H. Hagenguth, and J. G. Anderson, "Lightning to the Empire State Building," pt. 3, *AIEE Trans.*, vol. 71, (pt. 3), pp. 641–749, 1952.

E. P. Krider, and C. G. Ladd, "Upward Streamers in lightning discharges to mountainous terrain," *Weather*, vol. 30, pp. 77–81, 1975.

E. T. Pierce, "Triggered lightning and its application to rockets and aircraft," 1972 Lightning and Static Electricity Conference, AFAL-TR-72-325, Air Force Avionics Lab., Wright-Patterson Air Force Base, Ohio, 1972.

C. F. Wagner, "Relation between stroke current and velocity of the return stroke," *AIEE Trans.*, vol. 82, (pt. 3), pp. 606–617, 1963.

C. F. Wagner, "Lightning and transmission lines," *J. Franklin Inst.*, vol. 283, pp. 558–594, 1967.

E. Return Stroke Measured Fields

C. E. Baum, E. L. Breen, J. P. O'Neill, C. B. Moore, and D. L. Hall, "Measurement of electromagnetic properties of lightning with 10 nanosecond resolution, in Lightning Technology, NASA Conf. Publ. 2128, FAA-RD-80-30, Apr. 1980, pp. 39–84.

R. K. Baum, "Airborn lightning characterization, in Lightning Technology," NASA Conf. Publ. 2128, FAA-RD-80-30, Apr. 1980, pp. 153–172.

M. Brook and N. Kitagawa, "Radiation from lightning discharges in the frequency range 400 to 1000 Mc/s," *J. Geophys. Res.*, vol. 69, pp. 2431–2434, 1964.

R. J. Fisher, and M. A. Uman, "Measured electric fields risetimes for first and subsequent lightning return strokes," *J. Geophy. Res.* vol. 77, pp. 399–407, 1972.

S. N. Gupta, "Distribution of peaks in atmospheric radio noise," *IEEE Trans. Electromag. Compat.*, EMC-15, pp. 100–103, 1973.

J. E. Hart, "V.L.F. radiation from multiple stroke lightning," *J. Atmos. Terr. Phys.*, vol. 29, pp. 1011–1014, 1967.

R. O. Himley, "VLF radiation from subsequent return strokes in multiple stroke lightning," *J. Atmos. Terr. Phys.*, vol. 31, pp. 749–753, 1969.

D. M. Levine, and E. P. Krider, "The temporal structure of HF and VHF radiation during Florida Lightning return strokes," *Geophys. Res. Lett.*, vol. 4, pp. 13–16, 1977.

Y. T. Lin, M. A. Uman, J. A. Tiller, R. D. Brantley, E. P. Krider, and C. D. Weidman, "Characterization of lightning return stroke electric and magnetic fields from simultaneous two-station measurements," *J. Geophys. Res.*, vol. 84, pp. 6307–6314, 1979.

T. Nakai, "On the time and amplitude properties of electric field near sources of lightning in the VLF, LF, and HF bands," *Radio Sci.*, vol. 12, pp. 389–396, 1977.

J. E. Nanevicz, and E. F. Vance, "Analysis of electrical transients created by lightning," SRI4026, NASA Contractor Rep. 159308, July 1980.

F. L. Pitts, "Electromagnetic measurement of lightning strikes to aircraft," presented at AIAA 19th Aerospace Sciences Meeting, St. Louis, Jan. 1981.

F. L. Pitts, and M. E. Thomas, 1980 direct strike lightning data, NASA Technical Memorandum 81946, Langley Res. Center, Hampton, VA., Feb. 1981.

M. Takagi, "VHF radiation from ground discharges," in *Planetary Electrodynamics*, S. C. Coroniti and J. Hughes, Eds. New York: Gordon and Breach, 1969, pp. 535–538.

J. A. Tiller, M. A. Uman, Y. T. Lin, R. D. Brantley, and E. P. Krider, "Electric field statistics for close lightning return strokes near Gainesville, Florida," *J. Geophys. Res.*, vol. 81, pp. 4430–4434, 1976.

M. A. Uman, R. D. Brantley, J. A. Tiller, Y. T. Lin, E. P. Krider, and D. K. McLain, "Correlated electric and magnetic fields from lightning return strokes," *J. Geophys. Res.*, vol. 80, pp. 373–376, 1976a.

M. A. Uman, D. K. McLain, R. J. Fisher, and E. P. Krider, "Electric field intensity of the lightning return stroke," *J. Geophys. Res.*, vol. 78, pp. 3523–3529, 1973.

M. A. Uman, C. E. Swanberg, J. A. Tiller, Y. T. Lin, and E. P. Krider, "Effects of 200 km propagation on Florida lightning return stroke electric fields," *Radio Sci.*, vol. 11, pp. 985–990, 1976b.

C. D. Weidman, and E. P. Krider, "The fine structure of lightning return stroke wave forms," *Geophys. Res.*, vol. 83, pp. 6239–6247, 1978.

C. D. Weidman, and E. P. Krider, "Submicrosecond risetimes in lightning return-stroke fields," *Geophys. Res. Lett.*, vol. 7, pp. 955–958, 1980a.

C. D. Weidman, and E. P. Krider, "Submicrosecond risetimes in lightning radiation fields," in Lightning Technology, NASA Conf. Publ. 2128, FAA-RD-80-30, Apr. 1980b, pp. 29–38.

F. Return Stroke Frequency Spectra

J. S. Barlow, G. W. Frey, Jr., and J. B. Newman, "Very low frequency noise power from the lightning discharge," *J. Franklin Inst.*, vol. 27, pp. 145–163, 1954.

P. A. Bradley, "The spectra of lightning discharges at very low frequencies," *J. Atmos. Terr. Phys.*, vol. 26, 1069–1073, 1964.

P. A. Bradley, "The VLF energy spectra of first and subsequent return strokes of multiple lightning discharges to ground," *J. Atmos. Terr. Phys.*, vol. 27, pp. 1045–1053, 1965.

D. L. Croom, "The spectra of atmospherics and propagation of very low frequency radio waves," Ph.D. thesis, University of Cambridge, England, 1961.

D. L. Croom, "The frequency spectra and attentuation of atmospherics in the range 1–15 kc/s," *J. Atmos. Terr. Phys.*, vol. 26, pp. 1015–1046, 1964.

A. S. Dennis, and E. T. Pierce, "The return stroke of the lightning flash to earth as a source of VLF atmospherics," *Radio Sci.*, vol. 68D, pp. 779–794, 1964.

J. Galejs, "Amplitude statistics of lightning discharge currents and ELF and VLF radio noise," *J. Geophys. Res.*, vol. 72, pp. 2943–2953, 1967.

J. Harwood, and B. N. Harden, "The measurement of atmospheric radio noise by an aural comparison method in the range 15–500 kc/s," *Proc. Inst. Elec. Eng.*, B107, pp. 53–59, 1960.

F. Horner, "Narrow band atmospherics from two local thunderstorms," *J. Atmos. Terr. Phys.*, vol. 21, pp. 13–25, 1961.

F. Horner, "Atmospherics of near lightning discharges," in *Radio Noise of Terrestrial Origin*, F. Horner, Ed. New York: American Elsevier Publishing Company, 1962, pp. 16–17.

F. Horner, "Radio noise from thunderstorms," in *Advances in Radio Research*, vol. 2, J. A. Saxton, Ed. New York: Academic Press, 1964, pp. 122–215.

F. Horner, and P. A. Bradley, "The spectra of atmospherics from near lightning," *J. Atmos. Terr. Phys.*, vol. 26, pp. 1155–1166, 1964.

D. Levine, and E. P. Krider, "The temporal structure of HF and VHF radiation during Florida lightning return strokes," *Geophys. Res. Lett.*, vol. 4, pp. 13–16, 1977.

G. O. Marney, and K. Shanmugam, "Effect on channel orientation on the frequency spectrum of lightning discharges," *J. Geophys. Res.*, vol. 76, pp. 4198–4202, 1971.

E. L. Maxwell, "Atmospheric noise from 20 Hz to 30 KHz," *Radio Sci.*, vol. 2, pp. 637–644, 1967.

E. L. Maxwell, and D. L. Stone, "Natural noise fields from 1 cps to 100 kc," *IEEE Trans. Antenna Propagat.*, AP-11, pp. 339–343, 1963.

T. Obayashi, "Measured frequency spectra of VLF atmospherics," *J. Res. Nat. Bureau Standards*, vol. 64D, pp. 41–48, 1960.

G. I. Serhan, M. A. Uman, D. G. Childers, and Y. T. Lin, "The RF spectra of first and subsequent lightning return strokes in the 1-200 km range," *Radio Sci.*, vol. 15, pp. 1089–1094, 1980.

B. J. Steptoe, "Some observations on the spectrum and propagation of atmospherics," Ph.D. thesis, University of London, London, England, 1958.

W. L. Taylor, "Radiation field characteristics of lightning discharges in the band 1 kc/s to 100 kc/s," *J. Res. Nat. Bureau Standards*, vol. 67D, pp. 539–550, 1963.

W. L. Taylor, and A. G. Jean, "Very low frequency radiation spectra of lightning discharges," *J. Res. Nat. Bureau Standards*, vol. 63D, pp. 199–204, 1959.

A. D. Watt, and E. L. Maxwell, "Characteristics of atmospherics noise from 1 to 100 kc," *Proc. Inst. Radio Eng.*, vol. 45, pp. 55–62, 1957.

C. D. Weidman, E. P. Krider, and M. A. Uman, "Lightning amplitude spectra in the interval from 100 kHz to 20 MHz," *Geophys. Res. Lett.*, vol. 8, pp. 931–934, 1981.

J. C. Williams, "Thunderstorms and VLF radio noise," Ph.D. thesis, Harvard University, Cambridge, MA, 1959.

G. Return Stroke Current

K. Berger, "Die messeinrichtungen fur die blitzforschung auf dem Monte San Salvatore," *Bull. Schweiz. Elektrotech. Ver.*, vol. 46, pp. 193–204, 1955.

K. Berger, "Extreme blitzstrome und blitzschutz," *Bull. ASE/USC*, vol.

71, pp. 460–464, 1980.

K. Berger, "Front duration and current steepness of lightning strokes to earth," *Gas Discharges and the Electricity Supply Industry*, J. S. Forrest, P. R. Howard, and D. J. Littler, eds. London: England, Butterworth, 1962, pp. 63–73.

K. Berger, "Gewitterforschung auf dem Monte San Salvatore, Elektrotechnik," Z-A, vol. 82, pp. 249–260, 1967.

K. Berger, "Lightning research in Switzerland," *Weather*, vol. 2, pp. 231–238, 1947.

K. Berger, "Methoden und resultate der blitzforschung auf dem Monte San Salvatore bei Lugano in dem jahren 1963–1972," *Bull. Schweiz. Elektrotech. Ver.*, vol. 63, pp. 1403–1422, 1972.

K. Berger, "Novel observations on lightning discharges: Results of research on Mount San Salvatore," *J. Franklin Inst.*, vol. 283, pp. 478–525, 1967.

K. Berger, "Resultate der blitzmessungen der jahre 1947–1954 auf dem Monte San Salvatore," *Bull. Schweiz. Elektrotech. Ver.*, vol. 46, pp. 405–424, 1955.

K. Berger, R. B. Anderson, and H. Kroninger, "Parameters of lightning flashes," *Electra*, vol. 80, pp. 23–37, 1975.

K. Berger, and E. Vogelsänger, "Messungen und resultate der blitzforschung der jahre 1955–1963, auf dem Monte San Salvatore," *Bull. Schweiz. Elektrotech. Ver.*, vol. 56, pp. 2–22, 1965.

K. Berger, and E. Vogelsanger, "New results of lightning observations," in S. C. Coroniti, and J. Hughes, ed., *Planetary Electrodynamics*, 1, New York: Gordon and Breach, pp. 489–510, 1969.

D. W. Clifford, E. P. Krider, and M. A. Uman, "A case for submicrosecond rise-time current pulses for use in aircraft induced coupling studies," 1979 IEEE-EMC Conf., IEEE Rep. 79CH–1383-9 EMC, Library of Congress Catalog no. 78-15514.

R. Davis, and W. G. Standring, "Discharge currents associated with kite balloons," *Proc. Roy. Soc. London*, A191, pp. 304–322, 1947.

A. J. Eriksson, "Lightning and tall structures," *Trans. South African IEE*, vol. 69, pt. 8, pp. 238–252, 1978.

R. P. Fieux, C. H. Gary, B. P. Hutzler, A. R. Eybert-Berard, P. L. Hubert, A. C. Meesters, P. H. Perroud, J. H. Hamelin, J. M. Person, "Research on artificially triggered lightning in France," *IEEE Trans. Power Apparatus Syst.*, PAS-97, pp. 725–733, 1978.

E. Garbagnati, E. Giudice, G. B. Lo Pipero, "Measurement of lightning currents in Italy—Results of a statistical evaluation," *Elektrotechnische Zeitschrift etz-a*," vol. 99, pp. 664–668, 1978.

E. Garbagnati, E. Giudice, B. Lo Piparo, and U. Magagnoli, "Relieve delle caratteristiche dei fulmini in Italia," in Risultati ottenuti negli anni 1970–1973, *L'Elettrotecnica*, LXII, pp. 237–249, 1975.

J. H. Gilchrist, and J. B. Thomas, "A model for the current pulses of cloud-to-ground lightning discharges," *J. Franklin Inst.*, vol. 299, pp. 199–210, 1975.

J. H. Hagenguth, and J. G. Anderson, "Lightning to the Empire State Building—Pt. 3," *AIEE*, vol. 71, pt. 3, pp. 641–649, 1952.

N. Hylten-Cavallius, and A. Stromberg, "The amplitude, time to half-value, and the steepness of the lightning currents," *ASEA J.*, vol. 29, pp. 129–134, 1956.

N. Hylten-Cavallius, and A. Stromberg, "Field measurement of lightning currents," *Elteknik*, vol. 2, pp. 109–113, 1959.

D. G. McCann, "The measurement of lightning currents in direct strokes," *Trans. AIEE*, vol. 63, pp. 1157–1164, 1944.

K. B. McEachron, "Lightning to the Empire State Building," *J. Franklin Inst.*, vol. 227, pp. 149–217, 1939.

K. B. McEachron, "Wave shapes of successive lightning current peaks," *Elec. World*, vol. 56, pp. 428–431, 1940.

K. B. McEachron, "Lightning to the Empire State Building," *Trans. AIEE*, vol. 60, pp. 885–889, 1941.

B. J. Petterson, and W. R. Wood, "Measurements of lightning strikes to aircraft," Rep. no. SC-M-67-549, Sandia Labs., Albuquerque, NM, 1968.

F. L. Pitts, "Electromagnetic measurement of lightning strikes to aircraft," presented at the AIAA 19th Aerospace Sciences Meeting, St. Louis, MO, Jan. 1981.

F. L. Pitts, and M. E. Thomas, 1980 direct strike lightning data, NASA Technical Memorandum 81946, Langley Research Center, Hampton, VA, Feb. 1981.

F. Popolansky, "Frequency distribution of amplitudes of lightning currents," *Electra*, vol. 22, pp. 139–147, 1972.

M. A. Sargent, "The frequency distribution of current magnitude of lightning strokes to tall structures," *IEEE Trans. Power App. Syst.*, PAS-91, pp. 2224–2229, 1972.

S Szpor, "Comparison of Polish to American lightning records," *IEEE Trans. Power App. Syst.*, PAS-88, pp. 646–652, 1969.

S. Szpor, "Courbes internationales des courants de foudre," *Acta Geophys. Polonica*, vol. 19, pp. 365–369, 1971.

M. A. Uman, D. K. McLain, R. J. Fisher, and E. P. Krider, "Currents in Florida lightning strokes," *J. Geophys. Res.*, vol. 78, pp. 3530–3537, 1973.

C. D. Weidman, and E. P. Krider, "The fine structure of lightning return stroke wave forms," *J. Geophys. Res.*, vol. 83, pp. 6239–6247, 1978.

C. D. Weidman, and E. P. Krider, "Submicrosecond risetimes in lightning return stroke fields," *Geophys. Res. Lett.*, vol. 7, pp. 955–958, 1980.

H. Return Stroke Velocity

J. S. Boyle, and R. E. Orville, "Return stroke velocity measurement in multistroke lightning flashes," *J. Geophys. Res.*, vol. 81, pp. 4461–4466, 1976.

J. H. Hagenguth, "Photographic study of lightning," *AIEE Trans.*, vol. 66, pp. 577–585, 1947.

P. Hubert, and G. Mouget, "Return stroke velocity measurements in two triggered lightning flashes," *J. Geophys. Res.*, vol. 86, pp. 5253–5261, 1981.

V. P. Idone, and R. E. Orville, "Lightning return stroke velocities in the Thunderstorm Research International Program (TRIP)," *J. Geophys. Res.*, vol. 87, 1982, to be published.

D. Jordan, and M. A. Uman, "Variations of light intensity with height and time from subsequent return strokes," *Trans. Amer. Geophys. Union*, vol. 61, p. 977, 1980.

R. E. Orville, G. G. Lala, and V. P. Idone, "Daylight time resolved photographs of lightning, Science," vol. 201, pp. 59–61, 1978.

G. J. Radda, and E. P. Krider, "Photoelectric measurements of lightning return stroke propagation speeds," *Trans. Amer. Geophys. Union*, vol. 56, pp. 1131, 1974; and private communication.

Saint-Privat D'Allier Research Group, "Research in artificially triggered lightning in France," presented at the *Proc. Third Symp. Electromag. Compat.*, Rotterdam, Netherlands, May 1-3, 1979, ed. T. Dvorak, ETH Zentrum, Zurich 8092.

B. F. J. Schonland, and H. Collens, "Progressive lightning," *Proc. Roy. Soc. London*, vol. A143, pp. 654–674, 1934.

B. F. J. Schonland, D. J. Malan, and H. Collens, Progressive Lightning II, *Proc. Roy. Soc. London*, vol. A152, pp. 595–625, 1935.

C. D. Weidman, and E. P. Krider, "Time and height resolved photoelectric measurements of lightning return strokes," *Trans. Amer. Geophys. Union*, vol. 61, p. 978, 1980.

I. Dart Leader

N. W. Albright, and D. A. Tidman, "Ionizing potential waves of high-voltage breakdown streamers," *Phys. Fluids*, vol. 15, pp. 86–90, 1972.

M. Brook, and N. Kitagawa, "Radiation from lightning discharges in the frequency range 400 to 1000 Mc/s," *J. Geophys. Res.*, vol. 69, pp. 2431–2434, 1964.

M. Brook, N. Kitagawa, and E. J. Workman, "Quantitative study of strokes and continuing currents in lightning discharges to ground," *J. Geophys. Res.*, vol. 67, pp. 649–659, 1962.

E. P. Krider, C. D. Weidman, and R. C. Noggle, "The electric fields produced by lightning stepped leaders," *J. Geophys. Res.*, vol. 82, pp. 951–960, 1977.

L. B. Loeb, "The mechanism of stepped and dart leaders in cloud-to-ground lightning strokes," *J. Geophys. Res.*, vol. 71, pp. 4711–4721, 1966.

L. B. Loeb, "Ionizing waves of potential gradient," *Science*, vol. 148, pp. 1417–1426, 1965.

L. B. Loeb, "The mechanisms of stepped and dart leaders in cloud-to-ground lightning strokes," *J. Geophys. Res.*, vol. 71, pp. 4711–4721, 1966

D. M. LeVine, and E. P. Krider, "The temporal structure of HF and VHF radiations during Florida lightning return strokes," *Geophys. Res. Lett.*, vol. 4, pp. 13–16, 1977.

D. J. Malan, and B. F. J. Schonland, "The distribution of electricity in thunderclouds," *Proc. Roy. Soc. London*, A209, pp. 158–177, 1951.

R. E. Orville, G. G. Lala, and V. P. Idone, "Daylight time-resolved photographs of lightning," *Science*, vol. 201, pp. 59–61, 1978.

D. E. Proctor, "A hyperbolic system for obtaining VHF radio pictures of lightning," *J. Geophys. Res.*, vol. 76, pp. 1478–1489, 1971.

D. E. Proctor, "A radio study of lightning," Ph.D. thesis, University of Witwatersrand, Johannesburg, S.A., 1976.

M. Rao, "The dependence of the dart leader velocity on the interstroke time interval in a lightning flash," *J. Geophys. Res.*, vol. 75, pp. 5868–5872, 1970.

P. L. Rustan, "Properties of lightning derived from time series analysis of VHF radiation data," Ph.D thesis, University of Florida, Gainesville, FL, 1979.

P. L. Rustan, M. A. Uman, D. G. Childers, W. H. Beasley, and C. L. Lennon, "Lightning source locations from VHF radiation data for a flash," presented at Kennedy Space Center, *J. Geophys. Res.*, vol. 85, pp. 4893-4903, 1980.

B. F. J. Schonland, "The lightning discharge," *Handbuch der Physik*, vol. 22, pp. 576-628, Berlin: Springer-Verlag, 1956.

B. F. J. Schonland, D. B. Hodges, and H. Collens, "Progressive lightning, pt. 5, a comparison of photographic and electrical studies of the discharge process," *Proc. Roy. Soc. London*, A166, pp. 56-75, 1938.

B. F. J. Schonland, D. J. Malan, and H. Collens, "Progressive lightning, pt. 2," *Proc. Roy. Soc. London*, vol. A152, pp. 595-625, 1935.

M. Takagi, "VHF Radiation from ground discharges, in Planetary Electrodynamics," S. C. Coroniti and J. Hughes, eds. New York: Gordon and Breach, 1969a, pp. 543-570.

M. Takagi, "VHF Radiation from ground discharges," *Proc. Res. Inst. Atmos.*, Nagoya Univ., Japan, pp. 163-168, 1969b.

W. P. Winn, "A laboratory analogy to the dart leader and return stroke of lightning," *J. Geophys. Res.*, vol. 70, pp. 3256-3270, 1965.

J. Continuing Current

K. Berger, and E. Vogelsanger, "Messungen und Rsultate der Blitzforschung der Jahre 1955-1963 auf dem Monte San Salvatore," *Bull. Schweiz. Elektrotech. Ver.*, vol. 56, pp. 2-22, 1965.

M. Brook, N. Kitagawa, and E. J. Workman, "Quantitative study of strokes and continuing currents in lightning discharges to ground," *J. Geophys. Res.*, vol. 67, pp. 649-659, 1962.

M. Brook, M. Nakano, P. Krehbiel, and T. Takeuti, "The electrical structure of the Hokuriku winter thunderstorms," *J. Geophys. Res.*, vol. 87, pp. 1207-1215, 1982.

D. M. Fuqua, R. G. Baughman, A. R. Taylor, and R. G. Hawe, "Characteristics of seven lightning discharges that caused forest fires," *J. Geophys. Res.*, vol. 73, pp. 1897-1906, 1968.

N. Kitagawa, M. Brook, and E. J. Workman, "Continuing currents in cloud-to-ground lightning discharges," *J. Geophys. Res.*, vol. 67, pp. 637-647, 1962.

N. Kitagawa, M. Brook, and E. J. Workman, "The role of continuous discharges in cloud-to-ground lightning," *J. Geophys. Res.*, vol. 65, p. 1965, 1960.

D. J. Lathan, "A channel model for long arcs in air," *Phys. Fluids*, vol. 23, pp. 1710-1715, 1980.

J. M. Livingston, and E. P. Krider, "Electric fields produced by Florida thunderstorms," *J. Geophys. Res.*, vol. 83, pp. 385-401, 1978.

D. J. Malan, "Les decharges orageuses intermittentes et continu de la colonne de charge negative," *Ann. Geophys.*, vol. 10 pp. 271-281, 1954.

E. M. Thomson, "Characteristics of Port Moresby ground flashes," *J. Geophys. Res.*, vol. 85, pp. 1027-1036, 1980.

D. P. Williams, and M. Brooks, "Fluxgate magnetometer measurements of continuous currents in lightning," *J. Geophys. Res.*, vol. 67, p. 1662, 1962.

D. P. Williams, and M. Brook, "Magnetic measurement of thunderstorm currents," 1. Continuing currents in lightning, *J. Geophys. Res.*, vol. 68, pp. 3243-3247, 1963.

K. J- and K-Changes in Discharges to Ground

H. R.Arnold, and E. T. Pierce, "Leader and junction processes in the lightning discharge as a source of VLF atmospherics," *Radio Sci.*, vol. 68D, pp. 771-776, 1964.

M. Brook, and N. Kitagawa, "Radiation from lightning discharges in the frequency range 400 to 1000 Mc/s," *J. Geophys. Res.*, vol. 69, pp. 2431-2434, 1964.

M. Brook, N. Kitagawa, and E. J. Workman, "Quantitative study of strokes and continuing currents in lightning discharges to ground," *J. Geophys. Res.*, vol. 67, pp. 649-659, 1962.

M. Brook, and B. Vonnegut, "Visual confirmation of the junction processes in lightning discharges," *J. Geophys. Res.*, vol. 65, pp. 1302-1303, 1960.

A. J. Illingworth, "Electric field recovery after lightning as the response of the conducting atmosphere to a field change," *Quart. J. Roy. Meteorol. Soc.*, vol. 98, pp. 604-616, 1971.

H. Ishikawa, "Nature of lightning discharges as origins of atmospherics," *Proc. Res. Inst. Atmos.*, Nagoya Univ., 8A, pp. 1-273, 1961.

N. Kitagawa, "On the mechanism of cloud flash and junction or final process in flash to ground," Papers *Meteorol. Geophys. Tokyo*, vol. 7, pp. 415-424.

N. Kitagawa, "Types of lightning," in *Problems of Atmospheric and Space Electricity*, S. C. Coroniti, ed. New York: American Elsevier Publishing Company, 1965, pp. 337-348.

N. Kitagawa, and M. Kobayashi, "Field changes and variations of luminosity due to lightning flashes, in *Rec. Adv. Atmospheric Electricity*. London: Pergamon, 1959, pp. 485-501.

N. Kitagawa, and M. Brook, "A comparison of intracloud and cloud-to-ground lightning discharges," *J. Geophys. Res.*, vol. 65, pp. 1189-1201, 1960.

M. Kobayashi, N. Kitagawa, T. Ikeda, and Y. Sato, "Preliminary studies of variation of luminosity and field change due to lightning flashes," *Papers Meteorol. Geophys. Tokyo*, vol. 9, pp. 29-34, 1958.

P. R. Krehbiel, M. Brook, and R. McCrory, "Analysis of the charge structure of lightning discharges to ground," *J. Geophys. Res.*, vol. 84, pp. 2432-2456, 1979.

D. Mackerras, "A comparison of discharge processes in cloud and ground lightning flashes," *J. Geophys. Res.*, vol. 73, pp. 1175-1183, 1968.

D. J. Malan, "La distribution verticale de la charge negative orageuse," *Ann. Geophys.*, vol. 11, pp. 420-426, 1955.

D. J. Malan, and B. F. J. Schonland, "The electrical processes in the intervals between the strokes of a lightning discharge," *Proc. Roy. Soc. London*, vol. A206, pp. 145-163, 1951a.

D. J. Malan, and B. F. J. Schonland, "The distribution of electricity in thunder-clouds," *Proc. Roy. Soc. London*, A209, pp. 158-177, 1951b.

D. Muller-Hillebrand, "The magnetic field of the lightning discharge," in *Gas Discharges and the Electric Supply Industry*, J. S. Forrest, ed. London, England: Butterworth, pp. 89-111, 1968.

T. Ogawa, and M. Brook, "The mechanism of the intracloud lightning discharge," *J. Geophys. Res.*, vol. 69, pp. 5141-5150, 1964.

T. Ogawa, and M. Brook, "Charge distribution in thunderstorm clouds," *Quart. J. Roy. Meteorol. Soc.*, vol. 95, pp. 513-525, 1969.

P. L. Rustan, Jr., "Properties of Lightning Derived from Time Series Analysis of VHF Radiation Data," Ph.D. thesis, University of Florida, 1979.

P. L. Rustan, M. A. Uman, D. G. Childers, W. H. Beasley, and C. L. Lennon, "Lightning source locations from VHF radiation data for a flash at Kennedy Space Center," *J. Geophys. Res.*, vol. 85, pp. 4893-4903, 1980.

B. F. J. Schonland, "Progressive lightning, pt. 4, the discharge mechanism," *Proc. Roy. Soc. London*, A164, pp. 132-150, 1938.

B. J. Steptoe, "Some observations on the spectrum and propagation of atmospherics," Ph.D. thesis, University of London, England, 1958.

M. Takagi, "The mechanism of discharges in a thundercloud," *Proc. Res. Inst. Atmos.*, Nagoya Univ., 8B, pp. 1-105, 1961.

N. S. Wadhera, and B. A. P. Tantry, "VLF characteristics of K changes in lightning discharges," *Indian J. Pure Appl. Phys.*, vol. 5, pp. 447-449, 1967a.

N. S. Wadhera, and B. A. P. Tantry, "Audio frequency spectra of K changes in a lightning discharge," *J. Geomag. Geoelec.*, vol. 19, pp. 257-260, 1967b.

C. P. Wang, "Lightning discharges in the tropics-2. Component ground strokes and cloud dart streamer discharges," *J. Geophys. Res.*, vol. 68, pp. 1951-1958, 1963.

L. Cloud Discharges

J. I. Aina, "Lightning discharges studies in a tropical area," II. "Dischargers which do not reach the ground," *J. Geomag. Geoelec.*, vol. 23, pp. 359-368, 1971.

J. I. Aina, "Lightning discharges studies in a tropical area." III. "The profile of the electrostatic field changes due to nonground discharges," *J. Geomag. Geoelec.*, vol. 24, pp. 369-380, 1972.

M. Brook, and T. Ogawa, "The cloud discharge," in *Lightning*, I, *Physics of Lightning*, R. H. Golde, Ed. New York: Academic Press, 1977, pp. 191-230.

A. Huzita, and T. Ogawa, "Charge distribution in the average thunderstorm cloud," *J. Meteor. Soc. Japan*, vol. 54, pp. 285-288, 1976a.

A. Huzita, and T. Ogawa, "Electric field changes due to tilted streamers in the cloud discharge," *J. Meteor. Soc. Japan*, vol. 54, pp. 289-293, 1976b.

H. Ishikawa, "Nature of lightning discharges as origins of atmospherics," *Proc. Res. Inst. Atmos.*, Nagoya Univ., 8A, pp. 1-273, 1961.

H. Ishikawa, and T. Takeuchi, "Field changes due to lightning discharge," *Proc. Res. Inst. Atmos.*, Nagoya Univ., vol. 13, pp. 59–61, 1966.

S. R. Khastgir, and S. K. Saha, "On intracloud discharges and their accompanying electric field changes," *J. Atmospheric Terr. Phys.*, vol. 34, pp. 775–786, 1972.

N. Kitagawa, "On the mechanism of cloud flash and junction or final process in a flash to ground," *Papers Meteor. Geophys.*, vol. 7, pp. 415–424, 1957.

N. Kitagawa, and M. Kobayashi, "Field changes and variations of luminosity due to lightning flashes," in *Rec. Advanced Atmospheric Electricity.* London: Pergamon, 1959, pp. 485–501.

N. Kitagawa, and M. Brook, "A comparison of intracloud and cloud-to-ground lightning discharges," *J. Geophys. Res.*, vol. 65, pp. 1189–1201, 1960.

M. Kobayashi, N. Kitagawa, T. Ikeda, and Y. Sato, "Preliminary studies of variation of luminosity and field change due to lightning flashes," *Papers Meteor. Geophys.*, vol. 9, pp. 29–34, 1958.

E. P. Krider, G. J. Radda, and R. C. Noggle, "Regular radiation field pulses produced by intracloud discharges," *J. Geophys. Res.*, vol. 80, pp. 3801–3804, 1975.

E. P. Krider, C. D. Weidman, and D. M. LeVine, "The temporal structure of the HF and VHF radiation produced by intracloud lightning discharges," *J. Geophys. Res.*, vol. 74, pp. 5760–5762, 1979.

D. M. LeVine, "Sources of the strongest RF radiation from lightning," *J. J. Geophys. Res.*, vol. 85, pp. 4091–4095, 1980.

D. Mackerras, "A comparison of discharge processes in cloud and ground lightning flashes," *J. Geophys. Res.*, pp. 1175–1183, 1968.

M. Nakano, "The cloud discharge in winter thunderstorms of the Hokuriku coast," *J. Meteor. Soc. Japan*, vol. 57, pp. 444–445, 1979.

M. Nakano, "Initial streamer of the cloud discharge in winter thunderstorms of the Hokuriku coast," *J. Meteor. Soc. Japan*, vol. 57, pp. 452–458, 1979.

T. Ogawa, and M. Brook, "The mechanism of the intracloud lightning discharges," *J. Geophys. Res.*, vol. 69, pp. 514–5149, 1964.

B. J. Petterson, and W. R. Wood, "Measurements of lightning stroke to aircraft," Rep. SC-M-67-549 and DS-68-1 on Project 520-002-03X to Dept. of Transportation, FAA, Sandia Laboratory, Albuquerque, Jan. 1968.

S. A. Prentice, and D. Mackerras, "The ratio of cloud to cloud-to-ground lightning flashes in thunderstorms," *J. Appl. Meteor.*, vol. 16, pp. 545–550, 1977.

D. E. Proctor, "A radio study of lightning," Ph.D. thesis, Univ. of Witwatersrand, South Africa, 1976.

D. E. Proctor, "VHF radio pictures of cloud flashes," *J. Geophys. Res.*, vol. 86, pp. 4041–4071, 1981.

M. Rao, S. R. Khastgir, and H. Bhattacharya, "Electric field changes," *J. Atmospheric Terrest. Phys.*, vol. 24, pp. 989–990, 1962.

L. G. Smith, "Intracloud lightning discharges," *Quart. J. Roy. Meteorol. Soc.*, vol. 83, pp. 103–111, 1957.

M. Sourdillon, "Etude a la chambre de Boys de l'eclair dans l'air et du coup de foudre a cime horizontale," *Ann. Geophys.*, vol. 8, pp. 349–354, 1952.

M. Takagi, "The mechanism of discharges in a thundercloud," *Proc. Res. Atmos.*, Nagoya Univ., vol. 8B, pp. 1–105, 1961.

T. Takeuti, "Studies on thunderstorm electricity, 1, cloud discharges," *J. Geomag. Geoelec.*, vol. 17, pp. 59–68, 1965.

N. S. Wadhera, and B. A. P. Tantry, "VLF characteristics of K changes in lightning discharges," *Indian J. Pure Appl. Phys.*, vol. 5, pp. 447–449, 1967.

C. D. Weidman, and E. P. Krider, "The radiation fields wave forms produced by intracloud lightning discharge processes," *J. Geophys. Res.*, vol. 84, pp. 3157–3164, 1979.

C. D. Weidman, E. P. Krider, and M. A. Uman, "Lightning amplitude spectra in the interval from 100 kHz to 20 MHz," *Geophys. Res. Lett.*, vol. 8, pp. 931–934, 1981.

C. M. Wong, and K. K. Lin, "The inclination of intracloud lightning discharge," *J. Geophys. Res.*, vol. 83, pp. 1905–1912, 1978.

M. Flash Frequency Spectra Below 30–50 MHZ

S. V. C. Aiya, "Measurements of atmospheric noise interference to broadcasting," *J. Atmospheric Terrest. Phys.*, vol. 5, p. 230, 1954.

S. V. C. Aiya, "Noise power radiated by tropical thunderstorms," Proc. IRE, vol. 43, pp. 966–974, 1955.

S. V. C. Aiya, "Structure of atmosphere radio noise," *J. Science Ind. Res.*, vol. 21D, pp. 203–220, 1962.

S. V. C. Aiya, A. R. K. Sastri, and A. P. Shiraprasad, "Atmospheric

radio noise measured in India," *Indian J. Rad. Space Physics*, vol. 1, pp. 1–8, 1972.

S. V. C. Aiya, and B. S. Sonde, "Spring thunderstorms over Bangalore," *Proc. IEEE*, vol. 51, pp. 1493–1501, 1963.

P. A. Bradley, and C. Clarke, "Atmospheric radio noise and signal received on directional aerials at high frequencies," *Proc. Inst. Elec. Eng. London*, vol. 111, pp. 1534–1540, 1964.

J. Bhattacharya, "Amplitude and phase spectra of radio atmospherics," *J. Atmospheric Terrest. Phys.*, vol. 25, p. 445, 1963.

N. Cianos, G. N. Oetzel, and E. T. Pierce, "Structure of lightning noise; especially above HF, 1972 Lightning and Static Electricity Conference," AFAL-TR-72-325, Air Force Avionics Laboratory Wright-Patterson AFB, Ohio, 1972.

C. Clarke, "A study of atmospheric radio noise in a narrow width band at 11 Mc/s," *Proc. Inst. of Elec. Eng. London*, pp. 107, pt. B, 1960.

C. Clarke, "Atmospheric radio noise studies based on amplitude probability measurements at Slough, England, during the International Geophysical Year," Proc. IEE (London), p. 109, pt. B47, 1962.

C. Clarke, and P. A. Bradley, "Discussion on 1) Atmospheric radio noise and signals received on directional aerials at high frequencies and 2) Characteristics of atmospheric radio noise observed at Singapore," *Proc. Inst. Elec. Eng. London*, vol. 113, pp. 752–754, 1966.

C. Clarke, P. A. Bradley, and D. E. Mortimer, "Characteristics of atmospheric radio noise at Singapore," *Proc. Inst. Elec. Eng. London*, vol. 112, pp. 849–860, 1965.

F. Horner, "Narrow band atmospherics from two local thunderstorms," *J. Atmospheric Terr. Phys.*, vol. 21, pp. 13–25, 1961.

F. Horner, "Radio noise from thunderstorms," in Saxton, J. A. (ed.) *Advances in Radio Research*, vol. 2, New York: Academic Press, 1964, pp. 121–124.

F. Horner, and C. Clarke, "Radio noise form lightning discharges," *Nature*, vol. 181, pp. 688–690, 1958.

A. Iwata, and M. Kanada, "On the nature of the frequency spectrum of atmospheric source signal," *Proc. Res. Inst. Atmos.*, vol. 14, pp. 1–6, 1967.

D. M. Levine, "The effect of pulse interval statistics on the spectrum of radiation from lightning," *J. Geophys. Res.*, vol. 82, pp. 1773–1778, 1977.

M. A. Lind, J. S. Hartman, E. S. Takle, and J. L. Stanford, "Radio noise studies of several severe weather events in Iowa in 1971," *J. Atmos. Sci.*, vol. 29, pp. 1220–1223, 1972.

D. J. Malan, "Radiation from lightning discharges and its relation to the discharge process," *Recent Advances in Atmospheric Electricity.* L. G. Smith, ed. New York: Pergamon Press, 1958, pp. 557–563.

T. Nakai, "The frequency spectrum of atmospherics," *Proc. Res. Inst. Atmos.*, Nagoya Univ., vol. 3, 1955.

T. Nakai, "On the time and amplitude properties of electric fields near sources of lightning in the VLF, HF, and LF bands," *Radio Sci.*, vol. 12, pp. 389–396, 1977.

M. Takagi, "VHF radiation from ground discharges," *Proc. Res. Inst. Atmos.*, Nagoya Univ., vol. 16, pp. 163–168, 1969a.

M. Takagi, "VHF radiation from ground discharges," in *Planetary Electrodynamics*, vol. 1, S. C. Coroniti and J. Hughes, Ed. New Gordon and Breach, 1969b, pp. 535–538.

M. Takagi, and T. Takeuti, "Atmospherics radiation from lightning discharge," *Proc. Res. Inst. Atmos.*, Nagoya Univ., vol. 10, p. 1, 1963.

H. A. Thomas, and R. E. Burgess, "Survey of existing information and data on radio noise over frequency range 1–30 Mc/s," *Radio Res. Special Rep.*, #15, 1947.

N. Flash Frequency Spectra Above 30–50 MHZ

D. Atlas, "Radar lightning echoes and atmospherics in vertical cross-section," in Recent Advances in Atmospheric Electricity, L. G. Smith ed. New York: Pergamon Press, 1959, pp. 441–458.

M. Brook, and N. Kitagawa, "Radiation from lightning discharges in the frequency range 400 to 1000 Mc/s," *J. Geophys. Res.*, vol. 69, pp. 2431–2434, 1964.

R. E. Hallgren, and R. B. McDonald, "Atmospherics from lightning from 100 to 600 MHz," Rep. no. 63-538-89, IBM Federal Systems Division, 1963.

F. J. Hewitt, "Radar echoes from interstroke process in lightning," *Proc. Phys. Soc. London*, vol. 70, pp. 961–979, 1957.

F. Horner, and P. A. Bradley, "The spectra of atmospherics from near lightning discharges," *J. Atmospheric Terr. Phys.*, vol. 26, pp. 1155–1166, 1964.

E. L. Kosarev, V. G. Zatsepin, and A. V. Mitrofanov, "Ultra high frequency radiation from lightnings,'' *J. Geophy. Res.*, vol. 75, pp. 7524–7530, 1970.

J. L. Pawsey, ''Radar observations of lightning,'' *J. Atmospheric Terr. Phys.*, vol. 11, pp. 289–290, 1957.

W. D. Rust, P. R. Krehbiel, and A. Shlanta, ''Measurements of radiation from lightning at 2200 MHz,'' *Geophys. Res. Lett.*, vol. 6, 1979.

J. P. Shafer, and W. M. Goodall, ''Peak field strengths of atmospherics due to local thunderstorms at 150 Megacycles,'' Proc. IRE, vol. 27, pp. 202–207, 1939.

M. Takagi, and T. Takeuti, ''Atmospherics radiation from lightning discharge,'' *Proc. Res. Inst. Atmos.*, Nagoya Univ., vol. 10, pp. xxx–xxx, 1963.

O. Frequency Spectra, Review Papers

H. L. Jones, R. L. Calkins, and W. L. Hughes, ''A review of the frequency spectrum of cloud-to-ground and cloud-to-cloud lightning,'' *IEEE Trans. Geosci. Elec.*, GE-5, vol. 1 pp. 26–30, 1967.

A. Kimpara, ''Electromagnetic energy radiated from lightning,'' in *Problems of Atmospheric and Space Electricity*, S. C. Coroniti ed. New York: American Elesevier Publishing Company, 1965, pp. 352–365.

G. N. Oetzel, E. T. Pierce, ''Radio emissions from close lightning, in *Planetary Electrodynamics*, vol. 1, S. C. Coroniti and J. Hughes, Eds. New York: Gordon and Breach, 1969, pp. 543–570.

L. L. Oh, ''Measured and calculated spectral amplitude distribution of lightning sferics,'' *IEE Trans. Electromag. Compat.*, EMC-11, pp. 125–130, 1969.

E. T. Pierce, ''Spherics (sferics),'' in Encyclopedia of Atmospheric Sciences and Astrogeology, R. W. Fairbridge, Ed., Reinhold Publishing Co., 1967a, pp. 935–939.

——, ''Atmospherics–their characteristics at the source and propagation,'' Progress Radio Sci., 1963–1966, pt. 1, Inter. Scientific Radio Union, Berkeley, CA, pp. 987–1039, 1967b.

——, ''Atmospherics and radio noise, in Lightning,'' 1, Physics of Lightning, R. H. Golde, ed. New York: Academic Press, 1977, pp. 351–384.

Section III

V. G. Agapov, V. P. Larionov, and I. M. Sergievskaya, ''Appraisal of aircraft lightning protection reliability and test conditions,'' *Elec. Tech. USSR*, no. 3, 25–38, 1980. London, England: Pergamon Press, Ltd. (English translation of Elektrichestvo, no. 7, pp. 72–73, 1979.)

R. K. Baum, ''Airborne lightning characterization, Lightning Technology,'' Proc. Tech. Symp. NASA Langley Research Center, Hampton, VA, Apr. 22–24, 1980, Supplement to NASA Conf. Publication 2128, FAA-RD-80-30, pp. 1–19.

K. Berger, ''Methoden und resultate der blitzforschung auf dem Monte San Salvatore bei lugano in dem jahren 1963–1972,'' *Bull Schweiz. Elektrotech. Ver.*, vol.63, pp. 1403–1422, 1972.

K. Berger, and E. Vogelsanger, ''Messungen und resultate der blitzforschung der jahre 1955–1963 auf dem Monte San Salvatore,'' *Bull. Schweiz. Elektech. Ver.*, vol. 56, pp. 2–22, 1965.

W. G. Butters, D. W. Clifford, R. J. Lauber, and K. S. Zeisel. ''Evaluation of lightning-induced transients in aircraft using the *E*-dot, *V*-dot, and *I*-dot excitation techniques,'' Proc. of 1979 IEEE Int. Symp. Electromag. Compat., 9–11 Oct. 1979, San Diego, CA.

W. E. Cobb, and F. J. Holitza, ''A note on lightning strokes to aircraft,'' *Monthly Weather Rev.*, vol. 96, pp. 807–808, 1968.

A. J. Eriksson, ''A discussion on lightning and tall structures, CSIR Special Rep. ELEK 152, National Electrical Engineering Research Institute, Pretoria, S. A., July 1978; Lightning and tall structures, Trans. South African Institute of Electrical Engineers, vol. 69, pp. 2–16, Aug. 1978.

R. P. Fieux, C. H. Gary, B. P. Hutzler, A.R. Eybert-Berard, P. L. Hubert, A. C. Meesters, P. H. Perroud, J. H. Hamelin, J. M. Person, ''Research and artificially triggered lightning in France,'' *IEEE Trans. Power App. Syst.*, PAS-97, pp. 725–733, 1978.

D. R. Fitzgerald, ''Probable aircraft 'triggering' of lightning in certain thunderstorms,'' *Monthly Weather Rev.* vol. 95, pp. 835–842, 1967.

R. H. Golde, ''The frequency of occurence and the distribution of lightning flashes to transmission lines,'' *AIEE Trans.*, vol. 64, pp. 902–910, 1945.

R. H. Golde, ''Occurence of upward streamers in lightning discharges.'' *Nature*, vol. 160, pp. 395–396, 1947.

R. H. Golde, ''The attractive effects of a lightning conductor.'' *J. Inst. Elec. Eng. London*, vol. 9, pp. 212–213, 1963.

R. H. Golde, ''The lightning conductor,'' *J. Franklin Inst.*, vol. 283, pp. 451–477, 1967.

R. H. Golde, ''The lightning conductor,'' in *Lightning, Vol. II, Lightning Protection*, R. H. Golde, ed. New York: Academic Press, 1977, pp. 545–576.

R. H. Golde, ''Lightning and tall structures,'' *IEE Proc. Inst. Elec. Eng. London*, vol. 125, pp. 347–351, 1978.

J. H. Hagenguth, and J. G. Anderson, ''Lightning to the Empire State Building,'' pt. 3, *AIEE Trans.*, vol. 71, (pt. 3), pp. 641–749, 1952.

E. P. Krider, and C. G. Ladd, ''Upward Streamers in lightning discharges to mountainous terrain,'' *Weather*, vol. 30, pp. 77–81, 1975.

E. T. Pierce, ''Triggered lightning and its application to rockets and aircraft,'' 1972 Lightning and Static Electricity Conf. AFAL-TR-72-325, Air Force Avionics Lab., Wright-Patterson Air Force Base, OH, 1972.

F. L. Pitts, ''Electromagnetic measurement of lightning strikes to aircraft,'' AIAA 19th Aerospace Sciences Meeting, St. Louis, Jan. 1981.

F. L. Pitts, and M. E. Thomas, 1980 direct strike lightning data, NASA Technical Memorandum 81946, Langley Research Center, Hampton, VA, Feb. 1981.

B. Vonnegut, ''Electrical behavior of an airplane in a thunderstorm,'' A. L. Little, Inc., Cambridge, Mass., Feb. 1965, Defense Documentation Center, AD-614-914, Clearinghouse for Federal Scientific and Technical Information.

C. F. Wagner, ''Relation between stroke current and velocity of the return stroke,'' *AIEE Trans.*, vol. 82, (pt. 3), pp. 606–617, 1963.

C. F. Wagner, ''Lightning and transmission lines,'' *J. Franklin Inst.*, vol. 283, pp. 558–594, 1967.

Section IV

N. W. Albright, and D. A. Tidman, ''Ionizing potential waves and high voltage breakdown streams,'' *Phys. Fluids*, vol. 15, pp. 86–90, 1972.

C. E. R. Bruce, and R. H. Golde, ''The lightning discharge,'' *J. Inst. Elec. Eng. London*, vol. 88, pp. 487–520, 1941.

A. S. Dennis, and E. T. Pierce, ''The return stroke of the lightning flash to earth as a source of VLF atmospherics,'' *Radio Sci.*, vol. 68D, pp. 777–794, 1964.

L. P. Gorbachev, and V. F. Federov, ''Electromagnetic radiation from a return streamer of lightning,'' *Geomag. Aeron.*, vol. 17, pp. 1977.

E. L. Hill, ''Electromagnetic radiation from lightning strokes,'' *J. Franklin Inst.*, vol. 263, pp. 107–109, 1957.

R. D. Hill, ''Electromagnetic radiation from the return stroke of a lightning discharge,'' *J. Geophys. Res.*, vol. 71, pp. 1963–1967, 1966.

——, ''Analysis of irregular paths of lightning channels,'' *J. Geophys. Res.*, vol. 73, pp. 1897–1905, 1968.

——, ''Electromagnetic radiation from erratic paths of lightning strokes,'' *J. Geophys. Res.*, vol. 74, pp. 1922–1929, 1969.

R. O. Himley, ''VLF radiation from subsequent return strokes in multiple stroke lightning,'' *J. Atmos. Terr. Phys.*, vol. 31, pp. 749–753, 1969.

A. Iwata, ''Calculations of waveforms radiating from return strokes,'' *Proc. Res. Inst. Atmos.*, Nagoya Univ., vol. 17, pp. 115–123, 1970.

D. L. Jones, ''Electromagnetic radiation from multiple return strokes of lightning,'' *J. Atmospheric Terr. Phys.*, vol. 32, pp. 1077–1093, 1970.

R. D. Jones, and H. A. Watts, ''Close-in magnetic fields of a lightning return stroke,'' *Sandia Lab.*, SAND75-0114, 1975.

D. J. Jordan, and M. A. Uman, ''Variations of light intensity with height and time from subsequent return strokes,'' *Trans. Am. Geophys. Union*, vol. 61, p. 977, 1980.

R. E. Lefferts, ''A statistical simulation of ground-wave atmospherics generated by lightning return strokes,'' *Radio Sci.*, vol. 13, pp. 121–130, 1978.

——, ''Probabilistic model for the initial peaks of ground wave mospherics generated by lightning return strokes,'' *Radio Sci.*, vol. 14, pp. 1017–1026, 1979.

J. A. Leise, and W. L. Taylor, ''A transmission line model with general velocities for lightning,'' *J. Geophys. Res.*, vol. 82, pp. 391–396, 1977.

D. M. LeVine, and R. Meneghini, ''Simulation of radiation from lightning return strokes: The effects of tortuosity,'' *Radio Sci.*, vol. 13, pp. 801–809, 1978.

——, ''Electromagnetic fields radiated from a lightning return stroke: Application of an exact solution to Maxwell's equations,'' *J. Geophys. Res.*, vol. 83, pp. 2377–2384, 1978.

Y. T. Lin, M. A. Uman, and R. B. Standler, ''Lightning return stroke models,'' *J. Geophy. Res.*, vol. 85, pp. 1571–1583, 1980.

P. F. Little, "Transmission line representation of a lightning return stroke," *J. Phys. D. Applied Physics*, vol. 11, pp. 1893–1910, 1978.

——, "The effect of altitude on lightning hazards to aircraft," presented at the Proc. 15th European Conf. Lightning Protection, vol. 2, Institute of High Voltage Research, Uppsala University, Sweden, 1979.

M. J. Master, M. A. Uman, Y. T. Lin, and R. B. Standler, "Calculations of lightning return stroke electric and magnetic fields above ground," *Geophys. Res.*, vol. 86, pp. 12127–12132, 1981.

H. Norinder, "Lightning currents and their variations," *J. Franklin Inst.*, vol. 220, pp. 69–92, 1935.

J. Papet-Lepine, "Electromagnetic radiation and physical structure of lightning discharges," *Arkiv Geophysik*, vol. 3, pp. 391–400, 1961.

G. N. Oetzel, "Computation of the diameter of a lightning return stroke," *J. Geophys. Res.*, vol. 73, pp. 1889–1896, 1968.

R. A. Pearlman, "Lightning near fields generated by return stroke current," in *IEEE Int. Symp. EMC*, San Diego, CA, 1979, pp. 68–71.

E. T. Pierce, "Atmospherics from lightning flashes with multiple strokes," *J. Geophys. Res.*, vol. 65, pp. 1867–1871, 1960.

G. H. Price, and E. T. Pierce, "The modeling of channel current in the lightning return stroke," *Radio Sci.*, vol. 12, pp. 381–388, 1977.

J. Rai, "Current and velocity of the return lightning stroke," *J. Atmos. Terr. Phys.*, vol. 40, pp. 1275–1285, 1978.

J. Rai, and P. K. Bhattacharya, "Impulse magnetic flux density close to the multiple return strokes of a lightning discharge," *J. Phys. D: Appl. Phys.*, vol. 4, pp. 1252–1256, 1971.

M. Rao, "Notes on the corona currents in a lightning discharge and the emission of ELF waves," *Radio Sci.*, vol. 2, p. 1394, 1967.

M. Rao, and H. Bhattacharya, "Lateral corona currents from the return stroke channel and slow field change after the return stroke in a lightning discharge," *J. Geophys. Res.*, vol. 71, pp. 2811–2814, 1967.

M. Rao, and S. R. Khastgir, "The physics of the return stroke and the time-variation of its current in a lightning discharge," *Trans. Bose Res. Inst.*, vol. 29, pp. 19–24, 1966.

B. F. J. Schonland, "The lightning discharges," *Handbuch der Physik*, vol. 22, pp. 576–628, OHG, Berlin, Germany: Springer-Verlag, 1956.

K. M. L. Srivastava, "Return stroke velocity of a lightning discharge," *J.*

Geophy. Res., vol. 71, pp. 1283–1286, 1966.

K. M. L. Srivastava, and B. A. P. Tantry, "VLF characteristic of electromagnetic radiation from the return stroke of lightning discharge," *Indian J. Pure Appl. Phys.*, vol. 4, pp. 272–275, 1966.

J. A. Stratton, *Electromagnetic Theory*, New York: McGraw-Hill Book Company, Inc., pp. 582–583, 1941.

D. F. Strawe, "Nonlinear modeling of lightning return strokes," in Proc. Fed. Aviation Admin./Florida Inst. Tech. Workshop Grounding Lightning Tech., Mar. 6–8, 1979, Melbourne, Florida, Rep. FAA-RD-79-6, pp. 9–15.

J. Suchocki, "Characteristic impedence of a lightning channel and results for protection of high structures," *Elektrotechnische Zeitschrift, etz-a*, 99, 1978.

M. A. Uman, and D. K. McLain, "Magnetic field of lightning return stroke," *J. Geophys. Res.*, vol. 74, pp. 6899–6910, 1969.

——, "Lightning return stroke current from magnetic and radiation field measurement," *J. Geophys. Res.*, vol. 75, pp. 5143–5147, 1970.

M. A. Uman, "The electromagnetic radiation from a finite antenna," *Am. J. Phys.*, vol. 43, pp. 33–38, 1975.

M. A. Uman, and E. P. Krider, "Lightning environment modeling, Vol. I of atmospheric electricity hazards, analytical model development, and application," Electro Magnetic Applications EMA-91-R-21, Apr. 1, 1981.

C. F. Wagner, "Determination of the wave front of lightning stroke currents from field measurements," AIEE Trans. (pt. 3), vol. 79, pp. 581–589, 1960.

——, "Relation between stroke current and velocity of the return stroke," *AIEE Trans.*, vol. 82, (pt. 3), pp. 606–617, 1963.

C. F. Wagner, and A. R. Hileman, "The lightning stroke," (1), *AIEE Trans.*, vol. 77, pt. 3, pp. 229–242, 1958.

——, "Surge impedance and its application to the lightning stroke," *AIEE Trans.*, vol. 80, (pt. 3), pp. 1011–1022, 1962.

——, "The lightning stroke," (2), *AIEE Trans.*, vol. 80, (pt. 3), pp. 623–642, 1961.

C. D. Weidman, and E. P. Krider, "The fine structure of lightning return stroke wave forms," *J. Geophys. Res.*, vol. 83, pp. 6239–6247, 1978.

——, "Submicrosecond risetimes in lightning return stroke fields," *Geophys. Res. Lett.*, vol. 7, pp. 955–958, 1980.

Name Index

(References for Appendix E are not included in Name Index.)

Subject Index

(References for Appendix E are not included in Subject Index.)

Air discharges, 2, 23, 34, 96, 99, 109
Aircraft, lightning to, 134, 252
Alti-electrograph, 67
ArI, 145, 146, 148
Avalanche, 205

B (breakdown) field change, 75–82, 105,
    237
Ball lightning, 11, 12, 241, 243–248
Balloons, lightning to, 135
Basalt, 114, 118
Bead lightning, 11, 12, 240–242
Blackbody radiation, 153, 155, 165, 172
Boltzmann statistics, 154
Boys camera, 5, 15–18
Branch component, 10, 28
Branching of lightning, 9, 26–28, 34, 39,
    198
Breakdown criterion, 206
Breakdown electric field, 204, 213
Breakdown voltage, 204
Bremsstrahlung, 165

C (continuing current) field change, 74,
    75, 87–89

Camera, Beckman and Whitley streak-
    ing, 150
  Boys, 5, 15–18
  for cloud discharges, 28
  image converter, 153, 236
  image intensifier, 153, 236
  streak, 5, 17
Charge, deposited on dart leader, 4, 9,
    81–85, 222
  deposited on stepped leader, down-
      ward-moving, 4, 6, 78–81, 213, 217
  upward-moving, 39
  distribution in thundercloud *(see*
      Thundercloud, electrical charges
      of)
  transferred by continuing current, 9,
      73, 88, 102, 133
  transferred by flash, 4, 5, 71, 73,
      102, 130–132
  transferred by return stroke, 4, 73,
      86, 87, 125, 131, 132
Clear air lightning, 1
Cloud-to-cloud lightning, 1, 34, 39, 96
Cloud lightning *(see* Intracloud lightning)
CN spectrum, 143, 162, 163